THE IDENTIFICATION OF THE CF (CYSTIC FIBROSIS) GENE

Recent Progress and New Research Strategies

ADVANCES IN EXPERIMENTAL MEDICINE AND BIOLOGY

Recent Volumes in this Series

Volume 284
ENZYMOLOGY AND MOLECULAR BIOLOGY OF CARBONYL METABOLISM 3
Edited by Henry Weiner, Bendicht Wermuth, and David W. Crabb

Volume 285
HYPERCHOLESTEROLEMIA, HYPOCHOLESTEROLEMIA, HYPERTRIGLYCERIDEMIA, IN VIVO KINETICS
Edited by Claude L. Malmendier, P. Alaupovic, and H. Bryan Brewer, Jr.

Volume 286
TEMPERATURE AND ENVIRONMENTAL EFFECTS ON THE TESTIS
Edited by Adrian W. Zorgniotti

Volume 287
NEURORECEPTOR MECHANISMS IN BRAIN
Edited by Shozo Kito, Tomio Segawa, and Richard W. Olsen

Volume 288
DRUGS OF ABUSE, IMMUNITY, AND IMMUNODEFICIENCY
Edited by H. Friedman

Volume 289
NUTRITIONAL AND TOXICOLOGICAL CONSEQUENCES OF FOOD PROCESSING
Edited by Mendel Friedman

Volume 290
THE IDENTIFICATION OF THE CF (CYSTIC FIBROSIS) GENE: Recent Progress and New Research Strategies
Edited by Lap-Chee Tsui, Giovanni Romeo, Rainer Greger, and Sergio Gorini

Volume 291
FUEL HOMEOSTASIS AND THE NERVOUS SYSTEM
Edited by Mladen Vranic, Suad Efendic, and Charles H. Hollenberg

Volume 292
MECHANISMS OF LYMPHOCYTE ACTIVATION AND IMMUNE REGULATION III
Edited by Sudhir Gupta, William E. Paul, Max D. Cooper, and Ellen V. Rothenberg

---

A Continuation Order Plan is available for this series. A continuation order will bring delivery of each new volume immediately upon publication. Volumes are billed only upon actual shipment. For further information please contact the publisher.

THE IDENTIFICATION OF THE CF (CYSTIC FIBROSIS) GENE

Recent Progress and New Research Strategies

Edited by

Lap-Chee Tsui
The Hospital for Sick Children
Toronto, Ontario, Canada

Giovanni Romeo
Istituto Giannina Gaslini
Genoa, Italy

Rainer Greger
Albert-Ludwigs-Universität
Freiburg, Germany

and

Sergio Gorini
Fondazione Internazionale Menarini
Florence, Italy

PLENUM PRESS • NEW YORK AND LONDON

Library of Congress Cataloging-in-Publication Data

The Identification of the CF (cystic fibrosis) gene : recent progress
and new research strategies / edited by Lap-Chee Tsui ... [et al.].
p. cm. -- (Advances in experimental medicine and biology ; v.
290)
Proceedings of an international workshop held April 9-11, 1990, in
Sestri Levante (Genoa), Italy.
Includes bibliographical references and index.
ISBN 0-306-43909-3
1. Cystic fibrosis gene--Congresses. 2. Cystic fibrosis--Genetic
aspects--Congresses. I. Tsui, Lap-Chee. II. Series.
[DNLM: 1. Cystic Fibrosis--genetics--congresses. W1 AD559 v. 290
/ WI 820 I19 1990]
RC858.C95I34 1991
616.3'7--dc20
DNLM/DLC
for Library of Congress 91-3859
CIP

Proceedings of an international workshop on the Identification of the
CF Gene: Recent Progress and New Research Strategies,
held April 9–11, 1990, in Sestri Levante (Genoa), Italy

ISBN 0-306-43909-3

A Division of Plenum Publishing Corporation
233 Spring Street, New York, N.Y. 10013

Printed in the United States of America

PREFACE

Seven months after the report of the cloning of the CF gene by three North American groups, a workshop scheduled before that report was held in Sestri Levante, a splendid city on the Ligurian coast. For three days almost 100 scientists from Europe and North America had the opportunity to discuss the perspectives and the new research strategies opened up by the cloning of the gene. The issue of publishing these proceedings was discussed at length among the invited speakers of the workshop and the prevailing opinion was that they would be worthwhile if accompanied by the transcripts of the discussions which followed each presentation. Thanks to the generous suppport of the Menarini Foundation, represented at the meeting by Dr. Sergio Gorini, and thanks to the reviewing work done by the Chairpersons of each session, by many participants, by Dr. Donata Brugioni, by Mrs. Katherine Loparco and by my collaborators Drs. Marcella Devoto and Luis J.V. Galietta, that idea has generated this book, which should be a helpful tool at least for some years for all those interested in understanding how the CFTR gene works.

Prof. G. Romeo

CONTENTS

CYSTIC FIBROSIS - A STRATEGY FOR THE FUTURE

Robert Williamson

Department of Molecular Genetics
St. Mary's Hospital Medical School
Imperial College London, England

SUMMARY

The identification of the cystic fibrosis locus must be regarded as a major triumph for human molecular genetics. In less than five years, it has been possible to move from the designation of the chromosomal location of CF to the identification of the previously unknown protein which is mutated. This work is a credit to all of the groups which participated in it, and particularly to the Toronto and Ann Arbor groups, which identified the sequence coding for the cystic fibrosis transmembrane regulator as the mutated gene. However, as powerful as molecular genetics is, it can only provide the first step towards understanding the correlation between structure and function for CFTR, analysing the clinical correlates between gene and pathology, gaining knowledge of the population genetics of CF (including the numbers of mutations, as well as the question of whether or not there is a carrier advantage), and progressing to prevention through carrier testing in the community and to treatment through the development of pharmaceuticals and (ultimately) somatic gene therapy. The solutions to these problems will require the recruitment of new groups, with different techniques, to the CF research community.

Early Molecular Genetic Studies

Cystic fibrosis research has just taken a great step forward with the discovery of the gene that is mutated to cause the disease (Rommens et al., 1989; Riordan et al., 1989; Kerem et al., 1989). The use of the techniques of "reverse genetics" to find linkage (and chromosomal location), followed by movement from linkage to locus, has been fully justified. However, ten years ago, this approach was seen as, at best, a futuristic dream; at worst, a positive indication of mad hubris and over-optimism.

The Identification of the CF (Cystic Fibrosis) Gene
Edited by L.-C. Tsui *et al.*, Plenum Press, New York, 1991

In 1980, the thalassaemias had just been analysed in detail, and the first DNA diagnoses were being carried out on chorion. Many groups, including our own, had been able to contribute to the study of thalassaemia by cloning the human globin genes and determining the mutations which cause the disease (Little et al. 1978). However, we were still blinkered by the perception that it was necessary to know the nature of the protein which was affected before the gene could be cloned. When Kan and Dozy (1978) found that a marker restriction site outside the beta-globin gene is in linkage disequilibrium with the sickle cell mutation, Solomon and Bodmer (1979) immediately pointed out that this provided a new, reverse technique for gene mapping, making use of informative but anonymous markers. Our group, and others, quickly began to study diseases such as cystic fibrosis, where Mendelian inheritance was known to occur but the protein was not known (Davies et al. 1983; Williamson et al. 1984).

The first successes for reverse genetics did not occur for CF, but for Duchenne muscular dystrophy. Linkage was first demonstrated to anonymous probes located at Xp21 (Murray et al., 1982); then Kunkel and his colleagues isolated the gene and identified the altered protein dystrophin (Monaco et al., 1986). For Duchenne muscular dystrophy, the X-chromosome was known to be the one carrying the mutation by the pattern of inheritance, and there were translocation and deletion clues to the subchromosomal location. For CF, the only data available showed that the disease was due to single, defective autosomal gene locus, as proven by Romeo from pedigree studies (Romeo, 1984). Even the sites of tissue expression of the gene were controversial, with claims that the defect could be detected in leukocytes and fibroblasts. No carrier test for CF existed before DNA analysis, even in families with living affected individuals. Several methods for prenatal diagnosis were offered and then withdrawn (Nader and Walsh, 1980), until Brock devised a more or less reliable mid-trimester method, which still has a false positive and false negative rate of about 5% (Brock et al., 1987). It was within this unsure environment that several groups began to collect large families with multiple cases of CF for DNA linkage studies in the early 1980s (Tsui in Toronto, Williamson in London, Klinger at Case Western Reserve and White in Salt Lake City).

These groups first met at the Brighton International Cystic Fibrosis Conference (1984), and then at the Human Gene Mapping meeting in Helsinki in 1985. During the five years since 1980, the scientific atmosphere had changed, and reverse genetics had become an accepted technique for following genes and phenotypes by co-segregation. Between the two meetings, a most important finding was made by Hans Eiberg, in Copenhagen, who collected serum samples from several of the above CF groups as well as from CF families from Denmark, and showed significant segregation with a protein marker, the enzyme paraoxonase, which occurs in two allelic variants (Eiberg et al. 1985). The problem remained that no one had yet mapped the gene coding for this enzyme to a specific chromosome.

From Linkage to Gene

All of the molecular genetics groups greeted this finding with great interest. The data for all DNA and protein markers were pooled at Helsinki, and it became apparent that there were certain gaps in the map - I recall chromosomes 7, 8 and 18 as being particularly likely regions. Each team using segregation to find linkage approached the problem with their own probes and families. Linkage to three markers (MET, DocRI and J3.11) was published simultaneously by three groups (Salt Lake City, Toronto and London) towards the end of 1985 (Tsui et al., 1985; Wainwright et al., 1985; White et al., 1985).

A massive international collaboration to confirm the linkage, to determine whether CF was to one side of the close markers MET and J3.11 or between them, and to start applying the data to families by offering carrier testing and prenatal diagnosis. Throughout the entire period of this research programme, DNA from families was always shared, and probes usually were distributed between all the CF research groups after characterisation. Although there were exceptions, by and large the story of CF research was one of collaboration and mutual respect between the groups.

The first step was to confirm the linkage and to determine the positions of each probe relative to each other and to CF: the mutation proved to be between J3.11 and MET, and approximately one centimorgan from each (Beaudet et al., 1986). New probes were isolated, and a pulse field map of the region was made which showed that the physical distance between J3.11 and MET is approximately 1.5 megabase pairs (Poustka et al., 1988). At this point, the genetic and physical maps were defined as accurately as the technology permitted.

In 1987, our group cloned a close marker by cloning CPG-RICH sequences between J3.11 and MET; this was at the site of the HTF island associated with a protein related to the mouse growth factor gene INT (and was named IRP, for INT-related protein) (Estivill et al., 1987). Within four months, we reported with disappointment that this candidate gene was not in fact the gene itself, but a very close marker in disequilibrium with CF. With hindsight, we placed too much emphasis on this single neighboring gene in our research strategy, and were not as flexible in our approach as the Toronto and Michigan groups. However, the IRP markers KM-19 and CS7 were distributed to over 400 groups, and were (and still are) useful for population linkage studies as well as prenatal diagnosis and carrier assessment.

The Cystic Fibrosis Transmembrane Regulator

Lap-Chee Tsui and Francis Collins continued to isolate probes from the CF interval, and at this time began to use the powerful technique of chromosome jumping, which enabled them to move over certain sequences which were very difficult to clone using conventional methods. Their success, which deserves every credit both for its scientific excellence and its thoroughness, is now well known, and will be discussed by them during this meeting.

From their data, we know that the CF gene is indeed located where both groups predicted, between the gene coding for IRP and J3.11. It is a large gene, coding for a large mRNA and protein. While one mutation makes up 70% of the North American, and 80% of the London, CF defects, there are many other mutations as well. There is homology to a family of transmembrane proteins which include the MDR proteins found in many species, including bacteria and man.

The Questions that Remain

Many questions remain. The most obvious include the number of mutations; while approximately 30 have been identified to date, it appears that at least one major mutation (the common one in Southern Europe) is still to be identified. Associated with this is the question of the association between specific mutations and clinical course. How is it possible that the particular range of mutations which is seen causes the range of physiological and pathological changes which occur? Some patients seem to have termination mutations on both of their chromosomes; does this imply that the CF protein can be completely absent, and still give rise to classical CF? Why, indeed, do a proportion of CF patients have liver involvement, arthritis, severe pancreatic or lung symptoms, sweat electrolytes elevated to a greater or lesser extent, or meconium ileus? On a more theoretical level, why do there appear to be more CF carriers among men than women, in a ratio of 5:4? And in practical terms, can we offer carrier testing to the community, and what use will they make of this information if education and counselling are not available widely (Williamson, 1989).

For protein chemists, the problems are also immediate: in particular, what is the substance (ion or molecule) which is transported through the cell membrane by the CFTR? It does not have the appearance expected of a chloride channel, and indeed differs from at least one such channel which has been cloned. Perhaps it transports a molecule which regulates chloride transport? In which tissues, and at what time in development, is it expressed? Can site-directed mutagenesis, aided by knowledge of the natural mutations causing CF, give clues as to the structure/function relationship for this protein? The study of the protein is essential for understanding the pathophysiology of the disease, but it is also possible that therapy with the protein will succeed.

Therapy is, indeed, one of the major hopes and objectives of this research. While it is exciting to discuss somatic gene therapy, it is also worth remembering that most successful therapies at present are pharmacological - which is another reason why the protein chemistry is so essential, since it represents an approach to mathematical modelling of drug/protein interactions. Cystic fibrosis will be one of the major targets for those who wish to introduce normal functional genes into cells during the next ten years as a form of therapy, and of course this exciting prospect is helped by the knowledge that carriers (who, I anticipate, have a half-level of CFTR expression) are completely normal. It must be apparent, however, that the major problem in contemplating somatic gene therapy for CF is not only the absence of appropriate vectors to take genes into any human stem cells, but also our ignorance of the lung as a developing and dynamic tissue.

Personally, working on cystic fibrosis has been a pleasure and a privilege. The only negatives were brought about not by my colleagues (amongst whom I number all of the scientists and clinicians who are here today), but the press, and particularly the scientific journalists who were prepared to distort consistently the nature of the interactions and commitments between those who were involved in this scientific enterprise. Scientific journals, and their writers, have a special responsibility to our community, and I feel that in this case they let us down badly. While I will continue to have an interest and involvement in CF, it will not remain primarily a problem for molecular geneticists. Others will be able to make use of the data that the genetics groups have provided to make their own contribution in future.

Acknowledgements

I thank the Cystic Fibrosis Research Trust, and in particular Ron Tucker and Barbara Bentley, and the M.R.C., for sustained support over the past decade, and my many friends and colleagues in the St. Mary's C.F. group, the international scientific community, and amongst CF clinicians and families for their consistent cooperation.

References

Beaudet, A., Bowcock, A., Buchwald, M., Cavalli-Sforza, L., Farrall, M., King, M.C., Klinger, K., Lalouel, J.-L., Lathrop, G., Naylor, S., Ott, J., Tsui, L.-C., Wainwright, B., Watkings, P., White, R., Williamson, R., 1986, Linkage of Cystic Fibrosis to Two Tightly Linked DNA Markers: Joint Report from a Collaborative Study, **Am. J. Hum. Genet.**, 39:681.

Brock, D.J.H., Clarke, H.A.K., Barron, L. 1987, Prenatal Diagnosis of Cystic Fibrosis by Microvillar Enzyme Assay on a Series of 258 Pregnancies, **Hum. Genet.**, 78:271.

Davies, K.E., Giniam, T.C., Williamson, R., 1983, Cystic fibrosis is not caused by a defect in the gene coding for human complement C3, **Mol. Biol. Med.**, 1:185.

Eiberg, H., Mohr, J., Schmiegelow, K., Nielsen, L.S., Williamson, R., 1985, Linkage Relationships of Paraoxonase with other Markers: Identification of PON-Cystic Fibrosis Synteny, **Clin. Genet.**, 28:265.

Estivill, X., Farrall, M., Scambler, P.J., Bell, G.M., Hawley, K.M.F., Lench, N.J., Bates, G.P., Kruyer, H.C., Frederick, P.A., Stanier, P., Watson, E.K., Williamson, R., Wainwright, B.J., 1987, A Candidate for the Cystic Fibrosis Locus Isolated by Selection for HTF Islands, **Nature**, 326:840.

Kan, Y.W., Dozy, A.M., 1978, Polymorphism of DNA Sequence Adjacent to Human β-Globin Structural Gene: Relationship to Sickle Mutation, **Proc. Natl. Acad. Sci. USA**, 75:5631.

Kerem, B., Rommens, J.M., Buchanan, J.A., Markiewicz, D., Cox, T.K., Chakravarti, A., Buchwald, M., Tsui, L.-C., 1989, Identification of the Cystic Fibrosis Gene: Genetic Analysis, **Science**, 245:1073.

Little, P.F.R., Curtis, P., Coutelle, C.H., Van den Berg, J., Dalgleish, R., Malcolm, S., Courtney, M., 1978, The Isolation and Partial Sequence of Recombinant Plasmids containing Human α-, β- and gamma-Globin cDNA Fragments, **Nature**, 273:640.

Monaco, A.P., Neve, R.L., Colletti-Feener, C., Bertelson, C.J., Kurnit, D.M., Kunkel, L.M., 1986, Isolation of Candidate cDNAs for Portions of the Duchenne muscular dystrophy Gene, **Nature**, 323:646.

Murray, J.M., Davies, K.E., Harper, P.S., Meredith, L., Muller, C., Williamson, R., 1982, Linkage Relationship of a Cloned DNA Sequence on the Short Arm of the X Chromosome to Duchenne Muscular Dystrophy, **Nature**, 300:69.

Nadler, H.L., Walsh, M.M.J., 1980, Intrauterine Detection of Cystic Fibrosis, **Pediatrics**, 66:690.

Poustka, A.M., Lehrach, H., Williamson, R., Bates, G.P., 1988, A Long Range Restriction Map Encompassing the CF Locus and its Closely Linked Genetic Markers, **Genomics**, 2:337.

Riordan, J.R., Rommens, J.M., Kerem, B., Alon, N., Rozmahel, R., Grzelczak, J., Lok, S., Plasvic, N., Chou, J.L., Drumm, M.L., Iannuzzi, M.C., Collins, F.S., Tsui, L.-C., 1989, Identification of the Cystic Fibrosis Gene: Cloning and Characterisation of Complementary DNA, **Science**, 245:1066.

Rommens, J.M., Iannuzzi, M.C., Kerem, B., Drumm, M.L., Melmer, G., Dean, M., Rozmahel, R., Cole, J.L., Kennedy, D., Idaka, N., Zsiga, M., Buchwald, M., Riordan, J.R., Tsui, L.-C., Collins, F.S., 1989, Identification of the Cystic Fibrosis Gene: Chromosome Walking and Jumping, **Science**, 245:1059.

Romeo, G., 1984, Cystic Fibrosis: A Single Locus Disease <u>in</u>: "Cystic Fibrosis: Horizons," D. Lawson, ed., Wiley, London.

Solomon, E., Bodmer, W.F., 1979, Evolution of Sickle Variant Gene, **Lancet**, i:923.

Talamo, R.C., Rosenstein, B.J., Berninger, R.W., 1983, Cystic Fibrosis, <u>in</u>: "The Metabolic Basis of Inherited Disease," D. Stanbury, ed., McGraw Hill, New York.

Tsui LC, Buchwald, M., Barker, D., Braman, J.C., Knowlton, R., Schumm, J.W., Eiberg, H., Mohr, J., Kennedy, D., Plavsik,

N., Zsiga, M., Markiewicz, D., Akots, G., Brown, V., Helms, C., Gravius, T., Parker, C., Rediker, K., Donis-Keller, H., 1985, Cystic Fibrosis Locus Defined by a Genetically Linked Polymorphic DNA Marker, Science 230:1054.

Wainwright, B.J., Scambler, P.J., Schmidtke, J., Watson, E.S., Law, H.Y., Farrall, M.J., Cooke, H.J., Eiberg, H., Williamson, R., 1985, Localisation of the Cystic Fibrosis Locus to Human Chromosome 7cen-7q22, **Nature**, 318:384.

White, R., Woodward, S., Leppert, M., O'Connell, P., Hoff, M., Herbst, J., Lalouel, J.M., 1985, A Closely Linked Genetic Marker for Cystic Fibrosis, **Nature**, 318:382.

Williamson, R., Gilliam, C.T., Blaxter, M., Scambler, P., Robbins, T.P., Davies, K.E., 1984, Gene cloning - a Tool to Find the Basic Defect in Cystic Fibrosis, in: "Cystic Fibrosis: Horizons," D. Lawson, ed., Wiley, London.

Williamson, R., Allison, M.E.D., Bentley, T.J., Lim, S.M.C., Watson, E., Chapple, J., Adam, S., 1989, Community Attitudes to Cystic Fibrosis Carrier Testing in England: A Pilot Study, **Prenatal Diag.**, 9:727.

MOLECULAR GENETICS OF CYSTIC FIBROSIS

L.-C. Tsui, J. Rommens, B. Kerem, R. Rozmahel,
J. Zielenski, D. Kennedy, D. Markiewicz, N. Plavsic,
J.-L. Chou, D. Bozon and M. Dobbs

Department of Genetics
The Hospital for Sick Children, and
Departments of Medical Genetic and Medical Biophysics
University of Toronto

The gene for cystic fibrosis (CF) has recently been isolated (Rommens et al., 1989a; Riordan et al., 1989; Kerem et al., 1989) rendering it possible for the first time to study the elusive symptoms of the disease. In this brief review, we try to summarize our current understanding of the CF gene and our attempts to study the gene product and its function. We have also included a short discussion on the correlation between genotype and phenotype in CF patients.

ISOLATION OF THE CF GENE

Based on the recessive inheritance of the CF trait in families, it was possible to perform linkage analysis to identify a genetic marker linked to the disease gene, thereby mapping it to a chromosomal region. In 1985, linkage studies in our laboratory as well as others had allowed the assignment of the CF gene to the long arm of chromosome 7, in band q31. Several DNA markers tightly linked to CF were identified (see Figure 1) and they permitted the use of various novel gene cloning strategies to approach the CF gene. Our systematic examination of a large number of DNA markers from a flow-sorted chromosome 7-specific

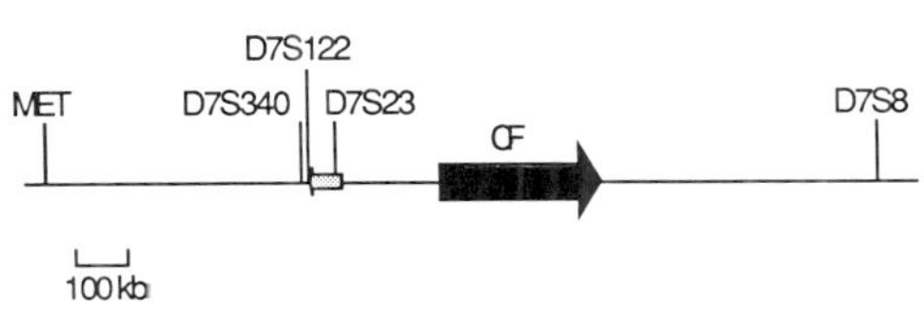

Fig. 1. Relation between CF and other DNA markers. The extent and direction of transcription of the CF gene is indicated by the filled arrow; the IRP locus is indicated by the small arrow.

The Identification of the CF (Cystic Fibrosis) Gene
Edited by L.-C. Tsui *et al.*, Plenum Press, New York, 1991

genomic DNA library had yielded two DNA markers (D7S340/D7S122) within close distance from the CF locus. (For general review, please see Boat et al., 1989).

Chromosome walking experiments were initiated from D7S340/ D7S122 and, in collaboration with Francis Collins' group (University of Michigan, Ann Arbor), the technique of chromosome jumping was also used. As a result, a total of 280 kb of contiguous DNA was isolated (Rommens et al., 1989a). DNA segments were examined for existence of candidate gene sequences and several transcribed sequences were detected. One of them, approximately 250 kb from D7S340, was found to be corresponding to the CF gene. The region which was subsequently determined to be the first exon of the CF gene was first identified by its ability to hybridize to DNA from a variety of animal species.

CHARACTERIZATION OF THE CF GENE

Extensive screening of cDNA libraries, in collaboration with Jack Riordan's group (Department of Biochemistry) of our Institute, resulted in the isolation of a single clone with an insert of 920 basepairs (bp) in length from a normal sweat gland library (Riordan et al., 1989). A transcript of 6.5 kb was detected in a variety of tissue known to be affected by CF. Additional cDNA clones were subsequently isolated from other libraries and a consensus sequence was derived from the overlapping clones. The length of the total cDNA sequence was found to be about 6.2 kb and to span approximately 250 kb of genomic DNA with 27 exons (see Figure 2). The cDNA sequence predicted a protein of 1,480 amino acids with molecular mass of about 170,000. This putative CF gene product (named cystic fibrosis transmembrane conductance regulator or CFTR) was found to contain two similar motifs, each with a domain having properties consistent with membrane association and a domain believed to be involved in ATP-binding.

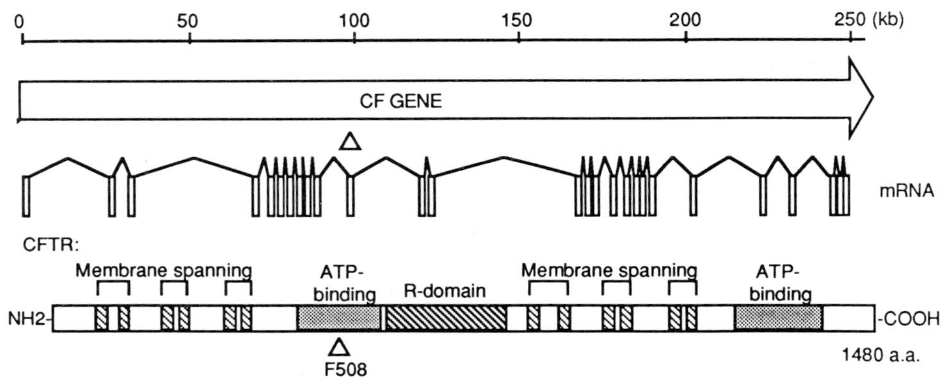

Fig. 2. Schematic diagram of the CF gene, with its mRNA and CFTR shown below. The location of the 3 bp deletion is also indicated (by an arrow).

DETECTION OF THE MAJOR MUTATION

Comparison of the cDNA sequences from an affected individual to that of a normal control revealed a deletion of 3 bp in exon 10; this deletion would result in the loss of a phenylalanine residue (position 508) at the center of the first putative ATP-binding site. Despite this and other highly suggestive biochemical properties of CFTR, however, confirmation for the CF gene was derived from genetic analysis of normal and CF chromosomes (Kerem et al., 1989). Population data showed that this deletion, deltaF508, was present in about 70-75% of the CF chromosomes studied in the North American populations (Kerem et al., 1989; Lemna et al., 1990). The 3 bp deletion was not found in any normal chromosomes, strongly arguing for the identity of the CF gene. Extended haplotype analysis based on DNA markers closely linked to the gene indicated that the deletion was not a tightly linked DNA polymorphism. The haplotype data also suggested that CF chromosomes carrying deltaF508 were all derived from a common origin of a single mutational event.

A total of over 13,000 CF chromosomes have been analyzed (as of April, 1990) in the world by many different CF research laboratories. Although the overall frequency of deltaF508 in this combined study is around .68, the proportion of the deltaF508 mutation is different among different populations. According to the report from the Consortium, the frequency of deltaF508 was only 40-50% of the CF chromosomes in some southern European countries, such as Italy and Spain, while some northern European countries, such as Denmark, were as high as 88%. This information is important when broad scale testing for CF carriers is to be implemented.

IDENTIFICATION OF ADDITIONAL MUTATIONS

Since each mutation marks a functionally important region of CFTR, identification of additional mutations in the CF gene should provide useful information about the structure and function of the protein; the knowledge would in turn be useful in understanding the pathophysiology of the disease and in development of pharmaceutical agents for better treatment of individuals with CF. Further, identification of CF mutations at the DNA level will provide the essential information for genetic testing.

The haplotype data suggested that the remainder of the CF mutant gene pool consisted of multiple, different mutations. A small set of these latter mutant alleles (about 8%) might confer residual pancreatic function in a subgroup of patients who are pancreatic sufficient (work done in collaboration with Peter Durie's group in the Department of Gastroenterology). More recent data derived from direct sequencing of genomic DNA amplified by the polymerase chain reaction revealed many additional

Table 1. Currently known mutations in the CF gene

Name	Amino Acid Change	Nucleotide Change	Exon	Reference
D110H	Asp→His at 110	G→C at 460	4	Dean et al 1990
R117H	Arg→His at 117	G→C at 482	4	Dean et al 1990
R347P	Arg→Pro at 347	C→G at 1173	7	Dean et al 1990
A455E	Ala→Glu at 455	C→A at 1496	9	Kerem et al 1990a
Q493X	Gin→Stop at 493	C→T at 1609	10	Kerem et al 1990a
deltaI507	deletion of Ile at 506/7	3 bp deletion	10	Kerem et al 1990a; Schwarz et al per.comm.
deltaF508	deletion of Phe at 508	3 bp deletion	10	Kerem et al 1989
1717-IG→A	splice mutation	G→A at 1717-1	intron 10	Kerem et al 1990a
G542X	Gly→Stop	G→T at 1756	11	Kerem et al 1990a
S549N	Ser→Asn at 549	G→A at 1778	11	Cutting et al 1990
S549I	Ser→Ile at 549	G→T at 1778	11	Kerem et al 1990a
S549R	Ser→Arg at 549	T→G at 1779	11	Kerem et al 1990a
G551D	Gly→Asp at 551	G→A at 1784	11	Cutting et al 1990
R553X	Arg→Stop at 553	C→T at 1789	11	Cutting et al 1990
A559T	Ala→Thr at 559	G→A at 1807	11	Cutting et al 1990
R560T	Arg→Thr at 560	C→G at 1811	11	Kerem et al 1990a
Y563N	Tyr→Asn at 563	T→A at 1819	12	Kerem et al 1990a
P574H	Pro→His at 574	C→A at 1853	12	Kerem et al 1990a
2566insAT	frameshift	AT insertion after 2566	13	White et al 1990
3659delC	frameshift	deletion of C3659/60	19	Kerem et al 1990a
W1282	Trp→Stop at 1282	G→A at 3978	20	Vidaud et al 1990

mutant alleles. We have so far identified 18 mutations in various locations of the CF gene.

A list of 21 currently known mutations is given in Table 1 (Cutting et al., 1990; Dean et al., 1990; Kerem et al., 1990a; Vidaud et al., 1990; White et al., 1990). The majority (12/21) of the changes are interpreted as missense mutations (D110H, R117H, R347P, A455E, S549N, S549I, S549R, G551D, A559T, R560T, Y563N, P574H), with 7 of them affecting highly conserved amino acid residues (A455, S549, G551, R560) in the first NBF among different ATP-binding proteins (Riordan et al., 1989). D110, R117 and R347 have also been reported to be conserved in several other species (Dean et al., 1990). The two other missense mutations (Y563N and P574H) are expected to cause a drastic change of the encoded amino acids. There are four nonsense mutations (Q493X, G542X, R553X, W128X) and two frameshift mutations (2566insAT and 3659delC) both of which would be predicted to produce truncated proteins. The most interesting one is a 3 bp deletion (deltaI507) removing a single amino acid, isoleucine, immediately upstream of F508 which is deleted in the most common mutation, deltaF508 (deltaI507 was initially thought to be a recurrent deltaF508 mutation). Finally, two mutations are found to reside in the splice junction: one (1717-IG->A) switched the highly conserved AG to AA in intron 10; the other (R560T) which is also a missense mutation substituted AC for AG immediately upstream of the splice acceptor site, creating an alteration known to reduce efficiency of splicing.

Although the above described mutations have yet to be shown to produce a defective protein, none of them have been found on normal chromosomes, arguing for their involvenent in the disease. Further, each of these mutations showed a unique extended DNA marker haplotype, and chromosomes carrying the same mutation appeared to have the same haplotype, suggesting that they probably derived from a common origin.

Among the 21 known mutations, 15 are located in the first NBF, two in the second NBF and one in the R-domain. The asymmetric distribution of mutations between the two NBFs suggests that the first NBF is either more important for the function of CFTR or a mutational hotspot. But the significance of this observation should not be overemphasized at this time because the first NBF is the region that has been most heavily searched for mutations. Nevertheless, many of the missense mutations were found associated with the highly conserved amino acid residues in the two NBFs, suggesting a functional role for these residues. It is intriguing, however, that only a single mutation has been reported for exon 13, which has been thought to be a region important for regulation of CFTR function.

While deltaF508 constitutes about 70% of all CF chromosomes, most of the newly identified mutations appear to be rare. Preliminary data from the consortium reports indicate that three of the mutations (G542X, G551D and A455E) might be more prevalent than others in several populations but knowledge of their actual frequency awaits a more complete survey of all populations. Therefore, while the highly heterogeneous nature of the remaining CF mutations may provide important insights into the structure and function of the protein, it also suggests that DNA-based genetic screening for CF carrier status will be difficult.

In addition to the mutations described above, we have detected three DNA sequence variations that occur on normal chromosomes: two of them, at nucleotide position 1716 (G or A) and 2694 (T or G), neither of which appear to affect the encoded amino acids, but the third one, at position 1540 (A or G) results in either methionine or valine.

CORRELATION BETWEEN GENOTYPE AND PHENOTYPE

We have previously noted that patients homozygous for the deltaF508 mutation tend to have a more severe pancreatic disease (i.e. pancreatic insufficiency). As a result, we predicted that approximately 90% of the CF chromosomes in the general population carry "severe" mutations and 10% carry "mild" ones; patients who are homozygous for the "severe" alleles have a much higher chance of developing PI and those who are heterozygous or homozygous for the "mild" alleles are expected to be PS. The deltaF508 mutation is one of the "severe" alleles. To test this hypothesis, we analyzed 60 patient families in our original study and obtained statistically significant support. We have since increased our study population to more than 293 families and obtained further support for the hypothesis (Kerem et al. 1990b).

In addition, we have examined a number of other clinical parameters for possible correlation between genotype and phenotype. The data so far suggest that the clinical course in CF patients can be attributed to specific genotype at the CF locus and confirm that deltaF508 is a common mutation associated with more severe course of disease typified by pancreatic insufficiency. We emphasize here that the classification of "severe" and "mild" are based on clinical phenotypes and we do not know their correlation with the severity of the molecular defects at this time.

PROTEIN ANALYSIS

CFTR shows remarkable sequence and structural similarity to a variety of proteins involved in transport functions, eg. the mammalian P-glycoproteins, many bacterial transport proteins, the Drosophila eye pigment transporter encoded by the white locus, and the yeast STE6 protein for the transport of "a" mating peptide. In order to produce a large amount of CFTR for biochemical studies, we have reconstructed a full length cDNA (the coding region) under the control of a heterologous promoter in a vector that allows biochemical selection in mammalian cells. In this construct, we have altered nucleotide sequences within exon 6 to avoid frequent rearrangement associated with a 15 bp direct repeat in this region. Transfection studies with mouse L cells and monkey COS-1 cells are in progress.

In addition, using a trpE-construct containing the first nucleotide-binding fold (NBF) and the R-domain, we have been able to show binding of $[_{\alpha}{}^{-32}P]N_3$-ATP to the fusion protein. This result thus confirms the predicted ATP-binding property of CFTR. Binding has also been observed for a similar construct carrying the deltaF508 mutation, suggesting the Phe508 residue is not involved in ATP-binding. Experiments are in progress to study the effect of other known mutations already identified in this region.

CLOSELY RELATED GENE MEMBERS

Preliminary studies in our laboratory suggest that DNA sequences closely related to CFTR exist in the human genome. One of the DNA segments has been isolated from a genomic DNA library and it shows remarkable sequence identity with exon 9 of CFTR. Gel blot hybridization analysis of total human genomic DNA revealed additional bands when other regions of the CF gene are used as probe. They may represent a family of related genes or pseudogenes. Understanding the nature of these sequences is obviously important for the study of CFTR. It will also be crucial to determine the function, if any, of these sequences.

PROMOTER ANALYSIS

We have tested a large series of DNA fragments upstream of the CFTR transcription initiation site for promoter activity in transfection experiments with a CAT reporter gene. After testing many fragments up to 50 kb upstream, we have been able to demonstrate weak promoter activity with a small (500 bp) DNA fragment from a region upstream of the transcription initiation site. Larger fragments containing the 500 bp segment do not have any activity, suggesting that the CFTR promoter contains negative regulatory element(s) in the upstream region. The inclusion of a viral enhancer sequence in the plasmid construct, however, appears to be able to activate the promoter element. Experiments to define the positive regulatory element(s) within the 500 bp DNA fragment and the putative upstream negative regulatory sequence(s) are in progress.

GENERATION OF TRANSGENIC ANIMALS

Two approaches are currently being pursued in our laboratory to exploit the transgenic mouse as a model system to study CF. The first is a gene "knock-out" experiment where a biochemically selectable marker (the neomycin-resistance gene) is introduced into the coding region of the mouse CF gene via homologous recombination. The second experiment involves creating a mouse with the identical 3 bp deletion as the major human mutation. The mouse genomic DNA sequences corresponding to exon 10 of CFTR have been isolated and, in collaboration with Alex Joyner and Janet Rossant (Mount Sinai Hospital, Toronto), appropriate constructs are being introduced into mouse embryonic stem cells for generation of transgenic animals.

CONCLUSION

The identification of the CF gene has provided the first opportunity to understand the basic defect at the molecular level and the pathophysiology of the disease, from which improved means of treatment can be developed. Knowledge of the CF mutations at the DNA sequence level also allows more accurate genetic diagnosis and, for the first time, detection of carriers without previous family history.

Acknowledgement

We would like to acknowledge the research support from the Canadian Cystic Fibrosis Foundation, the Cystic Fibrosis Foundation (USA), the National Institutes of Health (USA) and the Sellers Fund (the Hospital for Sick Children). We are also grateful to the Cystic Fibrosis Foundation (USA) for their special support of the CF Genetic Analysis Consortium.

References

Cutting, G.R., Kasch, L.M., Rosenstein, B.J., Zielensky, J., Tsui, L.-C., Antonarakis, S.E., Kazakian, H.H. Jr., 1990, Characterization of a cluster of disease-producing mutations in the first nucleotide binding fold domain of the CF gene, **Nature**, 346:366.

Dean, M., White, M.B., Amos, J., Gerrard, B., Stewart, C., Khaw, K.-T., Leppert, M., 1990, Multiple mutations in highly conserved residues are found in mildly affected cystic fibrosis patients, **Cell**, 61:863.

Kerem, B., Buchanan, J.A., Durie, P., Corey, M., Levison, H., Buchwald, M., Tsui, L.-C., 1989, DNA marker haplotype association with pancreatic sufficiency in cystic fibrosis, **Am. J. Hum. Genet.**, 44:827.

Kerem, B., Rommens, J.R., Buchanan, J.A., Markierwicz, D., Cox, T.K., Charkravarti, A., Buchwald, M., Tsui, L.-C., 1989, Identification of the cystic fibrosis gene: genetic analysis, **Science**, 245:1073.

Kerem, B., Zielenski, J., Markeiwicz, D., Bozon, D., Gazit, E., Yahav, J., Kennedy, K., Riordan, J.R., Collins, F.S., Rommens, J.M., Tsui, L.-C., 1990a, Identification of mutations in regions corresponding to the two putative nucleotide (ATP)-binding folds of the cystic fibrosis gene, **Proc. Natl. Acad. Sci. USA** (in press).

Kerem, E., Corey, M., Kerem, B., Rommens, J., Markiewicz, D., Levison, H., Tsui, L.-C., Durie, P., 1990b, Association between the major mutation (deltaF508) and phenotypes in cystic fibrosis, **New Eng. J. Med.** (in press).

Lemna, W.K., Feldman, G.L., Kerem, B., Fernbach, S.D., Zevkovich, E.P., O'Brien, W.E., Collins, F.S., Tsui, L.- C., Beaudet, A.L., 1990, Mutation analysis for heterozygote detection and prenatal diagnosis of cystic fibrosis, **N. Eng. J. Med.**, 322:291.

Riordan, J.R., Rommens, J.M., Kerem, B., Alon, N., Rozmahel, R., Grzelczak, Z., Zielensky, J., Lok, S., Plavsic, N., Chou, J.-L., Drumm, M.L., Iannuzzi, M.C., Collins, F.S., Tsui, L.-C., 1989, Identification of the cystic fibrosis gene: cloning and characterization of complementary DNA, **Science**, 245:1066.

Rommens, J.M., Zengerling, S., Burns, J., Melmer, G., Kerem, B., Plavsic, N., Zsiga, M., Kennedy, D., Markiewicz, D., Rozmahel, R., Riordan, J.R., Buchwald, M., Tsui, L.-C., 1988, Identification and regional localization of DNA markers on

chromosome 7 for the cloning of the cystic fibrosis gene, **Am. J. Hum. Genet.**, 43:645.
Rommens, J.M., Zengerling-Lentes, S., Kerem, B., Melmer, G., Buchwald, M., Tsui, L.-C., 1989, Physical localization of two DNA markers closely linked to the cystic fibrosis gene by pulsed field gel electrophoresis, **Am. J. Hum. Genet.**, 45:932.
Rommens, J.M., Iannuzzi, M.C., Kerem, B., Melmer, G., Drumm, M.L., Melmer, G., Dean, M., Rozmahel, R., Cole, J.L., Kennedy, D., Hidaka, N., Zsiga, M., Buchwald, M., Riordan, J.R., Tsui, L.-C., Collins, F.S., 1989, Identification of the cystic fibrosis gene: chromosome walking and jumping, **Science**, 245:1059.
Rommens, J., Kerem, B., Greer, W., Chang, P., Tsui, L.-C., Ray, P., 1990, Rapid non-radioactive detection of the major CF mutation, **Am. J. Hum. Genet.**, 46:395.
Vidaud, M., Fanen, P., Martin, J., Ghanem, N., Nicolas, S., Goossens, M., 1990, Three mutations in the CFTR gene in French cystic fibrosis patients: identification by denaturing gradient gel electrophoresis, **Human Genetics** (in press).
White, M.B., Amos, J., Hsu, J.M.C., Gerrard, B., Finn, P., Dean, M., 1990, A frame-shift mutation in the cystic fibrosis gene, **Nature**, 344:665.
Worldwide survey of the deltaF508 mutation. Report from the Cystic Fibrosis Genetic Analysis Consortium, 1990, **Am. J. Hum. Genet.**, 47:354.

DISCUSSION

BEAUDET: Did you find that you walked from CFTR to clones on other autosomes by using intron probes to screen genomic libraries?

TSUI: Yes.

AL AWQATI: Have you checked whether CFTR is expressed in ES cells?

TSUI: No, we haven't done this experiment yet; the first time we tried we found that the ES cell RNA preparation was not good enough. I don't know whether other people have looked.

COLLINS: Do you have comparisons of the clonability of constructs that either are wild-type or have single nucleotide substitutions to try to make exon 6 more stable? Are you convinced that these substitutions are responsible for the acquisition of stability?

TSUI: I think that the growth properties of these clones become rather different. After changing the nucleotide sequence around the unstable region, the clones become more stable and the colony sizes change and become larger (and hence are easier to grow) and the migration of the DNA fragments is altered slightly. We don't understand the reason for this. It is known that kinks in the DNA structure can alter gel mobility on polyacrylamide gels, and perhaps the change in sequence has caused kink formation. However, there are no sequence data supporting this, so at present this is a poorly understood but confirmed finding.

THE CF GENE PRODUCT AS A MEMBER OF A MEMBRANE TRANSPORTER (TM6-NBF) SUPER FAMILY

John R. Riordan, Noa Alon, Zbyszko Grzelczak,
Steve Dubel and Shi-Zhang Sun

The Hospital for Sick Children and
The University of Toronto, Toronto, Canada

INTRODUCTION

Since the isolation of the CF gene, efforts have been underway to determine the specific function of its protein product. Properties of the molecule inferred from its predicted primary structure (1) provided relatively strong biases as to its likely function. Hence, the presence of several sequence segments, capable of forming hydrophobic alpha helices of appropriate length to cross a lipid bilayer, suggests strongly that it is an integral membrane protein. The existence of hydrophilic domains with similarity to nucleotide-binding sequences of other membrane proteins involved in transport implies that CFTR itself could be a transporter. Of course, at this time, such a suggestion is only conjecture. To reach beyond this stage, functional studies employing antibodies to the protein and expressible cDNAs must be carried out. In the meantime, however, it is instructive to evaluate CFTR from the points of view of the known functions of related members of the superfamily and its possible relation to current models of epithelial ion transport. The need to undertake the latter considerations is emphasized by the fact that it is strongly expressed in tissues which are highly specialized to perform active salt secretion.

The membrane transporter superfamily to which CFTR belongs is briefly outlined in Table I. The basis for inclusion in this group includes both restricted regions of amino acid sequence similarity, especially within the so-called nucleotide-binding folds (NBFs) and what can be termed a general architectural similarity reflected in the apposition of an NBF and a particular membrane-associated domain having six membrane spanning segments

The Identification of the CF (Cystic Fibrosis) Gene
Edited by L.-C. Tsui *et al.*, Plenum Press, New York, 1991

TABLE I

TM6-NBF membrane transporter super family

Bacterial periplasmic uptake systems
Bacterial polypeptide and glycan exporters
Bacterial arsenical efflux pump
Drosophila eye pigment transporters
P-glycoprotein drug transporter
Yeast "a" mating factor exporter
CFTR
Others?

(Figure 1). Each of the members are relatively recently characterized, all having been discovered primarily by genetic means rather than from biochemical and physiological studies of the transport events which they mediate. As a grouping of related families, the classification as a superfamily seems appropriate.

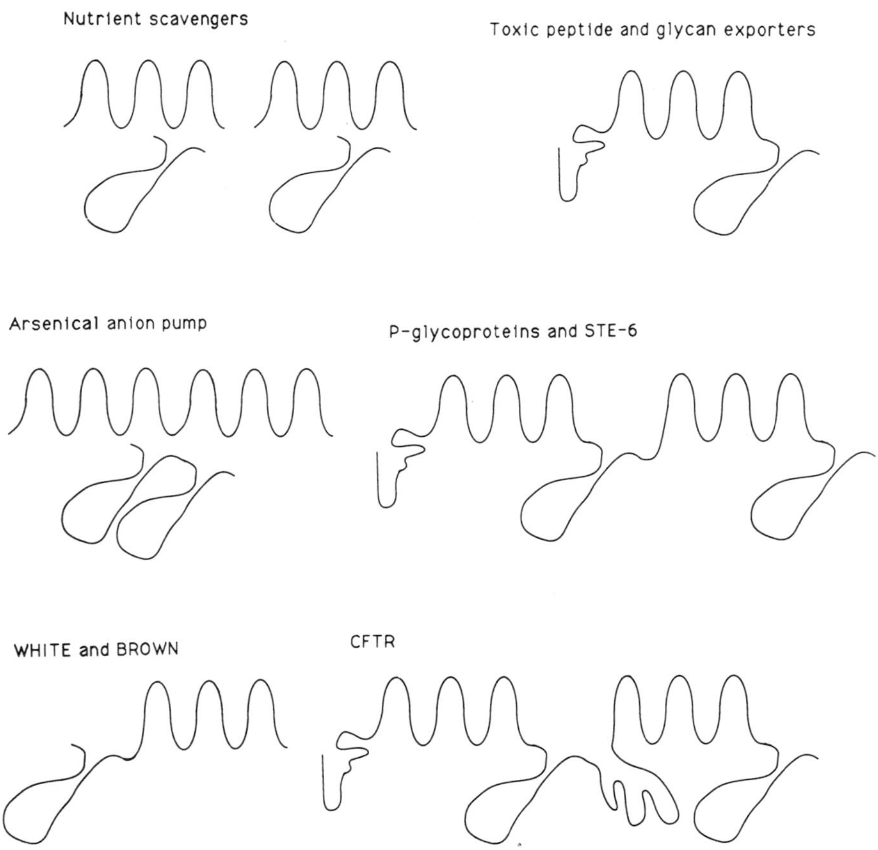

Fig. 1. Schematic depiction of members of the TM6-NBF transporter super family

Hence, the major families are 1.) the large number of periplasmic nutrient transporters of gram negative bacteria (2), 2.) the Drosophilia eye pigment transporter (3,4), 3.) the prokaryotic peptide and glycan exporters (5-10), 4.) the P-glycoprotein multidrug transporters (11) and 5.) CFTR (1). As yet there is no evidence for more than one CFTR gene.

TABLE II

Members of the TM6-NBF membrane transporter super family

Transporter	Substrate	Organism	Up-take	Ex-port	Additional Components Subst. binding	other
Nutrient scavengers	sugars, amino acids, peptides, vitamins, phosphate	E. coli	+	-	+	+
Toxic peptide exporters	hemolysin A	E. coli	-	+	-	+
	leukotoxin A	P.haemolytica	-	+	-	+
	colicin V	E. coli	-	+	-	+
	adenylate cyclase toxin	B.pertussis	-	+	-	+
Cyclic glycan exporters						
ndvA	β-1-2-glucan	Rhizobium	-	+	-	+
chvA	β-1-2-glucan	Agrobacterium	-	+	-	+
Arsenical anion pump	arsenate and arsenite	E.coli	-	+	-	+
WHITE and BROWN gene products	eye pigments	Drosophila	?	?	?	?
P-glycoproteins	hydrophobic compounds including cancer drugs	mammals and malarial parasites	-	+	?	?
STE-6 gene product	"a" mating factor	yeast	-	+	-	+
CFTR	?	mammals and Chondrichthyes	?	?	?	?

CHARACTERISTICS OF RELATED TRANSPORTERS

All of these molecules consist of at least one domain containing six potential membrane spanning sequences (TMs) and an extended NBF. The first family of transporters with this make up are the so-called periplasmic nutrient uptake systems of gram negative bacteria (2). These are highly efficient multi-protein systems which bind their substrates in the periplasm and translocate them into the cytoplasm. Substrates of the different members of this group include sugars such as maltose (12-14) and ribose (15), amino acids such as leucine, isoleucine and valine (16), histidine (17), vitamin B12 (18), phosphate (19) and oligopeptides (20). Most of these systems consist of separate hydrophobic integral membrane polypeptides and hydrophilic nucleotide binding polypeptides. In some instances two copies of each of these polypeptides are thought to constitute the transporter. In others, either or both of the individual NBFs or membrane spanning regions may be fused. In general, however, the distinguishing features of this group as outlined in table I are the high affinity binding periplasmic component which delivers substrate to the bacterial inner membrane complex composed of the membrane spanning domain(s) and associated NBF(s) which, in concert with usually two other membrane-associated polypeptides (2), are responsible for the translocation. The energy for transport is believed to be derived from the hydrolysis of ATP bound at the NBF as both binding and ATPase activities have been observed for several of these transporters (21,22).

The second family contains widely disseminated bacterial cytotoxin secretion systems of which E.coli hemolysin (5) was the first characterized and hence serves as the prototype for the others which include the leukotoxin of P.haemolytica (6), colicin V of E.coli (7), the so-called adenylate cyclase toxin of B.pertussis (8) and many others which have not yet been completely characterized. Like the periplasmic uptake systems, these transporters also consist of several protein components which cooperate functionally. One of these components which is believed to be directly involved in the transport event is the structural counterpart of CFTR, having a single N-terminal TM6 hydrophobic domain and a C-terminal NBF. In the case of the haemolysin system this is hlyB, whereas the other components are hlyA, which is the lysin, hlyD, which is a second membrane protein, and hlyC, which plays a poorly defined role. The function of the latter three proteins is to successfully export hlyA from the bacteria. In total, then, it is an active energy driven peptide export system as are the other members of this group. The actual mechanism of peptide export is little understood at this point.

More recently, a related set of exporters has been identified which bring about the secretion of glycans from certain bacteria. These include the chvA and ndvA gene products of Agrobacterium tumefaciens and Rhizobium species respectively, which contribute to the secretion of cyclic beta-1,2-glucans involved in the attachment of these bacteria to plant cells (9-10). These membrane proteins have the same molecular architecture as hlyB and its other lytic peptide-secreting relatives. Furthermore, as with the toxic peptide exporters, two additional protein components contribute to the function of these polysaccharide transporters.

Particularly interesting from the point of view of the potential involvement of the CF gene product in some aspect of Cl transports is the arsenical anion pump of E.coli (23-25). It brings about the ATP driven efflux of arsenate and arsenite anions from bacteria with plasmids containing the ArsA and ArsB genes (24). ArsB codes for an integral membrane protein containing essentially the equivalent of two TM6 domains. In a similar vein, ArsA codes for a hydrophilic protein containing sequences coding for two NBFs. It is assumed that these two proteins associate and cooperate to bring about the active export of anions, presumably with the membrane spanning elements of ArsB in some manner providing the route across the lipid bilayer, with the energy transduction being accomplished by the ArsA protein. Interestingly, like CFTR, the amino acid content of ArsB would predict a highly basic protein with a preponderance of positively charged amino acid side chains both in the predicted transmembrane segments and in the cytoplasmic loops between them. While it is tempting to speculate that CFTR, by virtue of its possession of these basic structural attributes may also be able to contribute directly to anion translocation, it is important to keep in mind the fact that specific functions of different proteins with similar basic structural motifs may be entirely different. There is also a third component, ArsC, which is a small hydrophilic protein that serves as a modifier of the anion transport.

At this stage there is not a great deal to be learned about the functional properties of this super family of molecules from the WHITE (3) and BROWN (4) genes of Drosophila. The former, the product of an early studied genetic locus (26) has been known much longer than has any other member of the group. Only with their more recent cloning and sequencing was the relationship to the proposed eye pigment transporters discovered. A feature of these molecules which is unique among known members of the super family is that the relative positions of the TM6 and NBF domains are reversed; i.e., the former is C-terminal to the latter (fig. 1). This may imply that the combination of these two domains is sufficient to accomplish transport regardless of the order. This would be consistent with the fact that the two different domains need not even be in the same polypeptide, as is the case in several of the periplasmic uptake transporters (see above).

By far the most similar molecules to CFTR are the P-glycoproteins which were discovered as cancer drug exporters (27). They are rather symmetrical proteins in which there is a significant level of amino acid sequence similarity between their N- and C-terminal halves (28,29), each of which is composed of a TM6 domain followed by an NBF. In fact, this internal homology had led to the suggestion that the complete molecule may have arisen as a gene duplication event (29). However, more recent analysis including the detection of quite different intron/exon structures in the two halves is more consistent with the fusion of related but distinct genes (30). P-glycoprotein genes constitute small families with members ranging from two in man (30) to three in rodents (31) and larger numbers in other species (11). Those most extensively studied are active efflux pumps which export to a diverse range of hydrophobic compounds out of cells (27). The basis of their ability to transport compounds with different chemical structures is not well understood, but many of these compounds have been shown to bind specifically to the protein (32-35). Hence, despite some efforts to detect them, associated substrate-binding components of the type which provide specificity to the periplasmic uptake systems have not been observed. Similarly, there is, at least as yet, no evidence of additional membrane-associated molecules of the type exemplified by hemolysin D which contributes to the active export of substrates. Nevertheless, there is some commonality of substrates between these toxic peptide exporters and P-glycoproteins in the sense that both can transport polypeptides. Although it is not known if there are physiologically significant peptide substrates of P-glycoproteins, ionophoric peptides such as gramicidin D (27) and the immunosupressive cyclosporine A are transported (36). During the past year it was discovered that the lipophilic "a" mating factor peptide of yeast is secreted by the product of the STE6 gene which is quite similar to a P-glycoprotein, having an amino acid identity of approximately 40% in the hydrophilic domains and greater than 20% in the hydrophobic segments (37). The functional similarity of these molecules is also illustrated by the fact that they are both capable of handling the peptide antibiotic valinomycin (38). In general terms, this class of molecules apparently functions primarily to exclude from membranes lipophilic compounds which are capable of partitioning into them.

RELATION OF CFTR TO CURRENT MODELS OF EPITHELIAL ION TRANSPORT

On the basis of the above comparative considerations, it does not seem unreasonable to speculate that CFTR may be a transporter. All other known proteins which combine the TM6 and NBF domains do contribute to a transport function even though one or the other of these motifs are present in many proteins not involved in membrane transport. If CFTR is a transporter, the question of the identity of its substrate arises. Clearly, one possibility would be the Cl^- anion itself, in which case CFTR would be an anion pump. However, such a pump is not immediately accomodated by the model of transepithelial ion movement which is widely accepted (39-41). In its simplest form it involves only three essential elements to accomplish trancellular anion movement in either direction: the energy providing Na/K ATPase, the Na + K + 2Cl cotransporter and a chloride conductive pathway or channel. Only the Na pump enzyme can be absolutely excluded as not being equivalent to CFTR since its primary structure is known (42). However, features of the predicted CFTR structure would not be expected of either the cotransporter or the channel; for example, neither exhibits apparent sensitivity to ATP. Therefore, for CFTR to be directly involved in ion pumping, it presumably would have to serve a secondary role not recognized in the physiological studies on which the model is based. This could be the case, as it is clear that CFTR is strongly expressed in at least some of the principle tissues in which those studies were performed, such as the shark rectal gland (see below). Alternatively, however, if CFTR proves not to be directly involved in ion pumping, it should act on other substrates which could in turn influence transport. It is difficult to anticipate what these might be, and only speculations can be made at this time. One of these, based on analogy with the ability of P-glycoproteins to transport hydrophobic compounds including peptides, would be that an ionophoric compound might be the transported substrate. Such possibilities should become testable in the near future as it becomes possible to introduce expressible forms of CFTR cDNAs into cells.

CFTR IN TISSUES SPECIALIZED FOR SALT TRANSPORT

At this stage, the only reason to think that the CF gene product may be involved in anion transport rather than some other membrane function, is the independent evidence (43-48) that a chloride conductive pathway is defective in the same epithelial tissues of patients which express CFTR. In an attempt to indirectly assess the likelihood that CFTR is involved in the pathway responsible for active chloride secretion, we have monitored its expression in a tissue which is very highly specialized for the performance of this specific function. To do this it was necessary to turn to a more primitive species, the spiney dogfish shark, Squalis acanthias, which employs the rectal gland to secrete the excess chloride which enters the organism from sea water (49,50). To determine if a counterpart of CFTR existed in this organism, low stingency nucleic acid hybridization was performed followed by cDNA cloning (51). The results indicate a high level of expression of a transcript coding for a protein product which is very similar to human CFTR. Overall sequence identity is about 70% and similarity is approximately 80% allowing for conservative amino acid substitutions. All of the predicted CFTR domains are present including the R-domain which distinguishes CFTR from all other known members of the membrane transporter super family to which it belongs. These observations are notable in at least two respects. First, the CF gene is apparently an ancient member of the super family rather than a more recently evolved derivative. The very high level of amino acid sequence similarity over this evolutionary distance is remarkable, indicating that a fairly limited degree of structural variability is compatible with the maintenance of CFTR function. Second, it seems reasonable to deduce that the molecule performs a function which is important to the rectal gland. Since its primary role is the active secretion of salt, these findings are consistent with CFTR being directly involved in that process.

CONCLUSIONS

The CF gene product is a member of a super family of membrane transporters characterized by the combination of two structural domains which presumably link the energy of ATP hydrolysis to the translocation of solute across a membrane. CFTR contains an additional charged hydrophilic domain, rich in potential phosphorylation sites, which is not present in any other known members of the super family. This unique portion of the molecule, possibly as well as the presence of residues with charged side chains in the membrane spanning segments and a concentration of positive charges in the intervening cytoplasmic loops, may provide the molecule with the ability to handle anions directly. Alternatively, CFTR may transport small molecules which play a role in ion permeation. Correspondence between the expression of a highly conserved CFTR counterpart and chloride secretion in the shark rectal gland is supportive of a direct role of the protein in that function.

Acknowledgements

Supported by Canadian and U.S. CF Foundations and the NIH-NIDDK.

References

1. J.R. Riordan, J.M. Rommens, B.-S. Kerem, N. Alon, R. Rozmahel, Z. Grzelczak, J. Zielenski, S. Lok, N. Plavsic, J.-L. Chou, M.L. Drumm, M.C. Iannuzzi, F.S. Collins and L.-C. Tsui, Identification of the cystic fibrosis gene: cloning and characterization of the complementary DNA, **Science**, 245:1066 (1989).
2. G.F.-L. Ames, Bacterial periplasmic transport systems: structure, mechanism and evolution, **Ann. Rev. Biochem.**, 55:397 (1986).
3. K. O'Hare, C. Murphy, R. Levis and G.M. Rubin, DNA sequence of the white locus of Drosophila melanogaster, **J. Mol. Biol.**, 180:437 (1984).
4. T.D. Dreesen, D.H. Johnson and S. Henikoff, The brown protein of Drosophila melanogaster is similar to the white protein and to components of active transport complexes, **Mol. Cell. Biol.**, 8:5206 (1988).
5. T. Felmlee, S. Pellett and R.A. Welch, Nucleotide sequence of an Escherichia coli chromosomal hemolysin, **J. Bacteriol.**, 163:94 (1985).

6. C.A. Strathdee and R.Y-C Lo, Cloning, nucleotide sequence and characterization of genes encoding the secretion function of the Pasteurella haemolytica leukotoxin determinant, **J. Bacteriol.**, 171:916 1989).
7. L. Gilson, H.K. Mahanty and R. Kolter, Four plasmid genes are required for colicin V synthesis, export and immunity, **J. Bacteriol.**, 169:2466 (1987).
8. P. Glaster, H. Sakamoto, J. Bellalou, A. Ullmann and A. Danchin, Secretion of cyclolysin, the calmodulin-sensitive adenylate cyclase-haemolysin bifunctional protein of Bordetella pertussis, **EMJO Journal**, 7:3997 (1988).
9. S.W. Stanfield, L. Ielpi, D.O. Brochta, D.R. Helinski and G.S. Ditta, The ndr A gene product of Rhizobium meliloti is required for β-(1->2) glucan production and has homology to the ATP-binding export protein Hly B. **J. Bacteriol.**, 170:3523 (1988).
10. G.A. Cangelosi, G. Martinetti, J.A. Leigh, C.C. Lee, C. Theines and E.W. Nester, Role of Agrobacterium temefaciens ChrA protein in export of β-1,2-glucan, **J. Bacteriol.**, 171:1609.
11. J.A. Endicott and V. Ling, The biochemistry of P-glycoprotein-mediated multidrug resistance, **Ann. Rev. Biochem.**, 58:137 (1989).
12. E. Gilson, H. Nikaido and M. Hofnung, Sequence of the mal K gene in E.coli K12, **Nucleic Acids Res.**, 10:7449 (1982).
13. S. Frashauer and J. Beckwith, The nucleotide sequence of the gene for mal F protein, an inner membrane component of the maltose transport system of Escherichia coli, **J. Biol. Chem.**, 259:10896 (1984).
14. R. Hengge and W. Boos, Maltose and lactose transport in Escherichia coli: Examples of two different types of concentrative transport systems, **Biochim. Biophys. Acta**, 737:443 (1983).
15. A.W. Bell, S.D. Buckel, J.M. Groarke, J.N. Hope, D.H. Kingsley and M.A. Hermodson, The nucleotide sequences of the rbsD, rbsA and rbsC genes of Escherichia coli K12, **J. Biol. Chem.**, 261:7652 (1986).
16. R. Landick, D.L. Oxender and G.F.-L. Ames, Bacterial amino acid transport systems, **in:** The Enzymes of Biological Membranes, A.N. Martonosi, ed., Plenum, New York (1984).
17. C.F. Higgins, P.D. Hoag, K. Nikaido, F. Ardeshir, G. Garcia and G.F.-L. Ames, Complete nucleotide sequence and identification of membrane components of the histidine transport operon of S.typhimurium, **Nature**, 298:723 (1982).
18. M.J. Friedrich, L.C. De Veaux and R.J. Kadner, Nucleotide sequence of the btu CED genes involved in vitamine B12 transport in Escherichia coli and homology with components of periplasmic-binding protein dependent transport systems, **J. Bacteriol.**, 167: 928 (1986).
19. B.P. Surin, H. Rosenberg and G.B. Cox, Phosphate-specific transport system of Escherichia coli: nucleotide sequence and gene-polypeptide relationships, **J. Bacteriol.**, 161:189 (1985).

20. J.D. Hiles, M.P. Gallagher, D.H. Jamieson and C.F. Higgins, Molecular characterization of the oligopeptide permease of Salmonella typhimurium, **J. Mol. Biol.**, 195:125 (1987).
21. A.C. Hobson, R. Weatherwax and G.F.-L. Ames, ATP-binding sites in the membrane components of histidine permease, a periplasmic transport system, **Proc. Natl. Acad. Sci. USA**, 81:7333 (1984).
22. C.F. Higgins, I.D. Hiles, K. Whalley and D.J. Jamieson, Nucleotide binding by membrane components of bacterial periplasmic binding protein-dependent transport systems, **EMJO Journal**, 4:1033 (1985).
23. C.-M. Chen, T.K. Misra, S. Silver and B.P. Rosen, Nucleotide sequence of the structural genes for an anion pump. The plasmid-encoded arsenial resistance operon, **J. Biol. Chem.**, 261:15030 (1986).
24. C.M. Hsu and B.P. Rosen, Characterization of the catalytic subunit of an anion pump, **J. Biol. Chem.**, 264:17349 (1989).
25. C.-E. Karkaria, C.-M. Chen and B.P. Rosen, Mutagenesis of a nucleotide binding site of an anion-translocating ATPase, **J. Biol. Chem.**, 265:7832 (1990).
26. M. Collins and G.M. Rubin, High frequency precise excision of the Drosophila feedback transposable element, **Nature**, 3032:259 (1983).
27. J.R. Riordan and V. Ling, Genetic and biochemical characterization of multidrug resistance, **Pharmacol. Ther.**, 28:51 (1985).
28. P. Gros, J. Croop and D.E. Housman, Mammalian multidrug resistance gene: complete cDNA sequence indicates strong homology to bacterial transport proteins, **Cell**, 47:371 (1986).
29. C. Chen, J.E. Chin, K. Eueda, D.P. Clark, I. Pastan, M.M. Gottesman and I.B. Roninson, Internal duplication and homology with bacterial transport proteins in the mdr1 P-glycoprotein gene from multidrug-resistant human cells, **Cell**, 47:381 (1986).
30. C. Chen, D. Clark, K. Ueda, I Pastan, M.M. Gottesman and I.B. Roninson, Genomic organization and evolution of the human mdr1 (P-glycoprotein) gene, **J. Biol. Chem.**, 265:506.
31. A. Devault and P. Gros, Two members of the mouse mdr gene family confer multidrug resistance with overlapping but distinct drug specificities, **Mol. Cell Biol.**, 10:1652 (1990).
32. A.R. Safa, C.J. Glover, M.B. Meyers, J.L. Biedler and R.L. Felsted, Vinblastine photoaffinity labelling of a high molecular weight surface membrane glycoprotein specific for multidrug-resistant cells, **J. Biol. Chem.**, 261:6137 (1986).
33. A.R. Safa, Photoaffinity labelling of the multidrug-resistance related P-glycoprotein with photoactive analogs of verapamil, **Proc. Natl. Acad. Sci. USA**, 85:7187 (1988).
34. R. Busche, B. Tümmler, J.R. Riordan and D.F. Cano-Gauci, Preparation and utility of a radioiodinated analogue of daunomycin in the study of multidrug resistance, **Molec. Pharmacol.**, 35:414 (1987).
35. R. Busche, B. Tümmler, D.F. Cano-Gauci and J.R. Riordan, Equilibrium, kinetic and photoaffinity labelling studies of

daunomycin binding to P-glycoprotein-containing membranes of multidrug-resistant Chinese hamster ovary cells, **Eur. J. Biochem.**, 183:189 (1989).
36. L. Slater, P. Sweet, M. Stupecky and S. Gupta, Cyclosporin A reverses vincristine and daunomycin resistance in acute lymphatic leukemia in vitro, **J. Clin. Invest.**, 77:1405 (1986).
37. J.P. McGrath and A. Varshavsky, The yeast STE6 gene encodes a homologue of the mammalian multidrug resistance P-glycoprotein, **Nature**, 340:400 (1989).
38. K. Kuchler, R.E. Sterme and J. Thorner, Saccharomyces cerevisiae STE6 gene product: a novel pathway for protein export in eukaryotic cells, **EMJO Journal**, 8:3973 (1989).
39. R.A. Frizzell, M. Fidd and S.G. Schultz, Sodium-coupled chloride transport by epithelial tissues, **Am. J. Physiol.**, 236:F1 (1979).
40. M. Welsh, Electrolyte transport in thick ascending limb, distal convolution and collecting duct, **Physiol. Rev.**, 67:1143 (1987).
41. R. Gregor, Chloride transport in thick ascending limb, distal convolution and collecting duct, **Ann. Rev. Physiol.**, 50:111 (1988).
42. G.E. Shull, A. Schwartz and J.B. Lingren, Amino acid sequence of the catalytic subunit of the (Na^+ + K^+) ATPase deducted from a complementary DNA, **Nature**, 316:691 (1985).
43. P.M. Quinton, Chloride impermeability in cystic fibrosis, **Nature**, 301:421 (1983).
44. M.R. Knowles, M.J. Stutts, A. Spock, N. Fischer, J.T. Gatzy and R.C. Boucher, Abnormal ion permeation through cystic fibrosis respiratory epithelium, **Science**, 221:1967 (1983).
45. R.A. Schoumacher, R.L. Shoemaker, D.R. Halm, E.A. Tallant, R.W. Wallace and R.A. Frizzell, Phophorylation fails to activate chloride channels from cystic fibrosis airway cells, **Nature**, 330: 752 (1987).
46. M. Li, J.D. McCann, C.M. Liedtke, A.C. Nairn, P. Greengard and M.J. Welsh, Cyclic AMP-dependent protein kinase opens chloride channels in normal but not cystic fibrosis airway epithelium, **Nature**, 331:358 (1988).
47. T.C. Hwant, L. Lo, P.L. Weitlin, D.C. Gruenert, R. Huganir and W.B. Guggino, Cl^- channels in CF: lack of activation by protein kinase C and cAMP-dependent protein kinase, **Science**, 244:1351 (1989).
48. M. Li, J.D. McCann, M.P. Anderson, J.P. Clancy, C.M. Liedtke, A.C. Nairn, P. Greengard and M.J. Welsh, Regulation of chloride channels by protein kinase C in normal and cystic fibrosis airway epithelia, **Science**, 244:1353 (1989).
49. F.H. Epstein, J.S. Stoff and P. Silva, Chloride secretion by shark rectal gland, **in:** Chloride Transport Coupling in Biological Membranes and Epithelia, G.A. Gerencser, ed., Elsevier (1984).
50. R. Greger, E. Schlatter and H. Gögelein, Chloride channels in the luminal membrane of the rectal gland of the dogfish (Squalus acanthias), **Pflügers Arch.** 409:114 (1987).
51. Z. Grzelczak, N. Alon, S. Dubel and J.R. Riordan, unpublished observations.

GENETIC ANALYSIS OF CYSTIC FIBROSIS

Xavier Estivill, Teresa Casals and Virginia Nunes

Department of Medical and Molecular Genetics
Fundació d'Investigació Santa Creu i Sant Pau
Barcelona, Spain

Cystic fibrosis (CF) is the most common severe autosomal recessive disorder affecting Caucasian populations, ocurring in the range of 1:500 to 1:3,000 (Thompson, 1980).

Genetic linkage analysis using DNA markers assigned the CF locus to the long arm of chromosome 7 (q31) (Knowlton et al., 1985; Tsui et al., 1985; Wainwright et al., 1985; White et al., 1985; Scambler et al., 1986; Estivill et al., 1986). Two of these marker loci, D7S8 (probe pJ3.11) and MET, flank the CF locus at a genetic distance of less than 1 cM each (Beaudet et al., 1986; Lathrop et al. 1988).

Chromosome jumping and HTF island cloning allowed the isolation of a generation of markers (XV-2c, CS.7, KM.19 and MP6d-9) in linkage disequilibrium with the disease (Estivill et al., 1987a; 1987b; 1989). Linkage disequilibrium between alleles at those loci and CF improved the informativity of CF families for prenatal diagnosis (Beaudet et al., 1988) and carrier exclusion (Farrall et al., 1987; Krawczak et al., 1988), and suggested that approximately 85% of chromosomes carrying the CF mutation are from a single mutational event (Estivill et al., 1987a; 1987b; 1988; Beaudet et al., 1988). The differences in levels of disequilibrium found in some populations suggested that independent mutations might be present in Southern-European populations.

Many advances in the understanding of the genetics of CF have taken place during the last year. The mutated gene causing CF has finally been isolated and the major mutation identified. A three base-pair (bp) deletion (deltaF$_{508}$ mutation) in the cystic fibrosis transmembrane regulator (CFTR) gene (Riordan et al., 1989) has been found to be associated with the majority of CF cases in North American families (Kerem et al., 1989a; Rommens et al., 1989).

The isolation of the CFTR gene and the identification of the major mutation causing CF allows the direct analysis of the disease for prenatal diagnosis and carrier testing in the population. Haplotype analysis with markers flanking the CFTR gene and within the gene suggests that several additional different mutations, or mutational events, contribute to the pancreatic insufficient (PI) cases (Kerem et al., 1989a; 1989b). Thus, population studies should help to reveal other mutations, and also help in analyzing their effect on the pathophysiology of the disease.

Allelic heterogeneity for CF in Southern European populations has been postulated on the basis of haplotypes with markers closely linked to CF (Estivill et al., 1987b; 1988; 1989b). A preliminary analysis of Spanish and Italian CF families showed that the frequency of the deltaF_{508} mutation is only 46% in these populations (Estivill et al., 1989c). A more detailed analysis of 194 Spanish CF families for both the deltaF_{508} mutation, and the haplotype correlation with markers closely linked to the CFTR gene, (XV-2c/TaqI, pKM.19/PstI and pMP6d-9/MspI), confirms that the described mutation is found in 50% of Spanish CF chromosomes, and that at least four different additional mutation events contribute to the total patient population (Table 1).

Allelic association (linkage disequilibrium) was measured using the Yule's association coefficient: (A) = (ad - bc) / (ad + bc), where a, b, c, and d are the numbers of normal (N) chromosomes with marker allele 1, CF with 1, N with 2, and CF with 2, respectively (Kerem et al., 1989a) (Table 2). Disequilibrium was also measured using the standardised linkage coefficient (data not shown). The difference in frequency of the deltaF_{508} mutation between north American and Spanish population is highly significant ($X^2 = 17.7$, $P < 10^{-5}$).

Four haplotypes have been found associated with the deletion, and at least seven haplotypes do not have the deletion. 98% of chromosomes with the deltaF_{508} deletion are associated with haplotype pKM.19/PstI allele 2, pMP6d-9/MspI allele 2. The deltaF_{508} mutation is also associated with two other haplotypes, which probably derive from the common CF haplotype by a single recombinational event or point mutation.

The CF chromosomes that do not have the deltaF_{508} mutation are associated with four haplotype groups (pKM.19/PstI, pMP6d-9/MspI): 2 2, 1 1, 1 2, and 1 3, with frequencies of 54%, 41%, 4% and 1% respectively. These data suggest that the second major mutation in the Spanish population has occurred in the same relatively uncommon haplotype on normal chromosomes, in which the deltaF_{508} arose. This "ancestral" CF haplotype suggested a single mutational event accounting for the majority of CF chromosomes (about 75%) in the Southern European population (Estivill et al., 1987b; 1988; 1989b). The nature of the other mutation(s) associated with this haplotype remains to be discovered. For the 1 1 haplotype, which accounts for 41% of CF chromosomes without

Table I

DNA marker haplotypes for pXV-2c/TaqI, pKM.19/PstI, and pMP6d-9/MspI, and correlation with the deltaF$_{508}$ mutation.

HAPLOTYPES			CF		N
pXV-2c TaqI	pKM.19 PstI	pMP6d-9 MspI	DeltaF$_{508}$	OTHERS	
1	2	2	122	50	39
2	2	2	13	14	37
-	2	2	5	6	6
1	2	-	11	8	2
1	-	2	5	1	2
2	-	2		1	3
2	2	-		1	5
1	2	1	2		2
2	2	1			5
1	1	2	2	2	14
2	1	2		3	5
-	1	2		1	
1	1	1		23	75
2	1	1		28	91
-	1	1		6	11
2	1	-		4	6
1	1	-			12
1	-	1			1
2	-	1			4
2	1	3		1	1
	Subtotal		160	149	321
	Unclassified				
-	-	-	34	45	67
	Total		194	194	388

CF = cystic fibrosis; N = normal.

the deletion, several mutational events might be expected as it is the commonest haplotype in normal chromosomes. A rare allele for pMP6d-9/MspI was found in one normal and one CF chromosome, which do not have the deltaF$_{508}$ mutation (Nunes et al., 1989).

Strong linkage disequilibrium was found between each allelic system and CF for the chromosomes with the deletion, without the F$_{508}$ mutation and for the overall CF chromosomes. As expected, disequilibrium was strongest on chromosomes with the deletion.

Table 2. Association between DNA markers and CF

Allelic system	CF mutation		
	$deltaF_{508}$	Others	Total
pXV-2c/TaqI	0.85	0.28	0.58
pKM.19/PstI	0.99	0.44	0.76
pMP6d-9/MspI	0.98	0.42	0.74

CF = cystic fibrosis; Values are the absolute Yule's association coefficients (see text).

Data of pancreatic function was available in 79 cases. The proportion of CF-pancreatic sufficiency (CF-PS) patients was lower than in the Canadian population (Kerem et al., 1989a; 1989b). CF-PS patients were either homozygous or heterozygotes for the absence of the $deltaF_{508}$ mutation.

The data presented here has important implications for CF diagnosis, for carrier testing and prenatal diagnosis in the Spanish population. The genotype frequencies in the CF population (31% homozygotes and 38% heterozygotes for the $deltaF_{508}$ mutation) are not in the Hardy-Weinberg equilibrium ($X^2 = 10.90$; $P < 10^{-3}$). This is probably due to a sampling effect caused by regional variations in the populations that are either prevalent for the $deltaF_{508}$ or for other mutations. Before the identification of the F_{508} mutation we had performed 22 prenatal diagnosis of CF using closely linked DNA markers (Casals et al., 1990), and since then 4 further prenatal diagnoses were carried out. The RFLPs used for prenatal diagnosis are shown in Figure 1. For the $deltaF_{508}$ mutation, the samples were analysed on a 6% polyacrylamide gel after PCR (Figure 2). Enzymatic analyses were performed by G. Novelli (Roma).

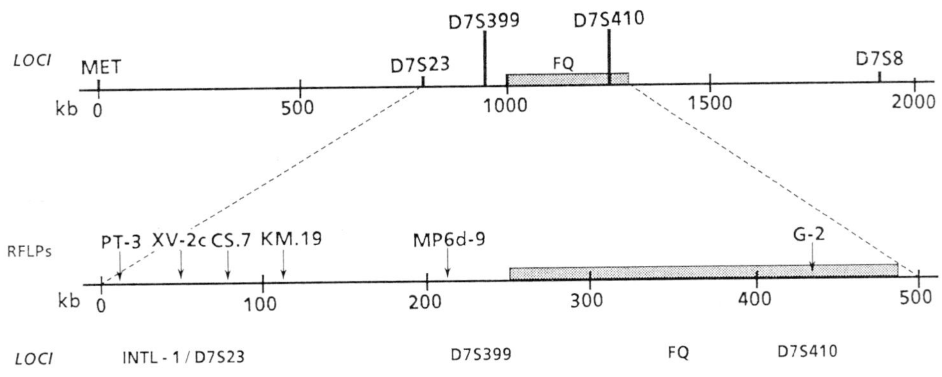

Fig. 1. Schematic representation of the CF locus with the RFLPs detected by probes flanking and within the CFTR gene.

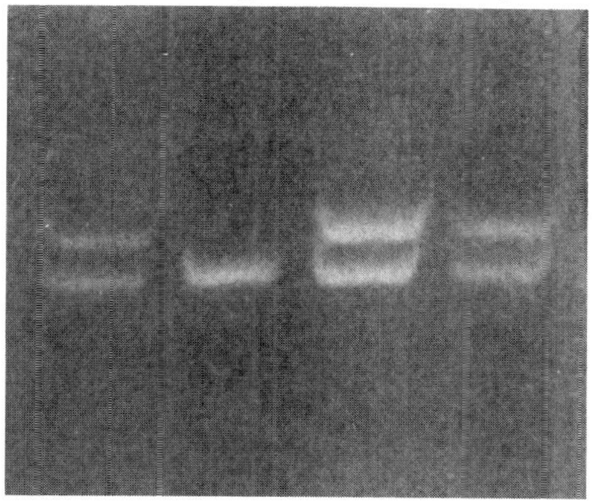

Fig. 2. Detection of the F_{508} mutation by direct visualisation of amplified DNA products on polyacrylamide gel electrophoresis. After amplification of 98/95 base pair (bp) fragments corresponding to the gene sequence containing the F_{508} mutation, 5 μl of formamide-dye mixture (95% formamide / 0.05% bromophenol blue / 0.05% xylene cyanol / 20 mM EDTA) were added to 15 μl of the PCR products. Samples were loaded on to a 1-mm thick, 20 cm x 20 cm, 8 M urea, 12% polyacrylamide gel in 1 x TBE buffer. Electrophoresis was performed at 500 V, for 2 hours.
1: chorionic villus sample; 2: cystic fibrosis; 3: father; 4: mother.

For all families in which there was a living CF child, prenatal diagnosis was possible with linked markers. Only in three families the diagnosis was performed with the enzymatic test; two of these families were 50% informative for the deltaF_{508} mutation such that a carrier or an affected child could not be distinguished; the third family was not informative for the mutation. In retrospect, for the families analysed with linked markers, 10 were fully informative for the mutation, and 9 were 50% informative. Thus, the deletion analysis by itself would only have allowed prenatal diagnosis in 58% of the families. However, the analysis performed in 194 Spanish CF families shows that only 50% of families will benefit from prenatal diagnosis by direct detection of the deltaF_{508} mutation.

Although much progress has been made in the understanding of the genetics of CF, only half of CF carriers can be detected in the Spanish population using the deletion test. Until the other mutations causing CF are identified, linkage analysis with closely linked markers should still be mandatory for diagnostic purposes in about half of the Spanish families. Thus, genetic screening in the general population should only be available for families that are particularly worried or at high risk for the disease. In addition, the haplotype data described here provide additional information about the origin and number of CF mutations, which should help in the molecular description of CF defects and contribute to the understanding of the pathophysiology of the disease.

Acknowledgments

This study was supported by the "Fondo de Investigaciones Sanitarias de la Seguridad Social" (89/0563 and F.J.G.), the "Direccion General de Investigación Científica y Técnica" (PB87-0074 and PB88-0340), and the "Institut Català de la Salut". X.E., and V.N. are investigators of the "Fundació d'Investigació Santa Creu i Sant Pau", Barcelona.

References

Beaudet, A., Bowcock, A., Buchwald, M., Cavalli-Sforza, L., Farrall, M., King, M.-C., Klinger, K., Lalouel, J.-L., Lathrop, G., Naylor, S., Ott, J., Tsui, L.-C., Wainwright, B., Watkings, P., White, R., Williamson, R., 1986, Linkage of cystic fibrosis to two tightly linked DNA markers: joint report from a collaborative study, **Am. J. Hum. Genet.**, 39:681.

Beaudet, A.L., Spence, J.L., Montes, M., O'Brien, M., Estivill, X., Farrall, M., Williamson, R., 1988, Experience with new DNA markers for diagnosis of cystic fibrosis, **New Engl. J. Med.**, 318:50.

Casals, T., Nunes, V., Giménez, J., Parra, J., Estivill, X., Diagnóstico prenatal de fibrosis quística en familias españolas utilizando marcadores del DNA: nuestra experiencia 1987-1989. **An. Esp. Ped.** (in press).

Estivill, X., Schmidtke, J., Williamson, R., Wainwright, B.J., 1986, Chromosome assignment and restriction fragment length polymorphism analysis of the anonymous DNA probe B79a at 7q22 (HGM8 assignment D7S13), **Hum. Genet.**, 74:320.

Estivill, X., Farrall, M., Scambler, P.J., Bell, G., Hawley, K.M.F., Lench, N.J., Bates, G.P., Kruyer, H.C., Frederick, P.A., Stanier, P., Watson, E.K., Williamson, R., Wainwright, B.J., 1987a, A candidate for the cystic fibrosis locus isolated by selection for methylation-free islands, **Nature**, 326:840.

Estivill, X., Scambler, P.J., Wainwright, B.J., Hawley, K., Frederick, P., Baiget, M., Kere, J., Williamson, R., Farrall, M., 1987b, Patterns of polymorphism and linkage disequilibrium for cystic fibrosis, **Genomics**, 1:257.

Estivill, X., Farrall, M., Williamson, R., Ferrari, M., Seia, M., Giunta, A.M., Novelli, G., Potenza, L., Dallapicolla, B., Borgo, G., Gasparini, P., Pignatti, P.F., De Benedetti, L., Vitale, E., Devoto, M., Romeo, G., 1988, Linkage disequilibrium between cystic fibrosis and linked DNA markers in Italian Families: a collaborative study, **Am. J. Human Genet.**, 43:23.

Estivill, X., McLean, C., Nunes, V., Casals, T., Gallano, P., Scambler, P., Williamson, R., 1989a, Isolation of a New DNA marker in linkage disequilibrium with cystic fibrosis situa-

ted between J3.11 (D7S8) and IRP, **Am. J. Human Genet.**, 44:704.
Estivill, X., Gasparini, P., Novelli, G., Casals, T., Nunes, V., Gallano, P., Savoia, A., Ruzzo, A., Dallapiccola, B., Pignatti, P.F., 1989b, Linkage disequilibrium for DNA haplotypes near the cystic fibrosis locus in two South European populations, **Hum. Genet.**, 83:175.
Estivill, X., Chillón. M., Casals, T., Bosch, A., Morral, N., Nunes, V., Gasparini, P., Seia, A., Pignatti, P.F., Novelli, G., Dallapiccola, B., Fernández, E., Benítez, J., Williamson, R., 1989c, deltaF$_{508}$ gene deletion in cystic fibrosis in Southern Europe, **Lancet**, II:1404.
Farrall, M., Estivill, X., Williamson, R., 1987, Cystic Fibrosis: Indirect Carrier Detection, **Lancet**, 2:156.
Kerem, B.-S., Rommens, J.M., Buchanan, J.A., Marckiewicz, D., Cox, T.K., Chakravarty, A., Buchwald, M., Tsui, L.-C., 1989a, Identification of the cystic fibrosis gene: genetic analysis, **Science**, 245:1073.
Kerem, B.-S., Buchanan, J.A., Durie, P., Corey, M.L., Levison, H., Rommens, J.M., Buchwald, M., Tsui, L.-C., 1989b, DNA marker haplotype association with pancreatic sufficiency in cystic fibrosis, **Am. J. Hum. Genet.**, 44:827.
Knowlton, R.G., Cohen-Haguenauer, O., Van Cong, N., Frazel, J., Brown, V.A., Barker, D., Braman, J.C., Schumm, J.W., Tsui, L.-C., Buchwald, M., Donis-Keller, H., 1985, A polymorphic DNA marker linked to cystic fibrosis is located on chromosome 7, **Nature**, 318:380.
Krawczak, M., Konecki, D.S., Schmidtke, J., Dück, M., Nützenadel, W., Trefz, F.K., 1988, Allelic association of the cystic fibrosis locus and two DNA markers, XV-2c and KM.1, in 55 German families, **Hum. Genet.**, 80:78.
Lathrop, G.M., Farrall, M., O'Connell, Wainwright, B., Leppert, M., Nakamura, Y., Lench, N., Kruyer, H., Dean, M., Park, M., Vande Woude, G., Lalouel, J.-M., Williamson, R., White, R., 1988, Refined linkage map of chromosome 7 in the region of the cystic fibrosis gene, **Am. J. Hum. Genet.**, 42:38.
Nunes, V., Casals, T., Gallano, P., Gimenez F.J., Kere, J., Williamson, R., Estivill, X., 1989, Detection of a rare allele with pMP6d-9/MspI RFLP near the cystic fibrosis locus, **Hum. Genet.**, 83:305.
Riordan, J.R., Rommens, J.M., Kerem, B.S., Alon, N., Rozmahel, R., Grzelczak, Z., Zielenski, J., Lok, S., Plasvic, N., Chou, J.L., Drumm, M.L., Iannuzzi, M.C., Collins, F.S., Tsui, L.-C., 1989, Identification of the cystic fibrosis gene: Cloning and characterization of complementary DNA, **Science**, 245:1066.
Rommens, J.M., Iannuzzi, M.C., Kerem, B.-S., Drumm, M.L., Melmer, G., Dean, M., Rozmahel, R., Cole, J.L., Kennedy, D., Idaka, N., Zsiga, M., Buchwald, M., Riordan, J.R., Tsui, L.-C., Collins, F.S., 1989, Identification of the cystic fibrosis gene: chromosome walking and jumping. **Science**, 245:1059.
Scambler, P.J., Wainwright, B.J., Watson E., Bates, G., Bell, Williamson, R., Farrall, M., 1986, Isolation of a further

anonymous informative DNA sequence from chromosome seven closely linked to cystic fibrosis, **Nucleic Acids Res.**, 14:1951.

Thompson, M.W., 1980, Genetics of cystic fibrosis, **In:** "Perspectives in cystic fibrosis," J.M. Sturgess, ed., Canadian Cystic Fibrosis Foundation, Toronto.

Tsui, L.-C., Buchwald, M., Barker, D., Braman, J.C., Knowlton, R., Schumm, J.W., Eiberg, H., Mohr, J., Kennedy, D., Plasvic, N., Zsiga, M., Markiewicz, D., Akots, G., Brown, V., Helms, C., Gravius, T., Parker, C., Rediker, K., Donis-Keller, H., 1985, Cystic fibrosis locus defined by a genetically linked polymorphic DNA marker, **Science**, 230:1054.

Wainwright, B.J., Scambler, P.J., Schmidtke, J., Watson, E.K., Law, H.-Y., Farrall, M., Cooke, H.J., Eiberg, H., Williamson, R., 1985, Localization of cystic fibrosis locus to human chromosome 7cen-q22, **Nature**, 318:384.

White, R., Woodward, S., Leppert, M., O'Connell, P., Hoff, M., Herbst, J., Lalouel, J.-M., Dean, M., Vande Woude, G.F., 1985, A closely linked marker for cystic fibrosis, **Nature** 318:382.

DISCUSSION

WILLIAMSON: I found it remarkable that there was equilibrium for those new polymorphisms that are right in the CFTR gene. Are they in disequilibrium with the deltaF508 mutation?

NUNES: We have just recently found these polymorphisms and we don't know yet.

COLLINS: The new ScrfI polymorphism you describe: if you study non-deltaF508 chromosomes, what is the comparison between normal and CF? Is the disequilibrium strong enough to be useful clinically?

NUNES: We are investigating this at present; ScrfI is always in high disequilibrium with CF chromosomes, whether or not they are deltaF508.

TSUI: Did you look at the ScrfI polymorphism in non-Spanish populations?

NUNES: We have also studied the British population with Michele Ramsay (St. Mary's London); the informativity and disequilibrium is similar.

MOLECULAR AND GENETIC ANALYSES AT THE CF LOCUS

K.W. Klinger, P. Stanislovitis, J. Merrill
and G.T. Horn

Integrated Genetics, Framingham, Mass USA

Cystic fibrosis (CF) is the most frequent lethal autosomal recessive disorder of caucasian populations. The disease incidence is commonly quoted as 1 in 2500 (Boat et al., 1989), but varies widely in different populations. Frequencies as low as 1:90,000 exist (Wright and Morton, 1968), and as high as 1:377 in Brittany (Bois et al., 1978), 1:313 in the Alberta Hutterites (Fujiwara et al., 1989), and 1:569 in the Ohio Old Order Amish (Klinger 1983). There has always been some question as to whether these diverse populations represented the same mutation.

The CF gene has now been isolated, and the major mutation shown to be a 3 base pair deletion in exon 10 (Rommens et al, 1989; Riordan et al, 1989; Kerem et al, 1989). This deletion, referred to as deltaF508, was originally reported to be carried on 68% of CF chromosomes (Kerem, 1989). More recent studies have indicated that as many as 75.8% of CF chromosomes may carry this mutation (Lemna, 1990). Similar results were obtained in our own analyses of unselected CF populations (see Table 1.) In these and other studies, the deltaF508 mutation occurred mainly on chromosomes with the XV2c/KM 19 haplotype 'B' (allele 1 at XV2c, allele 2 at KM19.) The data shown in Table I yield individual haplotype frequences of B/deltaF508 of 62.7%.

We have previously studied CF families in a number of inbred populations using a variety of RFLPs linked to the CF locus. For example, a Hutterite Brethren pedigree is shown in Fig 1. The marker haplotypes in many cases predicted that multiple mutations were segregating in these populations, and that at least some of these mutations might be rare in other populations.

In these families CF was associated with the "B", "C" and "D" XV2c/KM19 haplotype (Fujiwara et al., 1989). We predicted that three different CF chromosomes were present in the

TABLE I

HAPLOTYPE	CFTR*	#	Fraction
BB	FF	24	39.3%
BB	NF	18	29.5%
AB	NF	6	9.8%
BB	NN	4	6.6%
BD	FF	3	4.9%
BC	NF	3	4.9%
BD	NF	2	3.3%
AA	NF	1	1.6%

* "F" = the deltaF508 mutation, "N" = an unknown mutation. These data were derived from 61 phase known CF cases, for a total of 122 CF chromosomes tested.

Hutterite population, since examination of a 13 marker extended haplotype showed that it was unlikely that the "C" chromosome was derived by recombination subsequent to the founding of the Hutterite population. The discovery of the CF gene and the deltaF508 mutation has allowed us to test these hypotheses.

DNA amplification by the polymerase chain reaction (PCR) and analysis using allele-specific oligonucleotides (ASOs) were used to test genomic DNA for the presence of deltaF508. Primer GH353 (5'-GACTTCACTTCTAATGATGATT-3') is located at positions 1524-1545 in the "+" strand of the cDNA and primer GH353 (5'TCTTCTAGTTGGC-ATGCTTT-3') is located at positions 1696-1715 in the "-" strand. These primers combine to produce a 192-bp PCR product.
The annealing step of PCR was carried out at 45°C. A specific amplification product was visible after 28 cycles of PCR with 1μg of genomic DNA and using standard PCR conditions (Saiki et

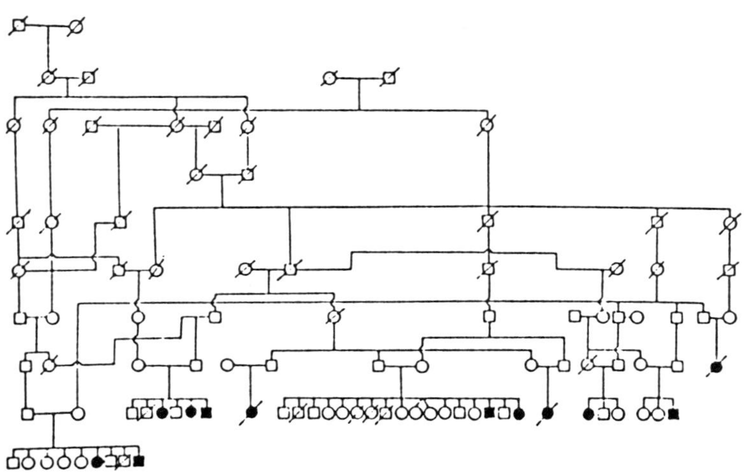

Fig. 1. Segregation of CF in a Hutterite Pedigree. Filled symbols represent affected individuals.

al., 1988). The synthetic ASO probes were based on the known sequence of the normal and deltaF508 alleles (Riordan et al., 1989). Probe GH356 (5'AAGAAAATATCATCTTTGGTGTTTC-3') is from the "+" strand of the normal cDNA, while probe GH357 (5'-AGGAAACACCAATGATATTTTCTT-3') is from the "-" strand of the mutant allele. The amplified DNA was analyzed in a dot-blot format using standard hybridization and washing conditions (Saiki et al. 1986), except that both probes were washed for 15 min. in 0.2% SSC/0.1% SDS at 44°C. Genomic DNA was analyzed from at least one CF child and at least one CF parent from each of the 10 Hutterite families shown in Fig. 1. One of the Hutterite CF haplotypes was the "B" haplotype commonly associated with deltaF508, was predicted to carry the common mutation, and indeed was shown to have deltaF508. The other two haplotypes lacked the deltaF508 mutation, as predicted from the haplotype data (Klinger et al. 1990). These CF chromosomes carry an as yet unidentified mutation.

In contrast, the situation is quite different in outbred patient populations. Our goal in these unselected CF populations is to both search for new mutations, and to screen these populations for sequence changes identified by members of the CF Genetic Analysis Consortium as potential mutations. In general, DNA amplification by PCR and analysis using ASOs were used to test genomic DNA for the presence of deltaF508 or other known sequence changes. 262 CF chromosomes from unselected North American CF families were tested. 186 were found to carry deltaF508, whereas 4 other mutations were found on a total of only 14 CF chromosomes.

We have been searching for additional mutations that account for CF in the following manner. Using synthetic primers within or flanking certain exons of the CFTR gene. PCR is used to amplify DNA from CF-affected individuals with one or two non-deltaF508 mutations. The amplified products are then cross-hybridized with labeled products from a non-CF individual, and heteroduplexes detected with a modified form of the hydroxylamine/osmium tetroxide chemical cleavage method (Cotton, 1988). At present we have searched exons 6, 10. 13, 17,and 19 for additional mutations. Although no coding sequence variations have yet been found, several polymorphisms and a potential splice site mutation were seen near exon 6. The PCR product from exon 6 was approximately 1 Kb larger than expected. We cloned and sequenced the exon-6 PCR products from several individuals, and found an IVS of approximately 1130 bp in length following nucleotide 875 of the cDNA sequence, separating what actually are exons 6A and 6B. The intron sequence and the various polymorphic changes are shown in Fig. 2.

The 3' end of the intron contains a small VNTR polymorphism, while the 5' end of the intron contains a diallelic polymorphism. We also detected what appeared to be a potential splice site mutation. Further studies suggest that this A→G variation is in fact a cloning artifact.

```
AlaLeuPheGlnAlaGlyLeuGlyArgMetMetMetLysTyrAr (g)
GCCCTTTTTCAGGCTGGGCTAGGGAGAATGATGATGAAGTACAGgtagcaacctattttca
                      (g)
taacttgaaagttttaaaaattatgttttcaaaagcccactttagtaaaaccaggactgc
                                      |----- Alu repeat --->
tctatgcatagaacagtggtcttcagtgtcattaaatttttttttttttttttttttttga
gacagagtctagatctgtcatcccaggctggagtgcagtggcacgattctggctcactgc
actgcaacttctgcctcccaggctcaagcaattctcctgcctacgcctccggagatgctg
ggattagaggcgacatgaccacccacacccagcaatcttttgtatttacgatagagacgg
gttacga ... cctagtcaagccacttcacctcactgagtcttgcttttcatctctaaa
tagagatacccaccgctcataggctgtcatcagggatagagatagcatatggaatgagtc
tgtacagcgtctggcacataggaggcatttaccaaacagtagttattattttgttacca
tctatttgataataaaataatgcccatctgttgaataaaagaaatatgacttaaaacctt
gagcagttcttaatagataatttgacttgtttttactattagattgattgattgattgat
                                        <--><--><--><--><-->
(      )     gAspGlnArgAlaGlyLysIleSerGluArgLeuValIleThrSer
gattgatttacagAGATC~GAGAGCTGGGAAGATCAGTGAAAGACTTGTGATTACCTCA
<--> <-->
```

Fig. 2. Sequence of the intron in 'exon 6', showing the various polymorphisms detected.

Acknowledgements

This work was supported in part by grants from the Cystic Fibrosis Foundation and NIH HD25348.

We thank M.A. Anderson, G. White. R. Schwartz, R.A. Doherty, S.R. Miller, N. Stuart, P.C. Watkins, M. Fujiwara, and K. Morgan for their contributions to these studies, and the clinicians and patients whose participation made this work possible.

References

Boat, T.F., Welsh, M.J., Beaudet, A.L., **in:** "The Metabolic Basis of Inherited Diseases," C.L. Scriver, A.L. Beaudet, W.S. Sly, D. Valle, eds., McGraw-Hill, New York.

Bois, E., Feingold, J., Demenais, F., Runavot, Y., Jehanne, M. and Toudic L., 1978, Cluster of cystic fibrosis cases in a limited area of Brittany (France), **Clinical Genetics**, 14:73.

Cotton, R.G.H., Rodrigues, M.R. and Campbell, R.D., 1988, Reactivity of cytosine and thymine in single-base pair mismatches with hydroxylamine and osmium tetroxide and its application to the study of mutations, **Proc. Natl. Acad. Sci. USA**, 85:4397.

Fujiwara, T.M., Morgan, K., Schwartz, R.H., Doherty, R.A., Miller, S.R., Klinger, K., Stanislovitis, Stuart, N. and Watkins, P.C., 1989, Genealogical Analysis of Cystic Fibrosis Families in the Hutterite Brethren, **Am. J. Hum. Genet.**, 44:327.

Kerem, B., Rommens, J.M., Buchanan, J.A., Markiewicz, D., Cox, T.K., Chakravarti, A., Buchwald, M. and Tsui, L.-C., 1989,

Identification of the Cystic Fibrosis Gene: Genetic Analysis, **Science**, 245:1073.
Klinger, K.W., 1983, Cystic Fibrosis in the Ohio Amish: Gene Frequency and Founder Effect, **Hum. Genet.**, 65:94.
Lemna, W.K., Feldman, G.L., Kerem, B., Fembach, S.D., Zevkovich, E.P., O'Brien, W.E., Riordan, J.R., Collins, F.S., Tsui, L.-C. and Beaudet, A.L., 1990, Mutation analysis for heterozygote detection and the prenatal diagnosis of cystic fibrosis, **New. Eng. J. Med.**, 322:291.
Riordan, J.R., Rommens, J.M., Kerem, B.-S., Alon, N., Rozmahel, R., Grzelczak, J., Lok, S., Plavsic, N., Chou, J.-L., Drumm, M.L., Ianuzzi, M.C., Collins, F.S., Tsui, L.-C., 1989, Identification of the Cystic Fibrosis Gene: Cloning and Characterization of Complementary DNA, **Science**, 245: 1066.
Rommens, J., Ianuzzi, M.C., Kerem, B.-S., Drumm, M.L., Melmer, G., Dean, M., Rozmahel, R., Cole, J.L., Kennedy, D., Hidaka, N., Zsiga, M., Buchwald, M., Riordan, J., Tsui, L.-C. and Collins, F.S., 1989, Identification of the Cystic Fibrosis Gene: Chromosome Walking and Jumping, **Science**, 245:1059.
Wright, S.W., Morton, N.E., 1968, Genetic Studies on Cystic Fibrosis in Hawaii, **Am. J. Hum. Genet.**, 20:157.

DISCUSSION

DEAN: I wonder if you could comment on the clinical characteristics of the Hutterite cases, and the correlation with the DNA mutation?

KLINGER: They look like classical CF, with high sweat electrolyte elevation, severe lung disease, and pancreatic insufficiency. Why they don't seem to be showing up any of the mutations one would expect is really not clear. They went through a bottleneck in their first round of immigration to North America, and we do not understand why they do not have one of the mutations which is common in central European CF chromosomes.

WILLIAMSON: What about your original Amish family. Do they have deltaF508 chromosomes, or are they another mutation?

KLINGER: The Old Order Amish of southern Ohio are almost 100% deltaF508. The Amish-Mennonite are more of a grab-bag, and have so many more ancestral chromosomes that it is not easy to make predictions about how many chromosomes there ought to be. They are segregating the deltaF508 mutation at about 60% or 70%. The Louisiana Cajuns are much the same, but in both cases the distribution of other chromosomes is quite skewed, with a lot of A-haplotypes.

KNOWLES: How precise can you be about the European geographical origin of the Hutterites?

KLINGER: Not as precise as I can be about the Amish, because I have better personal records on the Amish community, for whom I know which Canton in Switzerland each particular community came from. The Hutterites are mostly of German origin. When they left, there was a split, where a small percentage of the population went to Russia and then was forced to leave in the late 1800s, the final wave of immigration to the U.S. and Canada. I couldn't guarantee that there were not some non-German Russians that got recruited into the population and came with them in that last wave of immigration to the States.

COUTELLE: On the cloning problems, do you think that there is a structural problem in the DNA sequence, or is the protein toxic?

KLINGER: I think, as does Francis Collins, that the protein is toxic.

TSUI: We don't know whether it is. The cloning vectors we use are supposedly not expressing the protein, so I don't know whether there are any other cryptic promoters making a protein in these constructs.

KLINGER: If you look at the sequence of exon 6, there is quite a good consensus promoter sequence there. If you deliberately express the protein, it makes your bugs quite ill.

COLLINS: I agree, there is a nice prokaryotic promoter on exon 6. The other thing that leads us to think that this is protein mediated is the single nucleotide insertion at the beginning of exon 8 which also rescues toxicity. If this were DNA mediated I would think all the rescuing mutations would cluster in one "toxic" stretch of DNA; this one doesn't. We have also heard of another rescue mutation today from Transgene, which is even further out.

KLINGER: It seems like anything you do that interrupts correct expression gives a stable clone.

BEAUDET: Has anyone examined the complete sequence with deltaF508 to see if is toxic or not?

RIORDAN: This entire group of membrane proteins is known to be very toxic in bacteria - this is true for glucose transporters, lac transporters, and dozens of other family members. There are no exceptions from toxicity if they are deliberately expressed in E.coli. So my suspicion is that CFTR is toxic, whether it contains deltaF508 or not.

IDENTIFICATION OF CYSTIC FIBROSIS MUTATIONS

Michael Dean,(1) Bernard Gerrard,(1)
Claudia Stewart,(1) Leslie Krueger,(2)
Douglas Holsclaw,(2) Lynne Quittell,(3)
Vladislov Baranov,(4) Nicolai Kapranov,(5)
Mark Leppert,(6) Jean Amos,(7) and Marga White (3)

(1) Biological Carcinogenesis and Development Program
Program Resources, Inc., NCI-FCRF, Frederick, MD
(2) Hanneman University, Philadelphia, PA
(3) Columbia University Medical Center, New York, NY
(4) Institute of Obstetrics and Ginecology, Leningrad
(5) Institute of Pediatrics, Moscow
(6) Howard Hughes Medical Institute
University of Utah, Salt Lake City, UT
(7) Center for Human Genetics, Boston University
Boston, MA
(8) Laboratory of Viral Carcinogenesis
National Cancer Institute
Frederick Cancer Research Facility
Frederick, MD

Using oligonucleotide primers from the sequence of the cystic fibrosis transmembrane conductance regulator (CFTR) gene we have employed the polymerase chain reaction to amplify several coding exons. These regions have been examined from 110 patients that contain 127 chromosomes without the common CF mutation (DeltaF_{508}). Eight additional mutations have been identified in this group, in a total of four different exons. Most of the mutations were initially identified using an assay for single-stranded conformation polymorphisms. All mutations were subsequently characterized by direct sequencing of the amplified DNA, and can be assayed by restriction enzyme digestion or allele-specific oligonucleotide hybridization.

The mutations fall into two classes: (1) One or two nucleotide insertions and deletions that introduce termination codons into the gene and are predicted to result in severely truncated protein products; (2) Point mutations in the putative membrane-spanning domains that replace charged amino acids with non-polar residues. The clinical phenotypes of these patients and the ethnic distribution of the mutations provide insight into the origin and effect of CF mutations.

The Identification of the CF (Cystic Fibrosis) Gene
Edited by L.-C. Tsui *et al.*, Plenum Press, New York, 1991

INTRODUCTION

Cystic fibrosis (CF) was not clinically recognized as a distinct disease until the 1930s (Anderson 1938). Pedigree analysis established CF as an autosomal recessive disorder, and di Sant'Agnese (1953) found that the sweat of CF patients contains an excess of sodium and chloride ions. Defects in the regulation of chloride ion transport have been documented in CF epithelial cells (Knowles et al., 1981; Quinton and Bijman, 1983; Sato and Sato, 1984; Welsh and Liedkte, 1986). The chloride channel normally responds to β-adrenergic agents; however, CF cells are defective in this response (Quinton, 1983; Knowles et al., 1983; Sato and Sato, 1984). It has been proposed that the CF defect involves a pathway whereby cAMP regulates ion transport.

The clinical presentation of CF is heterogeneous between families and even within families (Rosenstein and Langbaum, 1984). Although most patients are diagnosed by the time they reach adolescence, a few remain undiagnosed until adulthood (di Sant'Agnese and Davis, 1979; Su and Beanblossom, 1989). Approximately 15% of CF patients display a milder form of the disease, with the principal feature being a reduced requirement for pancreatic enzymes (pancreatic sufficiency, PS). PS is usually concordant within families suggesting that these patients may have less severe mutations in the CF gene.

The cloning of the CF gene has considerably increased the potential research strategies for further understanding of the disorder. The gene encodes a 170 kd protein related to a family of membrane-bound active transport molecules (Rommens et al., 1989; Riordan et al., 1989; Kerem et al., 1989). A three-nucleotide deletion in a putative ATP binding domain has been found in 70% of CF chromosomes; this alteration removes a phenylalanine codon at position 508. To further understand the relationship between mutations in the gene and the phenotype of patients, we have examined a group of patients who do not contain the common mutation on one or both chromosomes.

RESULTS

While several methods have been developed for detecting small alterations in DNA sequence, they all involve considerable effort or expensive equipment. We initially attempted to identify CF mutations using an approach developed in the search for the gene (Dean et al., 1990). Specific regions of the CF gene were amplified by the polymerase chain reaction (PCR) and digested with restriction enzymes, and the DNA fragments resolved on high resolution acrylamide gels. Using this approach, a two nucleotide insertion was discovered in a single patient (CF Ins 2566, White et al., 1990). The mutation is carried by the mother and transmitted to each of her three offspring. The older sisters each inherited the father's normal chromosome and are free of disease. The affected child inherited a CF chromosome from the father that also does not contain the common mutation. The

CF Ins 2566 mutation is predicted to shift the reading frame of the protein and cause the introduction of a termination codon. This would lead to the production of a severely truncated protein. However, at present this child is over seven years of age and has a normal weight.

To search for additional mutations we sought a method that would be both sensitive and rapid. Orita et al. (1989) described a technique that allows point mutations to be detected in short stretches of DNA. The method detects single-stranded conformation polymorphisms (SSCP) by denaturing the DNA and resolving it on non-denaturing acrylamide gels. Each strand of the DNA fragment can form a unique conformation, and any mutation(s) within that segment can potentially affect its mobility.

We have used SSCP analysis to analyze 127 CF patients who have at least one CF chromosome that does not contain the common (DeltaF$_{508}$) mutation. Using the sequence of the CFTR gene, primers were chosen to amplify coding regions. We chose the larger exons of the gene for our initial analysis. Each patient that displays an aberrently migrating fragment on the SSCP gels is chosen for further analysis. Parents and siblings were also analyzed to follow the segregation of the variation, and then the sequence of the alteration was directly determined. Alterations are classified as CF mutations based on the following criteria. (1) The alteration shifts the reading frame and causes premature termination of the protein. (2) An amino acid is replaced with a dissimilar residue, and this alteration does not occur on a large number of normal chromosomes with the same haplotype.

We have used this approach to identify eight separate CF mutations, five of which are frame-shift mutations and three that replace amino acids (White et al., 1990; Dean et al., 1990a; White et al., Iannuzzi et al., manuscripts submitted). Each of the frame-shift mutations has been found in only a single family, whereas two of the three point mutations are found in multiple families. One of the point mutations, R347P, was traced back to the paternal grandfather. He has 12 brothers and sisters but no further history of CF.

The clinical characteristics of the individuals are summarized in Table 1. All individuals that we have examined that are homozygous for the DeltaF508 mutation are pancreatic insufficient and have moderate to severe disease. Most of the patients with the frame shift mutations are homozygous for the absence of the common mutation, and therefore must contain an additional unidentified mutation. These patients are clinically heterogeneous; they typically have moderate disease and can be either pancreatic insufficient (PI) or pancreatic sufficient (PS). The majority of the patients with CF point mutations have DeltaF508 on their other chromosome, are diagnosed as PS, and have mild disease. Therefore, the type of mutation at the CF locus appears to play an important role in the clinical presentation of the patient.

Table 1

Classes of CF Mutation and Clinical Outcome

Mutation	Abbrev.	Genotype	Pancreas	Lung
Frame-shift/ Termination	N	N/N	PI	SEVERE
Amino Acid		N/S*		
Deletion	S	S/S*		
Point Mutation	M	N/M*	PS	
		S/M*		
		M/M		MILD

The three classes of CF mutation--null (N), severe (S), and mild (M) are shown along with all of the possible combinations. The status of pancreatic function is shown for each class of patient, as well as the overall state of lung function. *, Combinations actually observed, see Table 2.

Table 2

Specific Mutation Combinations

Mutation	Pancreatic Status	Lung Disease
F/F	PI	Mod.-Severe
F/N	PI	Mod.
N/+	PI/PS	Mod.
F/M	PS	Mild-Mod.

Specific combinations of CF mutations are shown with the corresponding status of the pancreas as well as the severity of lung disease. DF refers to the DeltaF$_{508}$ mutation, and + is a non-DF mutation. Mod., moderate.

DISCUSSION

Several classes of mutation have been detected in the CF gene (Tables 1 and 2). These alterations include frame-shift and termination mutants, amino acid deletions, and point mutations. Since the most common mutation accounts for only 70% of CF chromosomes, 40%-50% of CF patients are compound heterozygotes, i.e., they have two different mutations in the gene. Therefore there is a large number of possible genotypes patients may have, and this appears to account, to a large extent, for the variation observed in the phenotype of patients.

The clearest correlation between genotype and phenotype is seen in the pancreas. Patients homozygous for the DeltaF508 deletion are typically PI. However, even in these patients, genetically identical at the CF locus, there is considerable variation in clinical outcome. This variation is expressed in the age of diagnosis, pulmonary function, and sweat chloride value.

Thus, additional environmental and genetic factors must play a role in these complications. In the lungs, the damage is principally caused by bacterial infection. These infections are secondary to the abnormal mucus present in patients and clearly have an environmental component. In addition, it is likely that the human lymphocyte antigens (HLA) and/or the T cell receptor locus could play a role in the susceptibility to bacterial infection.

In the compound heterozygote patients the correlation between genotype and phenotype is further complicated. Patients with null and/or DeltaF$_{508}$ mutations are PI, but show variation in other symptoms (Table 2). Patients who carry point mutations are mostly PS, and are among the mildest patients we have examined. However there is also variation between these patients, and among affected siblings in these families. While a patient's genotype at the CFTR locus will prove to be an important diagnostic indicator, the quality of life of individual patients remains only partially predictable.

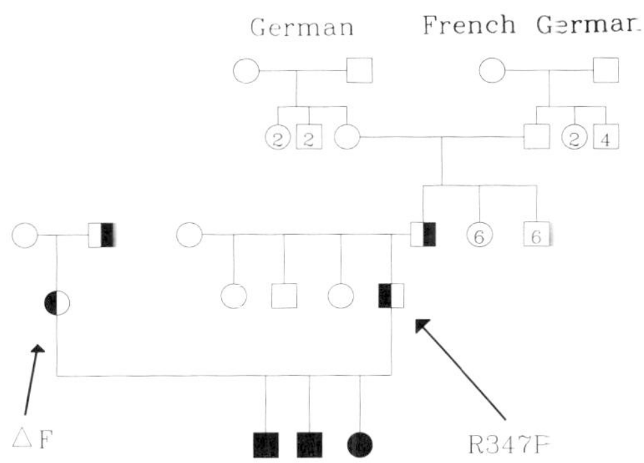

INHERITANCE OF THE R347P MUTATION

Fig. 1. Pedigree of the UT 1446 family. The family carries the R347P mutation on the paternal side and DeltaF$_{508}$ on the maternal side. We were able to further follow the inheritance of R347P to the paternal grandfather, whose ancestors are of German and French origin.

Acknowledgements

We thank the clinical personnel who provided the families, Bonnie Ramay and Donna Hedges for pedigree information on UT 1466, and Julie Cashour for preparing the manuscript.
"This project has been funded at least in part with Federal funds from the Department of Health and Human Services under contract number NO1-CO-74102 with Program Resources, Inc. The content of this publication does not necessarily reflect the views or policies of the Department of Health and Human Services, nor does mention of trade names, commercial products, or organizations imply endorsement by the U.S. Government."

References

Andersen, D.H., 1938, Cystic fibrosis of the pancreas and its relation to celiac disease, **Am. J. Dis. Child**, 56:344.

Dean, M., Drumm, M.L., Stewart, C., Gerrard, B., Perry, A., Hidaka, N., Cole, J.L., Collins, F.S., and Iannuzzi, M.C., 1990, Approaches to localizing disease genes as applied to cystic fibrosis, **Nucl. Acids. Res.**, 18:345.

Dean, M., White, M.B., Amos, J., Gerrard, B., Stewart, C., Khaw, K.-T., and Leppert, M., 1990a, Multiple mutations in highly conserved residues are found in mildly affected cystic fibrosis patients, **Cell**, 61:863.

di Sant'Agnese, P.A., Darling, R.C., Perea, G.A., and Shea, B.A., 1953, Abnormal electrolyte composition of sweat in cystic fibrosis of the pancreas: Clinical significance and relationship to the disease, **Pediatrics**, 12:549.

di Sant'Agnese, P.A., and Davis, P.B., 1979, Cystic fibrosis in adults, **Am. J. Med.**, 66:121.

Kerem, B.-S., Rommens, J.M., Buchanan, J.A., Markiewicz, D., Cox, T.K., Chakravarti, A., Buchwald, M., and Tsui, L.-C., 1989, Identification of the cystic fibrosis gene: Genetic analysis, **Science**, 245:1073.

Knowles, M., Gatzy, J., and Boucher, R., 1981, Increased bioelectric potential difference across respiratory epithelia in cystic fibrosis, **N. Eng. J. Med.**, 305:1489.

Orita, M., Suzuki, Y., Sekiya, T., and Hayashi, K., 1989, Rapid and sensitive detection of point mutations and DNA polymorphisms using the polymerase chain reaction, **Genomics**, 5:874.

Quinton,P.M., 1983, Chloride impermeability in cystic fibrosis, **Nature**, 301:421.

Quinton, P.M. and Bijman, J., 1983, Higher bioelectric potentials due to decreased chloride absorption in the sweat glands of patients with cystic fibrosis, **N. Engl. J. Med.**, 308:1185.

Riordan, J.R., Rommens, J.M., Kerem, B.-S., Alon, N., Rozmahel, R., Grzelczak, Z., Zielenski, J., Lok, S., Plavsic, N., Chou, J.-L., Drumm, M.L., Iannuzzi, M.C., Collins, F.S., and Tsui, L.-C., 1989, Identification of the cystic fibrosis gene: cloning and characterization of complementary DNA, **Science**, 245:1066.

Rommens, J.M., Iannuzzi, M.C., Kerem, B.-S., Drumm, M.L., Melmer, G., Dean, M., Rozmahel, r., Cole, J.L., Kennedy, D., Hidaka, N., Zsiga, M., Buchwald, M., Riordan, J.R., Tsui, L.-C., and Collins, F.S., 1989, Identification of the cystic fibrosis gene: Chromosome walking and jumping, **Science**, 245:1059.

Rosenstein, B.J. and Langbaum, T.S., 1984, Diagnosis, **in:** "Cystic Fibrosis," L.M. Taussig, ed., Thieme-Stratton, Inc., New York.

Sato, K. and Sato, F., 1984, Defective beta adrenergic response of cystic fibrosis sweat glands in vivo and in vitro, **J. Clin. Invest.**, 73:1763.

Su, C.T. and Beanblossom, B., 1989, Typical cystic fibrosis in an elderly woman, **Am. J. Med.**, 86:701.

Welsh, M.J. and Liedtke, C.M., 1986, Chloride and potassium channels in cystic fibrosis airway epithelia, **Nature**, 322:467.

White, M., Amos, J., Hsu, J.M.-C., Gerrard, B., Finn, P., and Dean, M., 1990, A frameshift mutation in the cystic fibrosis gene, **Nature**,.

MUTATION ANALYSIS FOR CYSTIC FIBROSIS

IN A NORTH AMERICAN POPULATION

A.L. Beaudet, G.L. Feldman, K. Kobayashi,
W.K. Lemna, S.D. Fernbach, M.R. Knowles,
R.C. Boucher and W.E. O'Brien

Howard Hughes Medical Institute and
Baylor College of Medicine, Houston, Texas, and
University of North Carolina, Chapel Hill, NC, USA

The common mutation causing cystic fibrosis (CF) is deletion of phenylalanine 508 (deltaF508). This mutation was analyzed in over 200 North American families using the polymerase chain reaction and hybridization with allele specific oligonucleotides. The deltaF508 mutation was present on 75.8% of non-Ashkenazic Caucasian chromosomes; 96% of the deltaF508 chromosomes carried the B haplotype for XV-2c and KM-19. Only 30.3% of Ashkenazic CF chromosomes carried the deltaF508 mutation, although the B haplotype for XV-2c and KM-19 was found on 97% of Ashkenazic CF chromosomes.

In this study, three "index parents" were identified who initially appeared homozygous for the deltaF508 mutation, although they were subsequently proven to represent compound heterozygotes. Cloning and sequencing of amplified genomic DNA identified a substitution of cysteine for phenylalanine 508 (F508C) and substitution of valine for isoleucine 506 (I506V). Two index parents were deltaF508/F508C compound heterozygotes and one was a deltaF508/I506V compound heterozygote. Detailed physiological studies of these three individuals were normal including pulmonary function, sweat chloride, sweat rate, sweat duct potential difference (PD), and detailed studies of nasal bioelectric parameters. These results indicate that the F508C and I506V mutations are benign, suggesting that amino acid substitutions are more benign than changes in the length of this portion of the putative nucleotide binding fold of the CF gene product.

Nasal scrapings from CF patients, heterozygotes and controls are being used to synthesize cDNA for amplification with the polymerase chain reaction. This proves to be a convenient method for analysis of CF mutations at a cDNA level. Using this method, a substitution of aspartic acid for glycine at position 1349 (G1349D) was identified in 2/85 non-deltaF508 CF chromosomes. Glycine 1349 is a highly conserved position in the nucleotide binding fold of this family of proteins. Another mutation, a substitution of lysine for asparagine at position 1313 (N1303K), was identified in 5/76 non-deltaF508 CF chromosomes. In addition, alternative splicing to include or exclude exon 9 was found in cDNA from normal and CF heterozygotes. The biological significance of the alternative splicing is uncertain at this time.

TOWARD A GEOGRAPHICAL HISTORY OF THE PREDOMINANT AND SECONDARY MUTATIONS IN EUROPE

J.L. Serre

I.N.S.E.R.M. (U 155), Paris, France

INTRODUCTION

One aim of population genetics studies is to account for observed genetic differentiation between populations by discriminating between factors which might have played a major role in the genetic features of human populations, mutation, migrations, genetic drift and natural selection.

Since the sixties numerous well documented studies, based on archeological data (Ammerman and Cavalli-Sforza, 1971) or on biochemical typing of large samples (Menozzi et al.,1978) have strengthened the hypothesis of a demic expansion of middle eastern farmers, during the neolithic period, that accounts for the diffusion process from which so many genetic clines are still observed in Europe.

More recently Piazza et al. (1988) could support the hypothesis that the genetic structure of Italy still reflects the ethnic stratification of pre-Roman times. This finding is in agreement with those of Rouche (1986) who found, on the basis of anthropological data, that the great invasions of the Middle-Ages might have led to a small admixture between populations in some parts of France. This fact, if strengthened, could be of interest for the CF population genetics.

The blood genetic markers are numerous, easy to type on large sample studies and display quite a good polymorphism as well as an informative differentiation between populations. On the contrary pathologic mutations are often too rare to provide any additional information in cluster analysis of human populations.

A first exception was the sickle-cell βS mutation due precisely to its high frequency in Africa. Pagnier et al. (1984)

The Identification of the CF (Cystic Fibrosis) Gene
Edited by L.-C. Tsui *et al.*, Plenum Press, New York, 1991

could evidence a multicentric origin of three independent and identical mutations in Senegal, Nigeria and Central Africa because each of them was highly associated with a specific RFLP haplotype defined for 11 endonuclease sites within 60 kb around the β locus.

A second exception will be the cystic fibrosis mutations in European populations because their absolute frequencies amount to 2.2% of the gene pool, because prenatal diagnosis as well as population screening will soon provide large scale sampling, and because the CF mutations are typed together with linked RFLPs. So we may ask:

- To what extent do the molecular data at the CF and linked RFLP loci provide, per se, anthropological information?
- To what extent can such an information be connected with previous studies of European populations ?

MATERIAL

The molecular typing of CF and normal chromosomes in every european population will afford three geographic maps of genetic polymorphism.

- First, a geographic map of relative frequencies of each mutation within the CF chromosomes.
- Second, a map of the relative frequencies of RFLP haplotypes associated with each characterized mutation.
- Third, a map of the relative frequencies of RFLP haplotypes within normal chromosomes.

These three maps are already available for most countries, but only for normal and DF508 chromosomes (European working group on CF genetics, 1990).

In this article we present only an analysis of our data from Spain (Jaume-Roig et al., 1990) and from two parts of France, Brittany (Ferec, personal communication) and Provence (Claustre, personal communication) where normal and DF508 chromosomes have been typed for XV2C, KM19, and MP6d-9 RFLP loci.

METHODS

Theoretical Equations

The high linkage disequilibrium between each mutation and a specific RFLP haplotype leads to the conclusion that they are single mutations. Each of these occurred once on a normal chromosome bearing the specific RFLP still associated with each mutation in the present times. But due to crossing-over within heterozygotes at the CF locus, each mutation could have been reassociated with other RFLP haplotypes, as reported in table 1.

TABLE I

Normal chromos. \ Chromos.			A (D_1 E_1 CF)	B (D_1 E_2 CF)	C (D_2 E_1 CF)	D (D_2 E_2 CF)
A	$(\theta-\psi)$	D_1 E_1 N	A	B	A	B
	ψ		A	A	A	A
B	$(\theta-\psi)$	D_1 E_2 N	A	B	A	B
	ψ		B	B	B	B
C	$(\theta-\psi)$	D_2 E_1 N	C	D	C	D
	ψ		C	C	C	C
D	$(\theta-\psi)$	D_2 E_2 N	C	D	C	D
	ψ		D	D	D	D

Considering this table, the frequencies of RFLP haplotypes associated with he CF mutation bear the following relationship to that in the preceding generation (i-1) :

(1)

$$\begin{aligned} h_{A,i} = {} & h_{A,i-1}\left[(1-\theta)+\theta.H_A+(\theta-\psi)H_B\right] \\ & +h_{B,i-1}\left[\psi.H_A\right] \\ & +h_{C,i-1}\left[\theta.H_A+(\theta-\psi)H_B\right] \\ & +h_{D,i-1}\left[\psi.H_A\right] \end{aligned}$$

(2)

$$\begin{aligned} h_{B,i} = {} & h_{B,i-1}\left[(1-\theta)+\theta.H_B+(\theta-\psi)H_A\right] \\ & +h_{A,i-1}\left[\psi.H_B\right] \\ & +h_{C,i-1}\left[\psi.H_B\right] \\ & +h_{D,i-1}\left[\theta.H_B+(\theta-\psi)H_A\right] \end{aligned}$$

(3)

$$\begin{aligned} h_{C,i} = {} & h_{C,i-1}\left[(1-\theta)+\theta.H_C+(\theta-\psi)H_D\right] \\ & +h_{A,i-1}\left[\theta.H_C+(\theta-\psi)H_D\right] \\ & +h_{B,i-1}\left[\psi.H_C\right] \\ & +h_{D,i-1}\left[\psi.H_C\right] \end{aligned}$$

(4)

$$\begin{aligned} h_{D,i} = {} & h_{D,i-1}\left[(1-\theta)+\theta.H_D+(\theta-\psi)H_C\right] \\ & +h_{A,i-1}\left[\psi.H_D\right] \\ & +h_{B,i-1}\left[\theta.H_D+(\theta-\psi)H_C\right] \\ & +h_{C,i-1}\left[\psi.H_D\right] \end{aligned}$$

where H_A, H_B, H_C, H_D, are the relative frequencies of A, B, C, D RFLP haplotypes on normal chromosomes and h_A, h_B, h_C, h_D, their relative frequencies on DF508 chromosomes.

With an additional marker (MP6d-9), there will be 8 different haplotypes named A1, A2, B1, B2, C1, C2, D1, D2, where 1 and 2 refer to the ± polymorphism at this third locus. Consequently there will be 8 recurrent equations with an additional recombination ratio.

Knowing the recombination ratio values, it would be possible to estimate the elapsed time since the mutation entered the gene pool of the population either by mutation or through migration, by iterating the calculation of frequencies from initial values to theoretical values which have the best fit with observed frequencies.

The fit can be estimated by a chi-square test between expected and observed frequencies.

Such an estimation of the age of the mutation is possible only if two conditions are assumed. First there have been no variations of RFLP frequencies within normal chromosomes, which seems to be strengthened by recent data (EWGCFG, 1990). Second there has been no genetic drift that could have disturbed the recurrent process of genetic recombination under study.

Application to our Data

We performed a set of computer iterations using the 8 recurrent equations in each population.

As the actual values of recombination ratio are still unknown, we took an array of two sets of values. In the first set the recombination values are proportional to physical distances on the DNA (Rommens et al., 1989). In a second set we assume a recombinational hot point between the XV2C and KM19 loci, as many authors have assumed in order to account for the higher relative frequency of the D haplotype associated with the DeltaF508 mutation, compared with A or C relative frequencies.

Since relative RFLP frequencies on DeltaF508 chromosomes for the first generation are also unknown, we performed two sets of simulations. A first calculation has been performed assuming that the mutation was exclusively associated with a B2 haplotype when it entered the gene pool of the population. If this first simulation could not give any fit between expected and observed RFLP frequencies, a second simulation has been processed assuming minimal values of D, C, and A frequencies in order to get the best fit between expected and observed RFLP frequencies.

This means that we calculated a maximal estimation of the age of the mutation by assuming a maximal association with the B RFLP, at the beginning of the recombination process in the population (generation i=0).

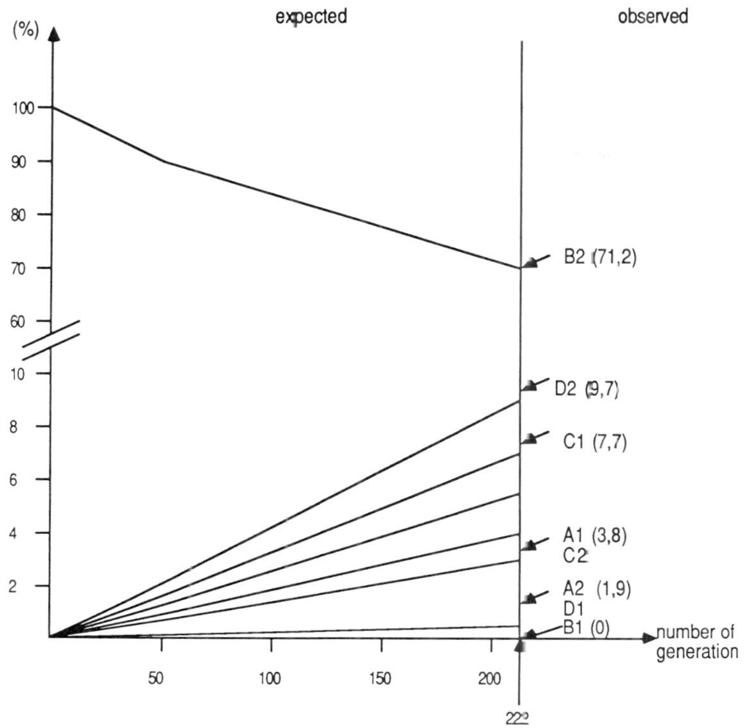

Fig. 1. Expected evolution of the frequencies of RFLP newly associated to the DeltaF508 mutation by means of crossover with respective recombination ratio values of 5.10-4, 3.10-4, 12.10-4 between XV2C/KM19/PM6d-9/CF loci. (From top, respective evolution lines of B2, D2, C1, A1, A2, C2, B1-D1 haplotypes

RESULTS

Spanish Sample

Figure 1 shows an example of simulation. If not equal, the expected and observed values have a good fit (X^2=2.54, with 7 d.f.). This means that the observed values are in agreement with the hypothesis that the maximal age of the mutation in this country ranges from 222 to 500 generations, depending on the actual recombination ratio values (table 2).

French Data

The same procedure, applied to samples from Brittany and Provence, gave significant departure between expected and

TABLE II

Recombination ratio values ($x10^{-4}$)			Mutation age estimated through recombination in		
a	b	c	Spain	French Brittany	French Provence
5	3	12	222 (1)	46 (1) / 20 (2)	36 (1) / 11 (2)
5	1	4	500 (1)	100 (1) / 44 (2)	80 (1) / 24 (2)

observed RFLP frequencies if we assumed that the mutation was exclusively associated with a B2 haplotype when it entered the gene pool of the population (respective X^2 values of 14.7 and 8 for 2 d.f.).

Such a departure may result from sampling variation and insufficient sampling in subpopulation where some associated RFLP are rare. It may also result from genetic drift or a different set of initial RFLP frequencies when the DeltaF508 mutation entered these populations.

A second simulation with an initial D relative frequency of 5% within DeltaF508 chromosomes gave a good fit in the two samples with a maximal estimated age of 20 (or 44) generations in Brittany and 11 (or 24) in Provence, which are obviously underestimated probably due either to sampling or to genetic drift.

DISCUSSION

The proposed method for analysing the CF data is too sensitive to recombination ratio values to give an estimation of the age of the predominant DeltaF508 mutation. However, whatever the actual values of these ratio, a set of different populations could be classified depending on the hierarchical relative ages of the mutation in each of them.

Such a classification is achieved assuming sufficient sampling, no genetic drift, and a maximal linkage disequilibrium between the mutation and its specific associated RFLP haplotype, at the beginning of the process (i=0). These two last conditions may be fulfilled only in some populations, resulting in a partially misleading view of the geographical history of the mutation.

So the obtained range of mutation ages in each country on the map of Europe must be connected with previous studies. In a first attempt, using data from the whole of Europe, but including only the CF and KM19 loci, we concluded that the observed data were in agreement with a neolithic origin in the Middle-East (Serre et al., 1990).

Recent data including RFLP haplotypes at the XV2C and KM19 loci in all the european populations are to be published in a special issue of Human Genetics. A first calculation leads to sone conclusions.

- First the relative ages of the DeltaF508 mutation, assuming a hot point for recombination between the XV2C and KM19 loci, range from 250 to 40 generations with a clear-cut south-north cline (may be due to insufficient sampling in Greece which plays a crucial part in determining whether there is a south-east/north-west cline or not).
- Second the mutation came probably to Europe already associated with other haplotypes, which means that it had occurred a long time ago in the ancestral population before these new associated RFLP could have emerged through crossover.
- Third there has been an initial founder effect because the C haplotype is underepresented in every country compared with the A one, whereas equal relative frequencies should have been expected from the recombination process, due to RFLP frequencies within normal chromosomes.

Of course it will be difficult to get any valuable information from those mutations which are very rare and so easily affected by genetic drift. Maybe there will be a more favourable outcome for this secondary mutation which is frequent enough to lead to a significant higher frequency of the B RFLP haplotype within the pool of secondary mutations, everywhere in Europe.

Acknowledgements

Work supported by grants from the Association Françoise de Lutte contre la Mucoviscidose.

References

Ammerman A.J., Cavalli-Sforza L.L., 1971, Measuring the rate of spread of early farming in Europe, **Man**, 5:674.

European Working Group on CF Genetics (EWGCFG), 1990, Gradiant of distribution in Europe of the major CF mutation and of its associated haplotypes, **Hum. Genet.**, special issue, in press.

Jaume-Roig B., Simon-Bouy B., Taillandier A., Serre J.L., Antich J., Bellon J., Boué J., Boué A., 1990, Genotyping of spanish cystic fibrosis population at the DeltaF508 mutation site and RFLP linked loci, **Hum. Genet.**, special issue, in press.

Menozzi P., Piazza A., Cavalli-Sforza L.L., 1978, Synthetic maps of human gene frequencies in Europeans, **Science**, 201:786.

Pagnier J., Mears J.G., Dunda-Belkhodja O., Schaefer-Rego K.E., Beldjord C., Nagel R.L., Labie D., (1984, Evidence for the multicentric origin of the sickle cell henoglobin gene in Africa, **Proc. Natl. Acad. Sci. USA**, 81:1771.

Piazza A., Cappello N., Olivetti E., Rendine S., 1988, A genetic history of Italy, **Ann. Hum. Genet.**, 52:203.

Rommens, J.M., Iannuzzi, M.C., Kerem, B., Drumm, M.L., Melmer, G., Dean, M., Rozmahel, R., Cole, J.F., Kennedy, D., Hidaka, N., Zsiga, M., Buchwald, M., Riordan, J.R., Tsui, L.C., Collins, F.S., 1989, Identification of the cystic fibrosis gene: chromosome walking and jumping, **Science**, 245:1059.

Rouche M., 1986, Les mélanges de population en Gaule (V-XI siècle), in: "Human Population Genetics," Ohayon E., ed., Cambon-Thomsen A., Paris : I.N.S.E.R.M.

Serre J.L., Simon-Bouy B., Mornet E., Jaume-Roig B., Balassopoulou A., Schwartz M., Taillandier A., Boué J., Boué A., 1990, Studies of RFLP closely linked to the Cystic Fibrosis locus throughout Europe lead to new considerations in population genetics, **Hum. Genet.**, 84:449.

ORIGIN AND DIFFUSION OF THE MAJOR CF MUTATION IN EUROPE

Data presented by Marcella Devoto in representation of
The European Working Group on CF Genetics (EWGCFG) (1)

(1) **M. Bonduelle, W. Lissens, I Dab**
(Dept. of Medical Genetics and Dept. of Pediatrics,
Academisch Ziekenhuis, Vrije Universiteit Brussel, Belgium)
J.J. Cassiman, H. Cuppens, P. Marynen
(Center for Human Genetics,
Catholic University of Leuven, Belgium)
P. Cochaux, J. Pencin, G. Vassart
(Service de Genetique, Hopital Erasme,
Universite Libre de Bruxelles, and
Universite de Liege, Belgium)
L. Kalaydjieva, J. Anton, J. Bronzova
(Laboratory of Molecular Pathology,
Medical Academy, Sofia, Bulgaria)
M. Macek Jr., V. Vavrova, M. Macek
(Dept. Clinical Biochemistry,
Fac. Medicine, Inst. Child Development Research and
Centre of Medical Genetics, Fac. Pediatrics,
Charles University, Praha, Czechoslovakia)
M. Schwartz, C. Koch, N. Sorensen
(Section of Clinical Genetics, Dept. of Pediatrics,
Righospitalet, Copenhagen, Denmark)
J. Kere, A. de la Chapelle, E. Savilahti
(Dept. of Medical Genetics, University of Helsinki, Finland)
J.C. Chomel, L. Tesson, A. Kitzis
(Laboratoire de Biochimie Genetique,
C.H.U. Cochin, Paris, France)
M. Claustres, H. Bellet, M. Desgeorges
(INSERM U 249, CRBM/CNRS, Montpellier, France)
C. Ferec, H. Guillermit
(Centre Departemental de Transfusion Sanguine,
Brest Cedex, France)
O. Pascal, J.P. Moisan
(INSERM U 211, Faculte' de Medicine, Nantes, France)
B. Simon-Bouy, E. Mornet, J.L. Serre
(INSERM U 73 and INSERM U 155, Chateau de Longchamp,
Paris, France)
M. Vidaud, M. Goossens
(INSERM U 91, Hopital Henri Mondor, Creteil, France)

The Identification of the CF (Cystic Fibrosis) Gene
Edited by L.-C. Tsui *et al.*, Plenum Press, New York, 1991

C. Coutelle, R. Szibor, J. Brock
(Akademie der Wissenschaften, Berlin-Buch, Medizinische Akademie, Magdeburg, and Wilhelm-Pieck-Universitat, Rostok, DDR)
J. Horst, C. Aulehla-Scholz, A. Eigel
(Institut fur Humangenetik, Universitat Munster, FRG)
M. Krawczak, J. Reiss, M. Wagner
(Institut fur Humangenetik, University of Gottingen, FRG)
K. Olek, R. Kaiser, J. Weber
(Institut fur Klinische Biochemie, Universitat Bonn, FRG)
M. Stuhrmann, A. Reis, J. Schmidtke
(Institut fur Humangenetik, Freie Universitat Berlin, FRG)
B. Tummler, S. Bremer, J. Hundrieser
(Abt. Biophysikalische Chemie, Medizinische Hochschule Hannover, FRG)
A. Balassopoulou, D. Loukopoulos
(First Dept. of Medicine, University of Athens, Greece)
M.A. De Arce, P. Humphries
(Dept. of Genetics, Trinity College, Dublin, Ireland)
D. Abeliovich, I. Lerer, A. Friedman
(Dept. Human Genetics, Hadassa University Hospital, Jerusalem, Israel)
N. Cappello, A. Piazza
(Dipartimento di Genetica, Biologia e Chimica Medica, Università di Torino, Italy)
O. Castiglione, F. Santamaria, G. Sebastio
(Dipartimento di Pediatria, II Facoltà di Medicina, Università di Napoli, Italy)
M. Devoto, P. Ronchetto, G. Romeo
(Laboratorio di Genetica Molecolare, Istituto G.Gaslini, Genova, Italy)
M. Ferrari, M. Seia, L. Piceni Sereni
(Ospedale San Raffaele, and Laboratorio, ICP, Milano, Italy)
G.B. Leoni, M.C. Rosatelli, P. Scarpa
(Istituto di Clinica e Biologia dell'Età Evolutiva, Università di Cagliari, Italy)
G. Novelli, F. Sangiuolo, B. Dallapiccola
(Dipartimento Sanità Pubblica e Biologia Cellulare, II Università di Roma, Italy)
G. Restagno, B. Santini, A. Carbonara
(Dipartimento di Genetica, Biologia e Chimica Medica, and Cattedra di Pediatria, Università di Torino, Italy)
D.J.J. Halley, B.A. Oostra, H.J. Veeze
(Depts of Clinical Genetics, Cell Biology and Genetics, and Pediatrics, Erasmus University, Rotterdam, The Netherlands)
H. Scheffer, G.J. te Meerman, C.H.C.M. Buys
(Dept. of Medical Genetics, University of Groningen, The Netherlands)
C.A. Graham, A.J.M. Hill, N.C. Nevin
(Dept. of Medical Genetics, The Queen University of Belfast, Belfast City Hospital, Northern Ireland)
A. Duarte, J. Lavinha
(Laboratorio de Genetica Humana, Instituto Nacional de Saude R.Jorge, Lisboa, Portugal)

E. Fernandez, J. Benitez
(Departamento de Genetica Molecular, Fundacion Jimenez-Diaz, Madrid, Spain)
B. Jaume-Roig, J. Antich, J. Bellon
(Servicio de Genetica, Hospital Son Dureta, and Conselleria de Sanidad, Palma de Mallorca, Spain)
V. Nunes, T. Casals, X. Estivill
(Molecular Genetics Unit, Hospital de la Santa Creu i Sant Pau, Barcelona, Spain)
C.A. Hardy, M.W. Kilpatrick, P.A. Farndon
(University of Birmingham, Dept. of Clinical Genetics, Birmingham Materniy Hospital, UK)
A. Harris, C. Mathew, F. Beards
(Pediatric Research Unit, United Medical and Dental School of Guy's and St.Thomas's Hospital, London, UK)
S. Malcolm, C. McMahon, S. Genet,
(Mothercare Dept. of Pediatric Genetics, Institute of Child Health, London, UK)
I. McIntosh, A. Curtis, D.J.H. Brock
(Human Genetics Unit, University of Edinburgh, UK)
G. Santis, L. Osborne, R. Knight
(Dept. of Cystic Fibrosis, Brompton Hospital, London, UK)
M.J. Schwarz, M. Super
(DNA Laboratory, Paediatric Genetics Unit, Royal Manchester Children's Hospital, UK)
A.J. Wallace, A. Ivinson, R.G. Elles
(Regional Molecular Genetics Laboratory, St.Mary's Hospital, Manchester, UK)
E.K. Watson, E.S. Mayall, R. Williamson
(Dept. of Biochemistry and Molecular Genetics, St.Mary's Hospital Medical School, London, UK)
V.N. Kalinin
(Laboratory of Molecular Genetics, Institute of Medical Genetics, Moscow, USSR)
G.V. Krainiaia, B.Y. Reznik
(Chair of Pediatrics, Odessa Medical School, USSR)
S.G. Vorsanova
(Institute of Pediatrics, Moscow, USSR)
B. Brukner Dabovic, D. Radojkovic, A. Savic
(Institute of Molecular Genetics and Genetic Engineering, Belgrade, Yugoslavia)
L. Simova, C. Williams, G.D. Efremov
(Regional DNA LAboratory, St.Mary's Hospital, London, UK, and Research Centre for New Technologies, Macedonian Academy of Sciences and Arts, Skopje, Yugoslavia)

After the localization of the CF locus on chromosome 7, several RFLPs have been isolated in strong linkage disequilibrium with the CF mutations (Estivill et al., 1987a; Estivill et al., 1987b; Rommens et al., 1988). However, the degree of disequilibrium and the distribution of the haplotypes defined by these polymorphisms were different in different populations. In particular, it soon becane evident that a more heterogenous distribution was present in Southern Europe with respect to Northern Europe, and therefore it was possible to hypothesize that a more heterogeneous distribution of mutations was present in Southern Europe (Estivill et al., 1988). At the same time, data from different groups showed that there was a correlation between presence of the most frequent haplotype and pancreatic insufficiency in CF patients, suggesting that the mutation associated with this haplotype was related to a severe phenotype in CF patients (Kerem et al., 1989a; Devoto et al., 1989; Ferrari et al., 1990).

All these hypotheses have been confirmed after the isolation of the CFTR gene and the identification of deltaF508 as the major mutation causing CF (Rommens et al., 1989; Riordan et al., 1989; Kerem et al., 1989b). DeltaF508 was found to represent about 70% of the CF mutations in the sample of Canadian patients studied by Kerem et al. (1989b). In addition, homozygosity for deltaF508 was always associated with pancreatic insufficiency.

However, when screenings of CF patients were carried out in different countries, it became evident that the frequency of this mutation was different in different populations, being higher in Northern Europe with respect to Southern Europe (McIntosh et al., 1990; Estivill et al., 1990).

Fig. 1. Linkage disequilibrium between XV-2c and KM.19 in normal chromosomes.

In order to establish the frequency distribution of deltaF508 in Europe, we have collected data on about 6000 CF and 4000 normal chromosomes from most of the European countries, including Turkey and Israel, and in particular on 1) the frequency of the haplotypes defined by the two polymorphic markers XV-2c and KM.19 in normal and CF chromosomes; 2) the frequency of the same haplotypes associated with deltaF508; 3) the distribution of genotypes in CF patients subdivided according to presence or absence of pancreatic insufficiency, and to presence or absence of meconium ileus.

The values of the standardized disequilibrium coefficient between XV-2c and KM.19 in normal chromosomes are reported in figure 1. The distribution of the haplotypes defined by XV-2c and KM.19 in normal chromosomes is quite similar in all the populations. A slight linkage disequilibrium between the two RFLPs results in an excess of the B and C haplotypes in most populations, but this is generally non significant. These observations lead to the conclusion that these two RFLPs are probably very ancient markers, because they have almost reached linkage equilibrium although their loci are only 60 Kb apart.

Fig. 2. Relative frequency of deltaF508.

Data on the relative frequency of the deltaF508 mutation in the different countries are reported in figure 2. These data indicate that this is the most frequent CF mutation in most European populations. Only in Finland, USSR, Turkey, and Israel is the frequency of deltaF508 below 50%. However, all these countries are represented by relatively small samples of patients. In particular for the USSR the sample is too small to be representative of the whole country, and for Israel the data used in the present analysis are those representing the patients of Ashkenazi descent only.

The data also show considerable variation in the frequency of deltaF508, with a minimum value of 30% observed in Turkey and a maximum of 88% observed in Denmark. Lower frequencies of the mutation are generally observed in the Southern populations, in agreement with the previously reported haplotype data.

Haplotype analysis with XV-2c and KM.19 in deltaF508 chromosomes is summarized in figure 3, which reports the frequency of the B haplotype in deltaF508 chromosomes. These data show that

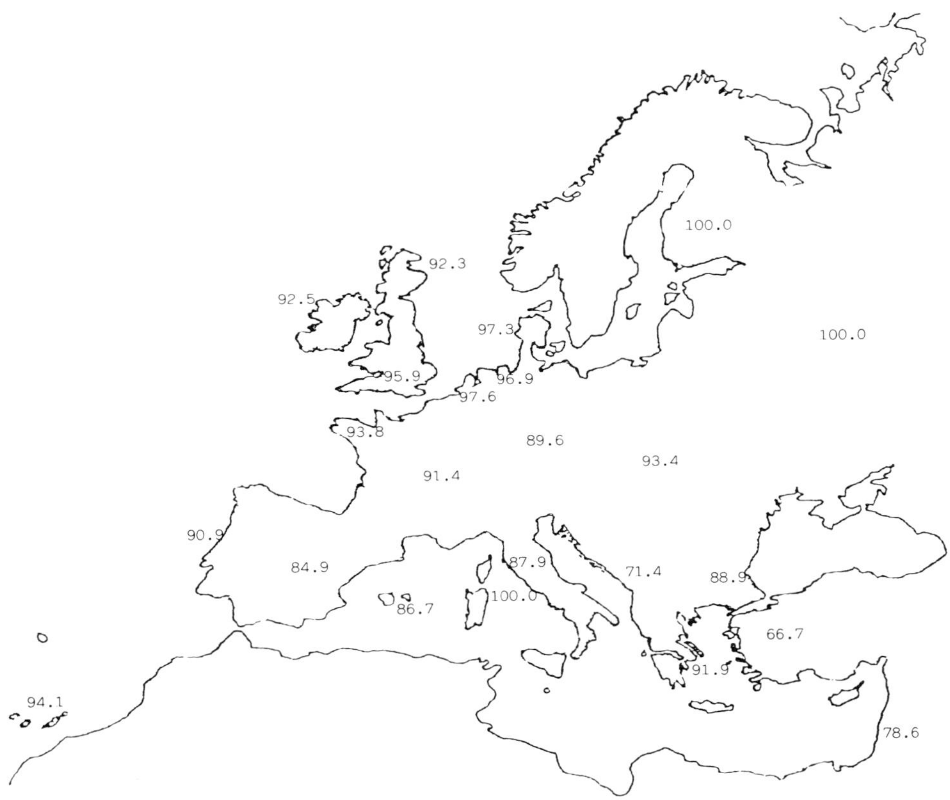

Fig. 3. Frequency of haplotype B in deltaF508 chromosomes.

in all the countries the great majority of chromosomes bearing the deltaF508 mutation is characterized by the B haplotype. However, there is a significant heterogeneity in the degree of association among the different populations. In particular, it appears that the degree of association is lower in the South European countries than in the North European ones. If the hypothesis of a single origin of the deltaF508 mutation is correct, and the mutation arose on a chromosome with a B haplotype, the association of deltaF508 with the other haplotypes can be explained by recombination events. The relatively higher frequency of deltaF508 chromosomes with A, C, and D haplotypes in Southern Europe might be due to a founder effect, or it might indicate that a longer time has passed since the mutation was first introduced in these populations, allowing more time for recombination, as already suggested by Serre et al. (1990).

The observation of deltaF508 in association with non B haplotypes is not unexpected, since approximately 200 Kb separate the two markers from the site of the mutation. In spite of the fact that the haplotype D is the rarest among normal chromosomes, its frequency is highest among deltaF508 non B haplotypes. On the contrary, the distribution of haplotypes in non

Fig. 4. Frequency of haplotype B in non-deltaF508 CF chromosomes. Frequency of haplotype B in normal chromosomes.

deltaF508 chromosomes shows that haplotype B is still the most frequent CF haplotype in most of the populations, in spite of the fact that it is the second less frequent haplotype among normal chromosomes. In figure 4, the frequencies of haplotype B in non deltaF508 CF chromosomes and in normal chromosomes are reported for each country. We now know that many other mutations associated with the same haplotype have been found in patients from different populations. The occurrence of most CF mutations on chromosomes characterized by the B haplotype, which represents only 16% of the normal haplotypes, is another interesting observation which might be explained by a selective advantage specific for mutations arising on this haplotype.

In conclusion, a South-East to North-West gradient of the frequency of the deltaF508 mutation in Europe represents the main finding of the present work. This gradient is similar to that observed for other allelic systems (Menozzi et al., 1978) and is in agreement with the hypothesis that spreading of the CF alleles throughout Europe is related to the migration of early farmers which took place during the Neolithic, starting from the Middle East and slowly progressing towards the North-West (Ammerman and Cavalli-Sforza, 1984).

References

Ammerman, A.J., Cavalli-Sforza, L.L., 1984, "The Neolithic Transition and the Genetics of Populations in Europe," Princeton University Press, Princeton, NJ.

Devoto, M., De Benedetti, L., Seia, M., Piceni Sereni, L., Ferrari, M., Bonduelle, M.L., Malfroot, A., Lissens, W., Balassopoulou, A., Adam, G., Loukopoulos, D., Cochaux, P., Vassart, G., Szibor, R., Hein, J., Grade, K., Berger, W., Wainwright, B., Romeo, G., 1989, Haplotypes in cystic fibrosis patients with or without pancreatic insufficiency from four European populations, **Genomics**, 5:894.

Estivill, X., Farrall, M., Scambler, P.J., Bell, G., Hawley, K.M.F., Lench, N.J., Bates, P., Kruyer, H.C., Frederick, P.A., Stainer, P., Watson, E.K., Williamson, R., Wainwright, B.J., 1987a, A candidate for the cystic fibrosis locus isolated by selection for the metilation-free islands, **Nature**, 326:840.

Estivill, X., Scambler, P.J., Wainwright, B.J., Hawley, K., Frederik, P., Baiget, M., Kere, J., Williamson, R., Farrall, M., 1987b, Patterns of polymorphism and linkage disequilibrium for cystic fibrosis, **Genomics**, 1:257.

Estivill, X., Farrall, M., Williamson, R., Ferrari, M., Seia, M., Giunta, A., Novelli, G., Potenza, L., Dallapiccola, B., Borgo, G., Gasparini, P., Pignatti, P.F., De Benedetti, L., Vitale, E., Devoto, M., Romeo, G., 1988, Linkage disequilibrium between cystic fibrosis and linked DNA markers in Italian families: a collaborative study, **Am. J. Hum. Genet.**, 43:23.

Estivill, X., Chillon, M., Casals, T., Bosch, A., Morral, N., Nunes, V., Gasparini, P., Seia, A., Pignatti, P.F., Novelli, G., Dallapiccola, B., Fernandez, E., Benitez, J., Williamson, R., 1989, DeltaF508 gene deletion in cystic fibrosis in Southern Europe, **Lancet**, II:1404.

Ferrari, M., Antonelli, M., Bellini, F., Borgo, G., Castiglione, O., Curcio, L., Dallapiccola, B., Devoto, M., Estivill, X., Gasparini, P., Giunta, A., Marianelli, L., Mastella, G., Novelli, G., Pignatti, P.F., Romano, L., Romeo, G., Seia, M., Williamson, R., 1990, Genetic differences in cystic fibrosis patients with and without pancreatic insufficiency. An Italian collaborative study, **Hum. Genet.**, (in press).

Kerem, B., Buchanan, J.A., Durie, P., Corey, M.L., Levison, H., Rommens, J.M., Buchwald, M., Tsui, L.C., 1989a, DNA marker haplotype association with pancreatic sufficiency in cystic fibrosis, **Am. J. Hum. Genet.**, 44:827.

Kerem, B.S., Rommens, J., Buchanan, J.A., Markiewicz, D., Cox, T.K., Chakravarti, A., Buchwald, M., Tsui, L.C., 1989b, Identification of the cystic fibrosis gene: genetic analysis, **Science**, 245:1073.

McIntosh, I., Lorenzo, M.L., Brock, D.J.H., 1989, Frequency of the deltaF508 mutation on cystic fibrosis chromosomes in UK, **Lancet**, ii:1404.

Riordan, J.R., Rommens, J., Kerem, B., Alon, N., Rozmahel, R., Grzelczak, Z., Zielenski, J., Lok, S., Plavsis, N., Chou, J.-L., Drumm, M.L., Iannuzzi, M.C., Collins, F.S., Tsui, L.-C., 1989, Identification of the cystic fibrosis gene: cloning and characterization of complementary DNA, **Science**, 245:1066.

Rommens, J.M., Zengerling, S., Burns, J., Melmer, G., Kerem, B., Plavsic, N., Zsiga, M., Kennedy, D.M., Markiewicz, D., Rozmahel, R., Riordan, J.R., Buchwald, M., Tsui, L.C., 1988, Identification and regional localization of DNA markers on chromosomes 7 for the cloning of the cystic fibrosis gene, **Am. J. Hum. Genet.**, 43:645.

Rommens, J., Iannuzzi, M.C., Kerem, B., Drumm, M.L., Melmer, G., Dean, M., Rozmahel, R., Cole, J.L., Kennedy, D., Hidaka, N., Zsiga, M., Buchwald, M., Riordan, J.R., Tsui, L.C., Collins, F.S., 1989, Identification of the cystic fibrosis gene: chromosome walking and jumping, **Science**, 245:1059.

Serre, J.L., Simon-Bouy, B., Mornet, E., Jaume-Roig, B., Balassopoulou, A., Schwartz, M., Thaillandier, A., Bouè, J., Bouè, A., 1989, Studies of RFLP closely linked to the cystic fibrosis locus throughout Europe lead to new considerations in population genetics, **Hum. Genet.**, (in press).

DISCUSSION

WILLIAMSON: I didn't understand the last point. I thought that it is believed that clines of this type start in the Middle East and then gradually move towards the west, but from your data and the data presented before you by Dr. Serre you are predicting the opposite, that the delta F508 mutations started in Central Europe and then migrated back to the Middle East.

DEVOTO: The migration of population which we know of, the migration of farmers, I mean, started from the Near East and slowly progressed towards the northwest, and this is a well-known fact. I do not mean by that, that the delta 508 followed the same course. What I personally believe is that maybe the delta 508 mutation was already present everywhere in Europe and this migration of population slowly introduced other mutations which were present in the near east. This is why you see this gradient in frequency of delta F508 mutation.

WILLIAMSON: That would involve a displacement of one mutant allele by another, which is different from a founder effect.

DEVOTO: Yes, it is not a founder effect, actually.

SERRE: It's like a diffusion process. You have a gradient, progressively enlarged on the whole of Europe.

COLLINS: I'm experiencing the same puzzlement that Bob is. Maybe I could just rephrase the question. If delta F508 originated in the Middle East - is that your supposition?

SERRE: No, no. It's just the opposite. All the others started there.

COLLINS: But your haplotype analysis claims that there are more non-B haplotypes on delta F508 chromosomes in the Middle East than elsewhere. Does that not, then, imply that the mutation is older?

DEVOTO: Yes, but this might have happened before the migration of the population from the near east.

COLLINS: So where do you think delta F508 originated?

DEVOTO: Really, I don't know. Probably in the south - but I repeat, probably that was before the migration of farmers which I just described. And then this migration introduced all over Europe other mutations, but these mutations, probably for many different reasons, are now more represented in the southern population than in the northern population. I believe there are two different processes, not only one.

COLLINS: I guess I'm still, then, puzzled by why the frequency of the allele is higher in the places where it ended up last, if you say it originated in the south...

DEVOTO: Because it was not replaced by other mutations. You also find other mutations which were introduced by this migration of population, or might have been introduced by this migration of population from the near east.

WILLIAMSON: But there is also a gradient of the absolute frequency of the gene, which is greater, in general, in North Europe.

SCHWARTZ: The incidence of cystic fibrosis in Denmark is only one out of 4,500 persons, and still we have the highest frequency of the deltaF508 mutation.

DEVOTO: As far as I know, in Italy we have the same incidence as you have in England.

SERRE: Yes, this is true for France, also. The incidence of CF doesn't seem to be very different among European countries, with the possible exception of Denmark.

DEVOTO: And of Finland.

SERRE: We can therefore say that there is no gradient in absolute frequencies of CF throughout Europe.

ROMEO: One other point that you might want to comment here, because it's been rather obscured by the question on the gradient, is that the B haplotype is still, among non deltaF, (that's what you both pointed out) the most frequent one. Would you like to comment on this?

SERRE: I have no comment about the presence of the B haplotype in the secondary mutation, but a lot of hypotheses can be made. At present we cannot choose among them. But what is clear from this data is that the B haplotype is clearly less associated in the south, which is a result of the recombination process. The mutation appears to be older in the south than in the north.

BEAUDET: Has anybody tried to calculate, if you throw out delta F, if this association of other mutations with the B haplotype is statistically significant or not?

DEVOTO: Yes, it is, very much.

ROMEO: I might remind this audience (this is something which turned up in the discussion when Luca Cavalli-Sforza was here last week) that a paper was published in the American Journal of Human Genetics in 1977 by Diane Wagener and Luca Cavalli-Sforza which deals with the effect of hitch-hiking and epistasis in the spreading of a disease gene. This might be one possible explanation for this interesting association between the B haplotype and different CF mutations.

BEAUDET: If you talk about hitch-hiking, why hasn't the B haplotype become itself much more enriched?

DEVOTO: It can be just the association between the CF mutation and this other gene associated to the B haplotype.

BEAUDET: But you have to hypothesize that whatever else is on that allele isn't itself a selective advantage but in combination with the CF gene is a selective advantage. That's a particularly complex version of hitch-hiking, not just a simple one.

ROMEO: That is exactly epistasis in addition to hitch-hiking. There are data which might be useful in discussing this hypothesis and they are now in press in Human Genetics. They come from the Hannover group (Tümmler et al.) and would indicate a role for epistasis.

GARDNER: I'm just puzzled about one thing. You postulated a gradient for deltaF508 which would have moved northward. You feel it started in the south because of the longer period of time allowed for recombination. Why don't you expect more CF in the south since other mutations have moved in? In other words, what causes replacement as opposed to addition?

SERRE: You mean that CF would be less in the north if you have migration? Because maybe there has been a selective effect, which has time to equalize the frequency of the CF mutations everywhere in Europe. If there is a selection effect (we must demonstrate it) there would have been enough time to equalize the frequencies in all of Europe.

WILLIAMSON: If you're only talking about a mutation that occurred 50 or 100 generations ago, as you are, you would have to have a major selective advantage, and no one has succeeded in showing such a selective advantage.

SERRE: Yes, I told you that this has been underestimated because certainly there has been some genetic drift (for example in Brittany) and secondly, there is not an estimation of the actual age of the mutation. It's just an estimate of the hierarchical relative age among populations. It's the same thing as the linkage disequilibrium, you know - it's another way to present the same data. But if you agree that the mutations enter Europe 4000 years B.C., there might be enough time for a selective effect.

BEAUDET: I certainly don't have much expertise in this area, but it troubles me to argue that the presence of the deltaF508 with other haplotypes at higher frequencies says anything about when it went where. If you believe there was a single origin, then every CF chromosome went through an equal number of generations from that origin, and you have to hypothesize that non-crossover people did migrate and crossover people didn't. This doesn't seem logical to me.

DEVOTO: It is much more likely that the mutation was introduced by migration in the North, always in association with the B haplotype because that association was by far the most frequent.

PRESENT STATE OF DNA-DIAGNOSIS FOR CYSTIC FIBROSIS AND ATTITUDES CONCERNING A HETEROZYGOTE SCREENING IN THE GDR

Ch. Coutelle (1), K. Grade (1), A. Speer (1), K. Will (1), Chr. Gille (1), R. Szibor (2), J. Gedschold (2), R. Brückner (2), V. Steinbicker (2), I. Bauer (3), K. Giermann (3), H. Gorki (3), S. Meyer (3), J. Brock (3), J. Hein (4) and U. Brell (5)

(1)Centr. Inst. Molec. Biol., Acad. Sci. GDR, Berlin
(2)Dept. Hum. Genet., Med. Acad., Magdeburg
(3)Inst. Biochem. & (4)Pediatr. Clin., WPU, Rostock
(5)lll. Pediatr. Clin., Berlin-Buch

Beginning with 1986 a nationwide system of genomic-diagnosis for PKU, Cystic fibrosis and Duchenne/Becker Muscular Dystrophy involving six DNA-laboratories has been introduced to the health care system of the German Democratic Republic. It is based on the long-standing system of genetic registers and the social and medical care available free of charge for the affected individuals and their families in this country [1].

Prenatal diagnosis for cystic fibrosis based on the determination of amniotic fluid enzymes and ultrasound was introduced by the Magdeburg group in 1984 [2]. This was a good preparation for the application of DNA-diagnosis for this disease, now performed routinely in Berlin, Magdeburg and Rostock. Presently about 900 children affected by cystic fibrosis are known among a population of more than 16 Mill.

DNA-diagnosis is offered to CF-affected families, giving them the opportunity to have a further nonaffected child. This offer was made in several talks to the parents/patients association and in individual counselling, explaining the scientific progress in CF-research as well as the growing diagnostic possibilities and their limitations. This information was always met with interest and sympathy. DNA-analysis was introduced using the flanking markers Met and pJ3.11 and crossover families were included in studies to determine probe localization [3,4]. Presently we are using the Delta F508 Mutation and RFLP analysis with XV2c, KM19, 7C22, D9, CS.7, MetD and MetH to gain full informativity for prenatal DNA diagnosis. Informed consent and freedom of choice concerning the application of this diagnosis and the termination of a pregnancy are fundamental principles of our work. Abortion on request is legal in the GDR until the 12th week of gestation.

By the end of 1989 34 prenatal diagnoses for CF had been performed and 8 pregnancies terminated because of the diagnosis of cystic fibrosis. 17 of the remaining fetus were CF-heterozygotes and 9 were homozygously nonaffected.

More than 160 CF-families have been investigated in a recent study on the occurrence of the Delta F S08 mutation and its relation to the tightly linked probes XV2c and KM19 [5]. The results shown in Table 1 indicate that 62% of all investigated CF-chromosomes carry the DeltaF508 mutation. 81% of the DeltaF505 and 56% of the NDF508 CF-chromosomes belong to haplotype B. This CF-deletion is linked in 92% to the KM19-allele 2. We also found a marked correlation of the meconium ileus syndrome with haplotype B and the Delta F508 deletion: More then 92% of the chromosomes from CF-patients with meconium ileus carry the deletion and 92% of them belong to haplotype B. (Table 1)

The high proportion of a single molecular defect causing CF has led to suggestions for the introduction of pilot studies in preparation of a heterozygote screening for this disease. Such studies should help to investigate the acceptance by the public of such a test, its ethical implications and also its logistics and costs. It should furthermore serve as a stimulus to develop and introduce new methods for the detection of the affected gene and to search for new mutations. Two of the involved centers in our country are preparing to offer it, on a strictly voluntary basis and after personal information, to a selected population of pregnant women as early as possible in gestation. This possibility is given by the present system of early reporting and monitoring of pregnancies in this country. Information would be given about the disease, its inheritance and frequency and the newly obtained possibility to avoid the birth of a substantial proportion of affected children. It would be explained why a blood sample or mouthwash from the mother and possibly also from the child's father has to been taken and that prenatal diagnosis is offered in case of detection of heterozygosity of both parents. The third of our centers is of the opinion that such pilot study should only be done when an at least 90% detection of heterozygosity can be reached, because of the possible unnecessary anxieties caused by the information, which can only be matched by an about 50% possibility to avoid the birth of an affected child. Complete agreement has been reached between the three centers and with the members of the CF-parents/ patients council to offer the DF508-test to members of CF-affected families.

TABLE 1

RELATION OF DF-MUTATION AND RFLP-HAPLOTYPES IN ALL CF-CHROMOSOMES AND IN CF-CHROMOSOMES FROM MECONIUM-ILEUS PATIENTS

	ALL PATIENTS						M.I.- PATIENTS					
	TOTAL	%	DF	%	NDF	%	TOTAL	%	DF	%	NDF	%
Nr.of DF/NDF CF-Chromos.	314		194	62	120	38	44		35	80	9	20
DF/NDF-Chromos. in phase with Haplotypes XV2c/KM19	205		134		71		27		25		2	
B 1 2		73		81		56		85		92		0
A 1 1		11		6		20		7		4		50
D 2 2		8		10		6		4		4		0
C 2 1		8		3		18		4		0		50
RFLP-Alleles in phase with DF/NDF-Alleles												
XV2c	215		142		73		28		26		2	
Allele 1		84		88		77		93		96		50
2		16		12		23		7		4		50
KM19	228		145		83		32		29		3	
Allele 1		18		8		35		13		6		66
2		82		92		6		87		84		34

One of the challenges connected with the introduction of screening programs is the development of techniques to easily process the large number of samples. To this end a theoretical and experimental study was performed to test the possibility of a mutiplex approach to heterozygote screening [6]. It is based on the electrophoretic detection of the heteroduplex formed by the 95bp and the 98bp PCR-product of the normal and the DF508 allele from DNA of a DeltaF508-CF heterozygous individual mixed together with DNA from non DeltaF508 individuals. The heteroduplex could easily be detected in our test-system at a constant amount of DNA of 500 ng and a ratio of DeltaF508-DNA to non DeltaF508-DNA of 1:49. The optimal number of samples from a randomly chosen population pooled together is reached, when the chance in this pool to contain or not to contain a heterozygote for the DeltaF508-mutation is equal. Assuming a general heterozygote frequency for CF of 1:22 and of the DeltaF508-mutation of 62% of all CF-chromosomes [5] an optimal pool size of 24 individuals was calculated. The number of tests for 1000 such individuals is reduced by this strategy to only 229. We did so far not succeed in applying this pooling strategy to a restriction based mutation test, but an oligonucleotide hybridization or an ARMS-PCR test should seem very promising.

References

1. Coutelle, Ch., Speer,A., 1990, Genomics in the GDR, **Genomics**, in press
2. Szibor, R., Weise, W., Gola, H., Roso, I., Gedschold, J., Steinbicker, V., Kropf, S., 1987, Pranatale Diagnostik der CF- Eine prospektive Studie auf der Basis der Bestimmung von Fruchtwasserenzymen und der Sonographie, **Z. schr. Klin. Med.**, 42:783.
3. Berger, W., Hein J., Gedschold, J., Bauer, I., Speer, A., Farrall, M., Williamson, R., Coutelle, Ch., 1987,Crossovers in two German cystic fibrosis families determine probe order for MET, 7C22,and XV2c/C5.7, **Human Genet.**, 77:187.
4. Estivill, X., Farrall, M, Scambler, P.J., Bell, G.M., Hawley, K.M.F., Lench, N.J., Bates, G.P., Kryer, H.C., Frederick, P.A., Stanier, P., Watson, E.K., Williamson, R., Wainwright, B.J., 1987, A candidate gene for the cystic fibrosis locus isolated by selection for methylation-free islands, **Nature**, 326:840.
5. Grade, K., Will, K., Szibor, R., Gedschold, J., Bruckner, J., Bauer, I., Giermann, K., Gorki, H., Hein, J., Brell, U., Coutelle, Ch., 1990, First analysis of the Phe508-Deletion in cystic fibrosis patients from the GDR, **Human Genet.**, in press.
6. Gille, C., Grade, K., Coutelle, Ch., A multiplex approach to heterozygote screening for the F508 cystic fibrosis mutation, **Human Genet.**, submitted.

DISCUSSION

WILLIAMSON: I want to present our experience in Britain briefly, and the reasons why we have come to a different conclusion from that which Coutelle has arrived at. In Britain we were faced with the same options, and as in the GDR, we have rejected the idea of neonatal screening and of screening children before the age of consent, in both cases for ethical reasons. However, when faced with a choice between screening at prenatal presentation and community screening, we have in general favoured the latter. We have a slight difference in practical terms because the time of presentation at prenatal clinics is Britain is considerably later than in the GDR, often 12 to 16 weeks and sometimes as late as 20 to 24 weeks. But that is not the basic reason we made the decision.

We made the decision, first, because there is a marked preference in favour of community screening prior to pregnancy on the part of the 2,500 people whom Eila Watson interviewed. There is a tradition of health care delivery by family practitioners in Britain, so most people look for advice from family practitioners rather than hospital obstetricians. Second, the key question of reproductive options - if you screen only when someone presents at prenatal clinic, you are only offering the options of continuing an affected pregnancy or terminating an affected pregnancy. We do not feel that deliberately limiting the range of reproductive options to these two (even though we believe that in our country the majority of people will choose termination of affected pregnancy) is correct.

Thirdly, and for me this is one of the most important points, when you offer screening in the community, you are offering screening to the individual. When you are offering screening in the prenatal clinic, you are offering screening to the couple. I don't know what it's like in the GDR, but in Britain now, 30 to 35% of pregnancies are to people who are unmarried and of those, approximately 15% are to people who are genuinely unmarried in the sense that there is no identifiable long-term partner. So we immediately have a major problem with respect to that. But I also think the whole concept of autonomy, which is a very important concept in medical care, means that it is better to give information on genetics and on reproduction to an individual, because an individual then has the right to make individual choices, as opposed to a couple.

The amount of anxiety involved in offering screening to a couple is a problem, and the question of education, counselling and time comes into it. Certainly in the context of the health service within which I work, the idea of having to offer counselling, to get the partner to come for testing, to offer education to large numbers of people, all in a few days at 12 to 13 weeks of pregnancy, fills most of us with trepidation. Because we are dealing, by and large, with a population that is interested in but has very little information about genetics, you will be approaching people at a very vulnerable time, and will have to give them a lot of information very quickly, and pregnancy, which is usually seen as a welcome event, suddenly becomes very traumatic for a large number of people. As compared to this, offering community screening pre-pregnancy, to individuals, where there is time for counselling, time for discussion, time for education - in Britain this seems to us to have great advantages, and we prefer trial schemes of this type to those based on the prenatal clinic.

SCHMIDTKE: Charles, I had the impression that your arguments against were mainly based on looking at the individual and the arguments in favor of screening were mainly based on looking at the society as a statistical argument. I think it's probably impossible to find a decision on the basis of weighing these two arguments. They are just incompatible in my view.

COUTELLE: There is something in it but I think you can't separate them both and at the end such a pilot study would have to show what it is. The big challenge to do it, is the present possibility that we could do it due to the well-organized situation. But I agree, the final decision will have to be taken at the end by the individual and we would hope that it would already come out during the consultation with the individual in the antenatal clinic, because there you're really dealing with individuals.

MACINTOSH: I would just like to point out that in Edinburgh we are going to look at heterozygote screening starting in the antenatal clinic. From past experience with establishing screening programs for neural tube defects and current screening also for Down's and so forth, this is taking place in the antenatal clinic. Obviously this is where you are just detecting an affected fetus rather than identifying carrier status, which requires looking at the partner and so forth. We believe that you just won't get the penetration into the community without having what are referred to as turnstiles by certain people.

COUTELLE: May I add two things? One thing is, of course, and there Jorge Schmidtke is quite right, the problem of how the screening has been thought through from the point of view of how you organize it. A different problem arises if you are offering it to a broader community: how do you get to them, how do you restrict what you can do? In an antenatal clinic you have already a very restricted area and you go in there and you can even select just by saying we will only take the very early pregnancies that we can select in this way. I must also admit that among our three centers we had the same discussions that we have here. Two of them are prepared to go this way, the other one is still reluctant and they are still thinking. So even in our society we are not completely through with it, that's fair to mention. The Rostock Clinic, I should identify it, is hesitant.

MEHTA: Bob, may I just ask you, are you hoping with your approach to sort of ride - to hitch-hike, to use an earlier term - on the screening approach by the general practitioner for other conditions. For example, most GP's now are going to be encouraged financially to screen for diseases X, Y and Z, so are you now going to say that CF counselling should be part of that process as and when the patient arrives, to make a turnstile at the GP, as it were?

WILLIAMSON: As of now, we're talking about pilot studies on carrier screening. Our group is working with general practitioners in one of the regions near London, and our idea is to make the counselling available through the G.P. service, partly because of the organization of the health service in Britain, where everyone is registered with a family practitioner and normally sees basic health advice as coming via that source. Of course it will be voluntary, and the invitation will go out from the GP to all his or her patients of reproductive age. We don't know whether 10% or 90% of people will come; we don't know exactly what will be required to get people who want to come to actually come. We are considering sending out a little tube and wooden spatula so that people can scrape the side of their cheek at home and pop it into the tube and send it to us via the G.P. in a stamped, addressed envelope. So far as uptake of screening in Britain is concerned, it varies from something on the order of 70% or so for cervical screening down to approximately 20% for breast cancer screening at the moment. This variation is partly due to financial incentives to G.P.s but it is probably due even more to the amount of publicity within the community and the number of prompts that exist to encourage people to come. We are dealing with very new questions, which have only been examined before in our community either very small populations (such as the Cypriots in London for thalassaemia, or the Askenazy Jews in the United States for Tay Sachs disease). From the point of view of most of the countries here, we are talking about something new, something we haven't tried before, so that's one reason why we're not sure.

BUCKWALD: It seems to me that to say that one chooses certain particular projects on ethical reasons is a little bit misleading. I mean, ethical issues are issues that one debates, and for which there may be quite valid different positions. Different individuals or different societies or different groups might make certain choices based on these discussions. It seems to me at this stage, since one really doesn't know what is going to work and what is not going to work, it would make a lot more sense to have as many different pilot projects as possible and evaluate them in terms of all the various issues that arise in relation to delivery of medical care, privacy, autonomy, etc.

SCHMIDTKE: I just want to add that this is not primarily a question of ethical concern. It is a question of the number of choices a person has. Certainly, when you offer the testing in pregnancy you have, as Bob pointed out quite rightly, you have only two choices, whereas if you do it much earlier, at a pre-pregnancy state, the number of choices is greater. So it is mainly a matter of strengthening the individual's autonomy.

BUCKWALD: The point is that it may turn out that by offering a particular program that offers the most autonomy you reach the fewest people.

PROS AND CONS OF NEONATAL SCREENING FOR CYSTIC FIBROSIS

Gerard J. te Meerman (1) and Jeannette E. Dankert-Roelse (2)

1 Department of Medical Genetics
University of Groningen, Groningen, The Netherlands
2 Department of Pediatrics
Free University of Amsterdam, Amsterdam

INTRODUCTION

In the past, several experimental schemes for neonatal screening for the presence of Cystic Fibrosis have been tried out. Several benefits of early detection have been claimed, but few studies are methodologically flawless, mainly because of the use of historical controls and/or a too short period of follow-up. At least initially, many programs suffered from the fact that after detection no systematic patient treatment scheme was available, and that no consensus was available on optimal treatment. The life expectancy of patients with cystic fibrosis has considerably improved over the last decades, mainly because better antibiotics became available and because treatment of complications has improved.

The quality of the tests that have been used for neonatal screening, such as the meconium (albumin) test and the IRT test, has not been optimal, especially because of a high number of false negatives.

Currently, controversies about optimal treatment are still present, but they center around (important) details of antibiotic treatment, rather than being differences of opinion on general aspects of management.

A new start with neonatal screening has to take into account all conceivable consequences, and it has to be shown that the negative aspects are outweighed by the positive ones. Alternatives, such as adult carrier screening, have to be considered, and neonatal screening has to prove to be the better choice. Several decision problems require empirical support. Before screening should start on a routine basis, experimental screen-

The Identification of the CF (Cystic Fibrosis) Gene
Edited by L.-C. Tsui *et al.*, Plenum Press, New York, 1991

ing projects, with a reasonable number of patients detected are to be carried out.

The purpose of the present contribution is to present a proposal for a neonatal screening program, and to discuss systematically the medical and ethical aspects.

Proposal for an Experimental Neonatal Screening Procedure

In most northern European countries the most prevalent deltaF508 mutation is so frequent that up to 79% of the CF chromosomes carry this mutation. This frequency would imply that only a fraction of .21x.21=4.3% of CF patients will not carry the delta mutation. The same applies to the parents of patients. From various sources it appears that the non-deltaF508 mutation carrying patients are in average less severely affected than other CF patients, implying that later detection may be less harmful than for other patients. This justifies an approach where only deltaF508 mutations are investigated in parents of newborn children. Selective PCR amplification with non-radioactive methods (Scheffer et. al. 1989), using material obtained from a mouthwash (Lench et. al., 1988) or urine (Gasparini et. al. 1989) is then able to detect 50% of the parents definitively at risk directly. A further 43% of parents at risk will belong to a group of parents where only one parent carries the delta mutation. More extensive DNA testing in this subgroup could reveal more parents definitively at risk. For children of parents at risk or with a higher probability of being at risk, sweat-testing of their children at two or three month will identify affected children.

Slightly inferior performance characteristics can be obtained from sweat-testing those children who themselves carry the delta mutation. The disadvantage of this method is that carrier status of the child would be known, which may not be in the best interests of the child, although the problem is not as important here as in the case of dominant diseases. The presence of an observation on the child which leads to further investigation, also could cause worry in the parents.

For parents not wishing to use a test where their carrier status is determined, an IRT test could be used, with a rather low threshold for further testing, and immediate DNA follow-up for suspect cases with a high probability of being affected, thus enabling either rapid diagnosis or further investigation on a rather low-risk population. Carrier status would not directly be communicated. Such an option for the parents would clearly indicate that finding the carrier status is a by-product of a method for determining CF affection in children, but that finding patients is the primary aim of a neonatal screening program.

The Organizational Consequences of Neonatal Screening

The primary health care system in the Netherlands provides regular checks on the medical status of a child in separate facilities (called "consultatiebureaus"), where children are seen by a physician and a registered nurse. Immunizations are given in well-organized centrally registered continuous campaigns, partly executed in the "consultatiebureaus" and in the schools. PKU screening shortly after birth is almost universal. Determining the carrier status of the parents will present a problem mainly if only one parent is present when the child is checked. The possibility of using urine or a mouthwash will probably make it easy to determine the carrier status of parents. Neonatal screening will require good cooperation between the primary health care system and specialized centers for diagnosis of CF, to carry out the sweat test. Sweat-testing the about 3% of the newborns where one of the parents is a carrier, would, in the Netherlands, amount to 5400 sweat-tests yearly. Liberal use of sweat-tests to detect CF patients among children presenting persistent pulmonary problens, using the facilities for sweat-testing necessary for the screening, could easily double this figure.

Treatment and follow up of detected patients has to be done in a systematic way. Our experience has been that regular support given from a CF center to local pediatricians, and regular visits to a CF center is effective and agreeable to all persons involved.

The overall effect of a neonatal screening program has to be evaluated from different viewpoints:

1. Positive and negative aspects of early detection for the detected patients.
2. Negative aspects of carrier and patient screening.
2.1 Confirmation of the diagnosis in carriers
2.2 False negatives
2.3 False positive cases
2.4 Assortive mating
3. Carrier detection and consequences for counseling
4. The burden on carrier families
5. The burden on families at risk for cystic fibrosis
6. Concluding ethical evaluation

1. Positive and negative aspects of early detection for the detected patients

The investigations of Dankert et. al. (1989) have shown that the life expectancy and clinical conditions of early detected patients are better than those of later detected patients, despite the presence of a bias in the other direction, because serious illness will be diagnosed earlier as cystic fibrosis. A screening programme should not be carried out before one is sure that the medical requirements for treatment, psychosocial counseling and prevention of permanent lung damage can be met. Detecting a disease may cause serious problems if there are no adequate insurance policies available for the screened patients.

From a medical point of view, patients are expected to gain advantages from neonatal screening. Their general well-being will be positively influenced by a better nutritional status and because infections are better treated. On the minus side, patients will be much earlier aware that they suffer from a serious condition with a strongly reduced life expectancy. This may cause serious problems in social adjustment, and lead to difficulties in their families. The psychological adjustment to life-threatening disease is manifold. Some may derive strength from a reduced life expectancy, others may become depressed. Early detection will on the other hand give a much longer period for parents and other family members to get acquainted with the disease, long before serious health problems develop. For the ethical evaluation it is especially important that some patients may suffer very strongly because of undertreatment as a consequence of late detection.

2. Negative aspects of patient and carrier screening

2.1 Diagnosis in children at elevated risk

About 95% of the patients will be detected because their parents appear initially as carriers of the DeltaF508 mutation. When both parents are carriers for a CF mutation, they have a 1 in 4 probability of being affected. A quick diagnosis is then possible. Children with only one parent as a proven carrier have a much smaller (about 1 in 400) probability of being affected. This small probability may however induce anxiety in parents and other family members. In previous investigations (Dankert et. al., 1983) it appeared that parents, after a false positive meconium test result, despite the anxiety which led in occasional cases to extended medical surveillance, were in favor of screening. From a public health view, one should be more pessimistic. The fact that no test on the child is carried out, will probably avoid too much worries.

2.2 False negatives

False negative cases are mainly possible because at present not all CF mutations can practically be detected, even if all of them could be known. As argued before, these cases will in average present with a milder form, for which later detection will not be as harmful as it could be in the severe form. False negative cases will also occur due to contamination or interchange of specimens. A strict system for quality control of the analysis is therefore mandatory. Current laboratory guidelines for similar analyses will apply, and no specific difficulties are to be expected.

Patients suffering from cystic fibrosis without a DeltaF508 mutation, will form a minority among the CF patients. If a screening programme is present, the diagnosis of CF will be more difficult, because the incidence of undetected cases will be very low. Probability diagnosis would then preclude detection of these cases. Detection of false negatives may however be improved because an efficient and easily accessible organization for sweat-testing will become available. The threshold for sweat-testing will therefore be lowered, and if doctors would still consider CF in the differential diagnosis of children with persistent pulmonary and/or digestive problems, diagnosis would still be made early.

2.3 False-positive cases

With the advent of DNA technology, false positive cases resulting from DNA analysis should be virtually absent, as a confirmation of the result will generally be carried out on the base material. False positive cases for the deletion will be virtually absent. False positive families will experience the same burden as those families in which a carrier is detected. Any false positive result from the DNA test, will therefore indicate a still low probability of being a patient. When appropriate precautions are taken, parents should be aware that the testing procedure will consist of a sweat-test in about 3% of the families, but that only a small fraction of the children where a sweat-test is necessary, will be affected. False positive results from a sweat-test at children age will generally occur rarely, but if they occur it will probably lead to closer medical surveillance over a few years, before the final diagnosis is made. The follow up of any experimental programme should look at these unavoidable side effects, to improve the quality of the detection process.

2.4 Assortive mating in carriers

Knowledge of carrier status before a partner is found, may induce assortive mating among other carriers, because of a common stigma. Assortive mating will most likely occur because those identified as non carriers would avoid mating with people identified as carriers. This is of course irrational from a genetic point of view, but misunderstanding of the nature of the CF gene could induce such processes in an important minority of the population. If carriership were interpreted as a stigma, then a carrier screening programme would quickly loose its impetus, and people will remain silent about their carrier status, with very unwanted side effects as a child with CF would result. This would result in exactly the opposite effect on the incidence of CF, as one would expect from carrier testing.

Curiously enough only a very minor assortive mating effect, could, given the fact that many parents at risk choose not to use prenatal diagnosis, lead to an increase in the frequency of cystic fibrosis, instead of to a decrease. If the probability for a carrier to mate with another carrier would triple, and only 50% of the couples at risk would use prenatal diagnosis, a 50% increase could result. Especially in countries where mating relations are unstable, both assortive mating and not using prenatal services could occur.

3. Carrier detection and consequences for counselling

Only the couples without one copy of the deltaF508 mutation would go undetected until they give birth to a child with CF. A neonatal screening programme where parents would seek counseling and further DNA diagnosis after having their carrier status determined, would therefore identify the large majority of the carriers in the population in the long run. It is quite conceivable that carrier detection through the described early detection programme, although as a side-effect, is very effective in detecting carriers, probably even more than carrier detection programs.

Counselling the parents at risk takes of course quite a bit of effort, but most couples would not be at risk, and need to be advised only of the probability of CF in other relations. Genetic counseling centers should be able to handle the extra load of work, which on the DNA level is low against the background of the effort required for routine testing. Most effort will be spent on advising carriers, especially brothers and sisters of carriers, especially when they do not have stable relations yet.

4. The burden on carrier families

On the basis of present knowledge there are no health risks involved in being a carrier. Theoretically even an evolutionary advantage may be present. Screening for autosomal recessive disease is quite different from screening for dominant or X-linked recessive diseases, because the consequences of carriership are of far less, close to zero, consequence for the carriers. Nevertheless, eugenic misunderstandings are still quite common, and knowledge of carriership of a serious disease could make people feel uncomfortable.

Before embarking on a carrier detection programme one should insure that any possible negative implications of carriership are absent. Preferably there should be legal provisions that knowledge about carriership is declared of nil value for insurances and that it can legally be withheld from insurance companies. Legal aid should be available to any person who might encounter difficulties because a carrier status has been determined.

5. The burden on families at risk for cystic fibrosis

Families having a 1 in 4 risk of getting a child with CF are in a difficult position. Most of these families will be detected without any suspicion that they might be at risk. It may be difficult to understand that, although they have perfectly healthy children, there is a risk of getting a child with CF. Whatever choice the parents make concerning prenatal diagnosis, subsequent pregnancies will not be without worries.

6. Concluding ethical evaluation

Screening programmes are results of institutional decisions made under the responsibility of the state. In some countries legislation will have to be made to carry out screening programmes. An ethical evaluation of screening programmes can meaningfully use the idea of "Justice as fairness" as set out by Rawls (1972). The primary question is then whether it is among the duties of the state to institute carrier screening and/or early patient detection. The first duty of the state is to protect the rights of the citizens. The principle that should be adhered to is that the good for society is not necessarily the highest average outcome. The fact that some patients suffer severely from too late diagnosis and subsequent adequate treatment, indicates that the state has a responsibility to protect patients, and neo natal screening is then an option. Promoting screening of heterozygotes as such, without direct medical benefit, would be a mandatory responsibility of the state only if it were impossible to exert the right to know one's genetic make-up otherwise. This is not the case, which leads to the conclusion that carrier screening as such is not an obligation for the state. The state may however institute carrier screening for utilistic arguments, such as high costs caused by disability to work. These utilistic arguments, should be carefully separated from individual benefits related to rights of individuals, to be derived from screening.

Carrier screening, in relation to a programme to detect and treat patients with cystic fibrosis, can be seen as a responsibility of the state, because it promotes important aspects of health in individuals, just as in the case of PKU screening. The obligation for the state with respect to carrier screening is that rights of individuals are not jeopardized by such a programme. This requires that besides carrier-detection based discovery of affected individuals, also a direct detection possibility exists, ensuring that detection as such does not depend on the willingness of parents to determine their carrier status.

Would people opt for participation in a screening programme if they were fully informed of all possible consequences? Advantages and disadvantages of a screening programme have to be clearly identified, before an individual can make a rational decision, especially about the choice between a diagnostic method with or without implied carrier determination.

Individuals should be free in using a screening programme, and should be free also in their subsequent decisions. Utilistic reasoning often relates these two forms of freedom. The argument is, for example, that it does not make any sense to possess knowledge if one is not inclined to use this knowledge. Thus, prenatal diagnosis should only be offered when parents intend to have an abortion when the expected child will be severely handicapped, and it is made available only to those who intend to use the information for a decision concerning an abortion of an affected child. Such a recommendation is actually made in connection with Huntington's disease by representatives of the International Huntington Association and the World Federation of Neurology (1990). A difference exists with cystic fibrosis, as children with CF will require treatment from birth onwards, while patients with Huntington's chorea will develop a serious condition only later in life. There is however an inconsistency in the actual policy with respect to Cystic Fibrosis and with respect to Huntington's chorea. One might object that the reasoning with respect to Huntington's chorea is utilistic, because the meaning of genetic information is not seen as informing people about their and their children's genetic make up, but merely as leading to use. The moral responsibility of the subject, which is to make a moral decision at the time when it has to be made, is ignored by asking for a premature decision. This responsibility to make a decision when it has to be made, is also a right: the right to postpone decisions to that moment when one knows exactly in which situation one is placed. One cannot make a decision to abort, unless one is pregnant. Even if one a priori objects to abortion, one may make another decision when all facts are there. The converse also applies: if one intends to abort for serious reasons, one may decide to refuse abortion. Requiring a premature decision to abort given a possible outcome of prenatal diagnosis is a fundamental violation of human rights: the right to make one's own moral decisions without interference; the right to freedom of conscience. (Rawls, 1972). One should never be forced to have an abortion, nor should one be forced to express opinions about it before the decision actually has to be made. The experience of Feldmann et al. (1989) shows that it quite often happens that a diagnosis of an affected child is not followed by an abortion. This should both be reassuring about the quality of the genetic counselling offered and about the moral complexity of the decision perceived by the parents. As long as parents can consider using genetic information, they should be allowed to obtain it, irrespective to what they intend to do with it. Of course the complexity of having certain knowledge should be clearly set out, but their responsibility is the final one.

Any screening programme should primarily be judged whether individuals may expect benefit from it. A limiting condition is that it respects basic rights such as the freedom of choice of people. The proposed neonatal detection programme should be compared to programmes for detection of carriers before any children are born, including those where prenatal tests are made available. The preventive aspect of adult screening may lead to serious consequences with respect to the acceptance of affected children: will there be no implicit or explicit societal attitude to blame the victim, or in any case the parents? Will there not be "wrongful-life suits" against parents not using carrier screening, or against doctors not advising to use carrier screening? In Europe this principle has not been accepted on any scale, but for other countries this is different. There exist strong psychological mechanisms to blame the victim anyhow, especially now more is known about the relation between life style and disease (smoking, obesity, alcohol and other drug consumption).

The advantage of a newborn CF detection programme is that it starts with a positive attitude towards the affected: the aim of the screening is primarily that the health and the quality of life of CF patients can be improved if adequate medical intervention is provided at the right time. A very important argument is also that some individuals may suffer extraordinarily if newborn screening is absent: it has been documented too often that children suffering from cystic fibrosis were undertreated, and reached a much worse clinical situation than would have been the case if they had been detected earlier and adequately treated. In the ethical analysis severe damage to some has to get more weight than relatively minor negative effects to many. We oppose utilistic reasoning where the suffering of a few people is averaged with minor disadvantages of many.

If carrier detection resulting from a neonatal detection programme should lead to preventive action on the part of parents and their family, it would be in another context, as mass screening for carriership would not be organized in the context of prevention of the birth of patients, but only related to children already born.

In conclusion there seem to be three options:

1. Only an IRT test is performed, and no carriership is determined, except when the child is affected.

2. In the context of IRT testing the parents get information on the possibility of carrier testing for one or both of them.

3. A DNA test on the parents is performed, and explicit carrier information is given on demand. Children of carriers are sweat-tested for possible CF at three months.

CONCLUSION

Newborn screening together with an option for carrier screening of parents is now the best option for using the present knowledge about the molecular origin of Cystic Fibrosis, rather than carrier screening in the population. Implementation of newborn screening requires, however, very careful medical follow-up of detected patients and very careful counseling of parents found to be at risk. The idea that CF is a preventable disease, for which parents are responsible in any way, should be avoided.

Acknowledgement

We would like to thank our colleagues and several participants in the discussion on the workshop "Recent Progress on the identification of the CF gene" Sestri Levante 9-12 April 1990, for their valuable contribution, without making them responsible for any opinion expressed in this paper.

References

Rawls, J., 1972, A Theory of Justice. Oxford University Press

Dankert-Roelse, J.E., G.J. te Meerman, A. Martijn, L.P. ten Kate, and K. Knol, 1989, Survival and clinical outcome in patients with cystic fibrosis, with and without neonatal screening, **J. Ped.**, 114:362.

Lench N, P. Stanier, R. Williamson, 1988, Simple non-invasive method to obtain DNA for gene analysis, **Lancet**,ii:1356.

Gasparini P, A. Savoia, P.F. Pignatti, B. Dallapiccola, G. Novelli, 1989, Amplification of DNA from epithelial cells in urine, **N. Eng. J. Med**, 12:809.

Feldman G.L., N. Lewiston, S.D. Fernbach, W.E. O'Brien, R. Williamson, B.J. Wainwright, and A.L. Beaudet, 1989, Prenatal diagnosis of cystic fibrosis by using linked DNA markers in 138 pregnancies at 1-in-4 risk, **Am. J. Med. Genet.**, 32:238.

Unnamed, Ethical issues policy statement on Huntington's disease molecular genetics predictive test, 1990, **J. Med. Genet.**, 27:34.

Scheffer H., E. Verlind, D. Penninga, G.J. te Meerman, L.P. ten Kate, and C. Buys, 1989, Rapid screening for deltaF508 deletion in cystic fibrosis, **Lancet**, ii,1345.

Bloch N, and N.R. Hayden, 1990, Predictive Testing for Huntington's Disease in Childhood: Challenges and Implications, 1990, **Am. J. Hum. Genet.**, 46:1.

DISCUSSION

SCHMIDTKE: Would not a test that would only detect the homozygotes be the optimal test in your view?

te MEERMAN: No, the test will detect heterozygotes and also homozygotes, of course. It can be technically done in many forms but my talk was not specific for the way you would detect it. The idea is that, however, you would do it. You would try to detect heterozygotes. You will automatically detect homozygotes and in these patients the diagnosis can be made very quickly.

SCHMIDTKE: If the main incentive for doing the screening is the benefit of the individual that gets better treatment by early diagnosis, the test would be optimal if it only detects the homozygotes (the affected ones). In this way you would not have the disadvantages of finding out genotype of those who would then know about their genotype without wanting to.

te MEERMAN: Yes, I agree that this is a disadvantage, but the point is that you could, of course, also choose just to look only for the patients - that's an option too - but I think that the carrier option is also reasonable. It's reasonable that people have this advantage because other people would have benefit from it. But I agree that you can have different opinions about that.

IANNUZZI: I wish you would comment on two issues. The first is, since I believe that 90% of CF patients are diagnosed by the age of two...

te MEERMAN: That is not true in the Netherlands.

IANNUZZI: Then what percentage of patients are diagnosed and by what age and what treatment is there to improve the patient's care administered, say, in the first few years of life? And the second issue is: since there probably isn't anything you are going to do which will improve treatment in the first year of life, how is informing the parents in the first six months of life going to interfere with maternal-infant bonding? This may have tremendous consequences in infancy. In other words, is there a treatment that benefits the patient more than not interfering with maternal-infant bonding?

te MEERMAN: Our experience is that when we get children very early we can prevent many complications and are able to keep them in much better condition than patients that are detected late, where you find many pulmonary complications, pseudomonas infections that all could have been prevented if they had been adequately treated. What typically happens is that you get young children that have contracted pneumonia and are under-treated, because they need much higher doses of antibiotics and we have had many cases where children were in a very bad state when we detected them finally at the age of 4 or 5 years. So I think there are serious reasons to believe that early detection is, for a few patients, very important.

KNIGHT: We've heard a lot today about the very wide variation in clinical expression of cystic fibrosis and we still have a sort of assumption that individual mutations in the gene are correlated with these variations. I'm not entirely convinced myself that they are. So can we in fact screen patients and tell them whether they are going to have mild disease or severe disease, or can't we? And if we can't, is there a justification for screening at all?

te MEERMAN: In my opinion, you can't say in advance whether it will be mild or severe. The point is that if cystic fibrosis is treated from the beginning on in a very systematic and well designed way, you can prevent many complications. If I tell you that the life expectancy of cystic fibrosis patients has, in the last twenty years, about tripled in the Netherlands, and that we have the idea that many of the patients dying at this moment could have lived much longer were they better treated at an earlier stage, I think that there is some reason to approach the disease more optimistically than in the past. In the past, the attitude was "It's cystic fibrosis - it's lethal - we can't do anything about it." But now we have the prospect that the life expectancy will be perhaps 30, 40 years, which would still enable people to have a productive life.

GOODCHILD: I'm interested in Dr. te Meerman's comments, which are based partly on the rather different findings of the Dutch community with respect to early diagnosis, compared with later diagnosis. But may I just say, in the United Kingdom, as a result of the Wales-West Midlands combined neonatal screening survey, up to the age of 4 1/2 years so far, we have not found any clinical differences whatsoever between those who were diagnosed early (that is at a mean of 9 weeks by screening) compared with those diagnosed later, and that is at a mean of 10 months, by non screening. In other words, these children have been diagnosed on the basis of their clinical symptoms. So we feel in the U.K. at the moment that the case has not been made for recognizing these children early, and therefore, I think as an elaboration of that, I don't think the case can be made at the moment, particularly with the development of all the various mutations, for screening early and for recognizing the cystic fibrosis babies early.

MEHTA: In relation to that, if you don't have the neonatal screening, what about other children that family might have? That you might have prevented if they wanted it. What do the parents feel about that?

GOODCHILD: I agree with you theoretically that I think it is the parents' right to know their genetic constitution and therefore to make informed plans about further pregnancies. But if you have a look at the data in Wales, for example, it's only something like 7% of pregnancies that occur within the 12 months of the last pregnancy. And therefore, if you look at the absolute numbers involved, I'm informed that there are only one or two families in whom this difference would emerge as a very real difference. Now I think even one case is individually important, but probably the people who work in the Welsh office wouldn't agree.

COLLINS: I'm concerned also about the strategy of screening those who are heterozygous for F508 by doing a sweat test at three months, for two reasons. One is that you are certainly going to induce a large amount of anxiety in the parents during that three month time interval, because presumably you're going to tell them that they need to come back for this test because there is some possibility their child might have CF. Secondly, that will certainly be the group where your ability to predict severity will be the worst, because if they do turn out to have a positive sweat test, you won't be able to say with nearly as great certainty as apparently you can with delta F508 homozygotes, what the course will be, and even in your study it may not be that early diagnosis for those children is beneficial. Do you think that this particular part of the screening makes sense in terms of the balance of these various factors?

te MEERMAN: I agree that this is a very difficult point. I'm even inclined to think that you should be able to call it off when this aspect appears to be very bad. But on the other hand, our experience clinically is that we have very large phenotypic variation which is not always associated with the presence of the deletion as such.

WILLIAMSON: I was concerned at your implication that the availability of carrier testing within the community necessarily results in stigma for carriers.

te MEERMAN: Not the availability. The promotion, by the state.

WILLIAMSON: Well, the making of something available, whether by the state or privately or any other way, doesn't necessarily cause stigma. The general experience with other diseases has been that stigma results when no other alternatives, no other options, are offered other than carrier testing. In the cases of thalassemia, Tay Sachs, various other conditions, as soon as other alternatives are offered, as soon as people are actually given the reproductive freedom to make their own choices, far from resulting in stigma, the increase in knowledge and the increase in autonomy usually seems to result in a very much better situation.

I'm also very concerned at the implication that the availability of prenatal diagnosis results in a negative attitude to patients. Certainly, in the case of thalassemia and other conditions, that has not been true. The availability and implementation of prenatal diagnosis by and large has gone hand in hand with a tremendous improvement in attitude to those who are affected by the disease. I think that to deny people the ability to make their own choices on reproduction in relation to what is by and large still a very serious disease, is to deprive people of a fundamental autonomy, and I think that the real argument for community screening is one of increased freedom and not one of decreased freedom.

te MEERMAN: That's too easy. The point is not that I am depriving people of any freedom. The only thing that I'm talking about is whether the state should promote it or not. It should be made available, but promotion is something completely different from making things available and making things publicly known. That is the point.

GENOTYPE-PHENOTYPE CORRELATIONS IN CYSTIC FIBROSIS PATIENTS

Manfred Stuhrmann (1), Thilo Dörk (2), Michael Krawczak (3), Margarete Dueck (4), Ulrich Banholzer (1), Jutta Domagk (5), Nicola Hoffknecht (6), Hans-Georg Posselt (7), Andre Reis (1), Manfred Schlösser (3), Fritz Trefz (4), Michael Wagner (3), Ulrich Wahn (8), Brigitte Wulf (2), Jörg Schmidtke (1), Jochen Reiss (3), and Burkhard Tümmler (2)

(1) Institut für Humangenetik
Freie Universität Berlin
(2) Abteilung Biophysikalische Chemie
Medizinische Hochschule Hannover
(3) Institut für Humangenetik, Universität Göttingen
(4) Kinderklinik, Universität Heidelberg
(5) Kinderklinik, Universität Göttingen
(6) Abteilung Humangenetik
Medizinische Hochschule Hannover
(7) Kinderklinik, Universität Frankfurt
(8) Kinderklinik, Freie Universität Berlin

SUMMARY

Genetic and biomedical data from 346 cystic fibrosis patients of German origin have been evaluated. We demonstrated an age dependent distribution of CFTR genotypes, and confirmed the previously reported association between the dF508 mutation in the CFTR gene and pancreatic insufficiency. However 3 out of 22 pancreatic sufficient patients were dF508 homozygous. When patients were grouped with respect to height development, significant differences were seen in the distribution of J3.11-MspI alleles. We conclude that genetic determinants in and around the CFTR gene contribute to the variability in the clinical course of the disease.

INTRODUCTION

The recent isolation of the cystic fibrosis transmembrane conductance regulator (CFTR) gene (Rommens et al., 1989; Riordan et al., 1989) and the identification of mutations within this

The Identification of the CF (Cystic Fibrosis) Gene
Edited by L.-C. Tsui *et al.*, Plenum Press, New York, 1991

gene associated with cystic fibrosis (Kerem et al., 1989; this conference) have opened up the prospects for a thorough understanding of the etiology and pathogenesis of this frequent genetic disease in Northern European populations. A striking feature of cystic fibrosis is the variability of its clinical course, ranging from early onset with severe symptoms to mild manifestation in later life. It has been speculated that both environmental and genetic factors may contribute to the clinical variability; genetic determinants may include allelic heterogeneity, modifying genes linked to CFTR, or, less specifically, the "genetic background" of the affected individual (Getschold et al., 1987; Knowles et al., 1989; Mornet et al., 1989; Tümmler et al., 1990).

We have initiated a multi-center collection of genetic and clinical data of cystic fibrosis patients of German origin, and present here a first attempt to correlate the distribution of alleles and genotypes in and around the CFTR gene with specific patterns of disease manifestation.

PATIENTS AND METHODS

Patients: Altogether 346 patients were ascertained either through laboratories involved in routine gene diagnosis or through cystic fibrosis outpatient clinics at Berlin, Frankfurt, Göttingen, Hannover and Heidelberg. As most of these centers regularly received diagnostic requests from outside, our sample is likely to form a representative survey for West Germany.

Genetic testing: The following genetic markers were tested: MetD-BanI, MetD-TaqI, MetH-MspI, XV2c(D7S23)-TaqI, KM19(D7S23)-PstI, dF508(CFTR), J3.11(D7S8)-MspI, using standard Southern blot and/or PCR methodology. For technical reasons not all patients could be tested for the complete set of markers.

Clinical data: The patients' records were evaluated using a standardized protocol. The following parameters were included: (1) sex, (2) age, (3) meconium ileus at birth, (4) pancreatic function (a patient was scored pancreatic sufficient if pancreatic enzyme substitution was not required up to the age of five years, and if fecal excretion of fat did not exceed 5g per day), (5) height, weight, percentage of weight predicted for height based on the latest clinical assessment, (6) sweat chloride concentration, (7) lung function, (8) X-ray score and (9) bacterial colonisation of the lung.

RESULTS AND DISCUSSION

The methodological procedure used was to classify the patients according to the biomedical parameters and to calculate the allelic and genotype distributions within these groups.

1. Distribution of the CFTR-dF508 mutation in cystic fibrosis patients of different age groups

Cystic fibrosis patients who reach adulthood are probably more likely to suffer from a milder type of the disease. We therefore subdivided our patients into four age groups, and calculated the distribution of the dF508 mutation in these groups. If the dF508 mutation represents a "severe" CF-mutation it is to be expected that its relative frequency decreases with age. Table 1 shows that a negative correlation between age and the frequency of the dF508 mutation does exist, but the differences between the groups are only slightly significant (e.g. age <10 years vs. age >20 years: chi-square=3.2; p=0.07). A similar correlation is observed with respect to genotypes (e.g. age <20 years vs. age >=20 years: chi-square=5.50; 2 d.f.; p=0.06). Therefore, it appears that a higher life expectancy in part depends on the presence of at least one "mild" allele.

Table 1

FREQUENCY OF DF508 MUTATION IN CYSTIC FIBROSIS PATIENTS OF VARIOUS AGE GROUPS

age	number	dF508/dF508	(%)	dF508/N	(%)	N/N	(%)	dF508 (%)
<10y	(78)	51	(65)	21	(27)	6	(6)	79
10-20y	(83)	52	(63)	21	(25)	10	(12)	75
20-30y	(67)	33	(48)	29	(43)	6	(9)	70
>=30y	(5)	3	(60)	1	(20)	1	(10)	70

dF508: dF508-CFTR allele
N: non-dF508 CFTR allele

2. Distribution of the CFTR-dF508 and flanking marker alleles among pancreatic sufficient and pancreatic insufficient patients

A highly significant correlation between the presence of the dF508 mutation and pancreatic insufficiency was observed (Table 2). This finding also confirms the observation of Kerem et al. (1989), but contrary to their assertion that presence of at least one "mild" allele is required for the manifestation of the pancreatic sufficient phenotype, we have observed homozygosity for the dF508 mutation in 3 out of 22 pancreatic sufficient patients. A similar observation was made in cystic fibrosis patients of Czech origin (Stuhrmann et al., 1990). The age of the three pancreatic sufficient patients was 3 9/12, 18, and 23 5/12 years, respectively, at the time of the last clinical assessment. However, pancreatic sufficiency mostly occurs in patients carrying other mutations. We have previously shown that one or more of these "mild" alleles may be associated with MetH-MspI allele 2 (Stuhrmann et al., 1990).

Table 2

CORRELATION BETWEEN THE DF508 MUTATION IN THE CFTR GENE AND PANCREATIC INSUFFICIENCY

	PI	PS
dF508 alleles	349	15
other CF alleles	97	29

n= 490; Fisher's exact test: p=4*10E-9

PI: pancreatic insufficient
PS: pancreatic sufficient

3. Correlation between J3.11 (D7S8)-MspI alleles and physical development

We recently reported on a relation between J3.11(D7S8)-MspI genotypes and the course of the disease (Schmidtke and Krawczak, 1989; Tümmler et al., 1990). In patients older than 18 years we had observed a marked sex difference for the MspI-RFLP detected by J3.11 (D7S8). The enlarged sample present in this study no longer supports such a sex difference. However, an association between J3.11-MspI alleles and development is still obvious from the data. We classified the patients into two groups, one below and one above the 50th height percentile. Comparing the two groups, a difference in the J3.11-MspI allele frequencies becomes apparent, with allele 1 occurring at a significantly higher frequency in the group of patients above the 50th percentile of height development (Table 3a). Similarly, the J3.11-MspI 1-1 genotype appears to be overrepresented among the patients with better height development while genotype 2-2 appears to be underrepresented. The group of patients with worse development shows a reverse effect. The heterozygotes occur at the expected frequencies in both groups (Table 3b). Haplotype analyses revealed that the above-mentioned effects are not due to allelic association between J3.11 and dF508.

Table 3a

CORRELATION BETWEEN J3.11(D7S8)-MSP ALLELE FREQUENCIES AND GROWTH

HEIGHT PERCENTILE	ALLELE 1	ALLELE 2
>50	125	69
<=50	162	136

n=492; chi-square=4.9; p=0.03

Allele 1: absence of MspI-restriction site
Allele 2: presence of MspI-restriction site

Table 3b

CORRELATION BETWEEN J3.11(D7S8)-MSP GENOTYPE FREQUENCIES AND GROWTH

HEIGHT PERCENTILE	1-1	1-2	2-2
>50	40	45	12
<=50	47	68	34

n=246; chi-square=4.99; 2 d.f.; p=0.08

Mornet et al. (1988) reported an association of J3.11-MspI alleles with meconium ileus. In our sample, which included 21 patients with and 225 patients without meconium ileus, we observed neither an association of meconium ileus with J3.11-MspI nor any other CF marker including dF508.

Our results suggest that both the allelic heterogeneity at the CFTR locus and genetic factors associated with flanking markers contribute to the variation in the clinical course of cystic fibrosis.

References

Getschold, J., Kropf, S., Szibor, R., et al., 1987, Cystic fibrosis - a single locus disease? Results of a population genetics study, **Hum. Genet.**, 75:277.

Kerem, B.S., Rommens, J.M., Buchanan, J.A., et al., 1989, Identification of the cystic fibrosis gene: Genetic analysis, **Science**, 245:1073.

Knowles M.J., Barnett, T.B., McConkie-Rosell, A., et al., 1989, Mild cystic fibrosis in a consanguineous family, **Ann. Intern. Med.**, 110:599.

Mornet, E., Simon-Bouy, B., Serre, J.L., et al., 1988, Genetic differences between cystic fibrosis with and without meconium ileus, **Lancet**, I:376.

Riordan, J.R., Rommens, J.M., Kerem, B.S., et al., 1989, Identification of the cystic fibrosis gene: Cloning and characterization of complementary DNA, **Science**, 245:1066.

Rommens, J.M., Iannuzzi, M.C., Kerem, B.S., et al., 1989, Identification of the cystic fibrosis gene: Chromosome walking and jumping, **Science**, 245:1059.

Schmidtke, J. and Krawczak, M., 1989, Sex difference in D7S8 marker allele distribution in adult cystic fibrosis patients, **Lancet**, I: 393.

Stuhrmann, M., Macek, M. Jr., Reis, A., et al., 1990, Genotype analysis of cystic fibrosis patients in relation to pancreatic sufficiency, **Lancet**, I:738.

Tümmler, B., Aschendorff, A., Darnedde, T., et al., 1990, Marker haplotype association with growth in German cystic fibrosis patients, **Hum. Genet.**, 84:267.

DISCUSSION

PIGNATTI: My question regards your last point about the sex difference of patients over 18 years old. We did answer your Lancet letter using a greater number of patients and we didn't find the effect. So on what number now do you not find your effect? Is it a larger number of patients?

SCHMIDTKE: Yes, the proportion of adult patients is much greater now than previously but I don't have the exact number here right now.. It is an interesting question, as to why we saw this effect, whether it was just chance or what. However, at that time, we probably had a very strongly biased sample because the families were not surveyed systematically. They had mainly asked for diagnosis themselves. So maybe the sample was biased towards a more severe cause of the disease, but we are not sure.

PIGNATTI: The second question is on PS-PI. You have an enormous percentage of delta F508 homozygotes in the PS class: 6 over 36. We don't find that. We just are coming out of a study with the CF center of Verona (Professor Mastella and Dr. Borgo) on a large number - there are 50 - very well characterized PS patients. I mean also with duodenal intubation (analysis of fat balance) and we find no - zero over 50 delta F508 homozygotes. So the question is, do you think there might be a problem in the diagnosis of those patients.

SCHMIDTKE: Well all I can say is that we have at least three patients 16 years and older that never required any pancreatic enzyme substitution and have a low fat excretion (below 5g/day). As I pointed out, these are the criteria we used in our study. We could not actually ask the patients to come to a clinic and undergo further diagnostic tests. So it may very well be a matter of defining pancreatic sufficiency why we find a higher proportion.

SANTIS: We are interested in your results. I am afraid we do not agree with your definition of pancreatic sufficiency on clinical grounds alone, particularly in adults. It is well known that adult patients can tolerate pancreatic insufficiency in the absence of steatorrhea without pancreatic enzyme supplementation. Having said that, we identified two adult patients in our group of 500 adults who were pancreatic sufficient on objective criteria, and were homozygous for F508 deletion, which was confirmed on direct sequencing of exon 10 in both patients. Of interest, one of those patients is a British black, of West Indian origin. So we think it is possible to be an adult, be homozygous for the 508 mutation and be pancreatic sufficient, although that is extremely unusual. This is two people out of 25, in our group of adult patients.

DALEMANS: One of the phenotypes you studied was the electrolyte concentration in the sweat. Did you have enough data to see if the homozygous deltaF508 versus the heterozygous with another form show any differences by these criteria?

SCHMIDTKE: We have not yet fully evaluated the data, but Dr. Liechti-Gallati may wish to comment. I think you have already evaluated your data on that parameter.

LIECHTE-GALLATI: Yes, we have looked at all our patients and have made the correlations, homozygous with delta F, compound heterozygous without any delta F and we didn't find any correlation. We were very disappointed because we hoped to find something.

ROMEO: Before ending the session, since the point regarding selective advantage was raised several times, and it is going to be relevant for the discussion of tomorrow, I would like just to point out something which has not been discussed at all this afternoon: the possibility that heterozygote advantage has taken place in a very specific form, as pointed out already a number of years ago in a paper by Paul Quinton (Quinton, P.M. in: "Fluid and Electrolyte Abnormalities in Exocrine Glands in CF," Quinton et al., eds., San Francisco Press pp. 53-176, 1982). In 1982, Paul was saying that "if the basic defect is expressed in the intestine as it is in the pancreas, in the form of a reduced capacity to secrete fluids, the CF gene might impart some physiological advantage in surviving cholera infections in some individuals. Hence, decreasing the permeability of an anion sensitive pathway should limit the maximal rate at which anions (chloride and bicarbonate) and therefore fluid, can be secreted. Cholera toxin activates intestinal fluid secretion by stimulating chloride secretion, which involves a passive movement, and this ion, through an anion pathway in the apical membrane of the enterocyte." The conclusion of these statements was that "the intestinal secretion of persons carrying the mutant CF gene may be less than that of the normal population when stimulated by cholera toxin during an episode of infection. Consequently this person would lose body fluids more slowly than normal and should be at some advantage in surviving an acute infection. At least until recent times, cholera may have been a disease of sufficient proportion and incidence to act as an agent capable of imparting a selective advantage to humans carrying the mutant gene." This hypothesis has been more recently discussed by our group together with the published indirect evidence supporting it (Romeo G., Devoto M., and Galietta J.L.V., Hum. Genet. 84, 1-5, 1984).

ION TRANSPORT IN NORMAL AND CF AIRWAY EPITHELIA

Richard C. Boucher, Thierry Chinet, Niels Willumsen,
Michael R. Knowles, and M. Jackson Stutts

University of North Carolina at Chapel Hill
Department of Medicine, Division of Pulmonary Diseases
Chapel Hill, North Carolina

INTRODUCTION

Cystic fibrosis (CF) of the lung is characterized by chronic suppurative airways infection in which two bacterial species, Pseudomonas aeruginosa and Staphylococcus aureus, predominate. Because no abnormalities in the migratory inflammatory cells or immunoglobulins that protect airway surfaces from infection have been detected in CF, most efforts to analyze the pathophysiology of CF airways disease have focused on elements of the mucociliary transport apparatus. This system is typically analyzed in terms of three components: 1) cilia/cilial activity; 2) mucus glyco-proteins; and, 3) the aqueous, periciliary layer. Most data indicate that cilial motility is normal in CF and cannot account for abnormalities and clearance of secretions. Compositional abnormalities, particularly relating to sulfate ions (1), may exist beween CF and normal mucins but the effects of these abnormalities on mucus clearance are not clear. Because bioelectric abnormalities have been detected routinely in CF airway epithelia, attention has focused on the electrogenic ion transport processes that regulate the depth and composition of the periciliary liquid layer.

As will be discussed below, abnormalities in two ion transport systems, sodium (Na^+) and chloride (Cl^-), have been detected in CF airways epithelium. However, it must be stressed that surprisingly little is known about the relationship of immunoregulatory compounds, e.g., interleukins, that interact directly with inflammatory cells, and/or compounds that exhibit direct antibacterial effects, e.g., maiginin-like molecules, to airways infection in CF. Only when a full description of the airways defense mechanisms expressed in airway surface liquid exists will a complete analysis of the failure of airways defense in CF be possible.

The Identification of the CF (Cystic Fibrosis) Gene
Edited by L.-C. Tsui *et al.*, Plenum Press, New York, 1991

Active Sodium (Na^+) Absorption in Normal and Cystic Fibrosis Airway Epithelia

Electrogenic, amiloride sensitive Na^+ absorption is the major active ion translocation exhibited by proximal human airway epithelia. This process is described by a mechanism similar to that originally described by Ussing forty years ago (4). Specifically, ouabain sensitive Na-K-ATPase pumps located in the basolateral membrane extrude Na^+ from the cell, generating a favorable electrochemical gradient for Na^+ to enter the cell across the apical membrane. Na^+ appears to enter the cell via an electrically conductive process that is inhibited by amiloride.

There is convincing evidence that the rate of active Na^+ transport is elevated in CF airways epithelium. Three independent lines of investigation support this conclusion:

1) In vivo studies of transepithelial electrical potential difference (PD): The basal transepithelial electric PD, which is proportional to the rate of active ion transport, is raised in CF. Further, the amiloride sensitive component of the PD is absolutely and proportionally increased in CF airways (5). These features characterize CF patients from the time of birth, indicating that they arise from a primary defect rather than a secondary abnormality.

2) Assessment of active ion transport rate in vitro: Studies of freshly excised airway epithelia from CF and normal patients have revealed that the rate of net transepithelial active ion transport [short circuit current, (Isc)] is increased in CF and that the Isc is an index of the rate of net Na^+ transport (6). More recently (Fig. 1) similar data have accrued from studies of

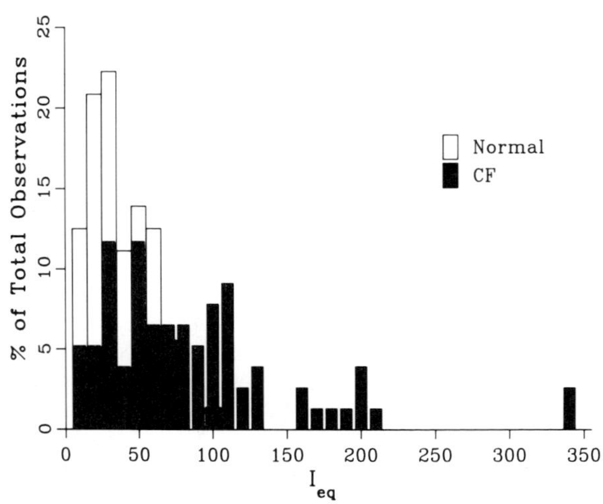

Fig. 1. Distribution of rate of Na^+ transport (Ieq; $\mu A \cdot cm^{-2}$) in preparations cultured from nasal epithelium of normal and CF subjects.

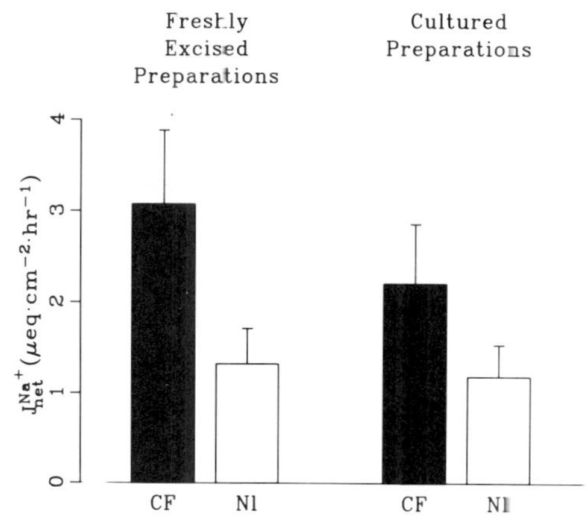

Fig. 2. Rate of net absorption of Na^+ by freshly excised and cultured normal and CF airway epithelia under open circuit conditions. N=8 or more/group.

normal and CF cells in culture, indicating that the abnormality is not secondary to a chronic in situ stimuli.

That the accelerated rate of Na^+ transport is expressed in vivo is made likely by the observations that isotopically measured Na^+ transport is accelerated in "open circuit" conditions. This is the experimental condition in which ion transport is measured where the open circuit potential is not nulled by a voltage clamp, so that the effect of the spontaneous PD, similar in magnitude to that observed in vivo, is exerted on the net movement of ions. As shown in Figure 2, the open circuit Na^+ absorptive fluxes in both freshly excised and cultured CF airway preparations are approximately double those observed in preparations obtained from normal subjects.

3) Direct evidence for an increase in the apical membrane sodium permeability (P_a^{Na+}) in cystic fibrosis cells: Theoretically, the increase in Na^+ transport in CF could reflect the absence of the apical membrane Cl^- conductance, perhaps by hyperpolarizing the apical membrane, or a direct increase in the apical membrane Na^+ permeability. To distinguish between these possibilities, direct measurements of the driving force for Na^+ movement across the apical membrane were made in cultured CF and normal airway epithelial cells with double barrelled Na^+ selective microelectrodes. Representative values from CF and normal cells are shown in Table 1. In brief, it can be seen that the apical membrane in CF cells is actually depolarized as compared to normal cells. The intracellular sodium activity (a_c^{Na+}) is similar in CF and normal cells. Utilizing these two parameters, the driving force for sodium entry across the apical membrane (DF_a^{Na+}) can be calculated and is indeed smaller in CF than normal cells. Simultaneous measurements of the rate of transepithe-

Table I

Transepithelial and Intracellular Bioelectric Parameters, and Intracellular Na^+ Activities in Cultured Normal and CF Airway Epithelia*

	Vt (mV)	Isc ($\mu A.cm^{-2}$)	Va (mV)	ac^{Na+} (mM)	DF_a^{Na+} (mV)	P_a^{Na+} ($cm.sec^{-1}$)
Normal	-11.8 ± 1.5	-38.7 ± 4.9	-25.9 ± 2.5	23 ±1	63.1	$1.6x10^{-6}$
CF	37.4 ±8.4	-93.1 ±26.8	- 8.2 ± 5.0	21 ±2	51.0	$6.2x10^{-6}$

* Mean + SEM (N=30 or greater/group)

lial Na^+ transport, Isc, in these preparations revealed that Isc was approximately three fold higher in CF than normal cells. The observation that the driving force for Na^+ entry across the apical membrane for CF is smaller, coupled with the larger net Na^+ transport rate, indicates that the apical membrane Na^+ permeability for CF must be increased. Direct quantitative estimates of the apical membrane Na^+ permeability can be made utilizing model analyses employing an equivalent circuit (7). These data also indicate that P_a^{Na+} is absolutely increased in CF (Table I). Thus, the simplest hypothesis is that the increase in Na^+ transport in CF is not secondary to the abnormality of the Cl^- channel but is a primary abnormality of the regulation of the apical membrane Na^+ permeability.

Two fundamental problems have plagued the study of the regulation of Na^+ transport in normal and CF airway epithelia. First, there is virtually no knowledge of endogenous regulators of Na^+ transport in airway cells. Agents that regulate the chronic rate of sodium transport in other tissues, e.g., ADH and aldosterone, appear to be ineffective in human airway epithelia. Further, agents that acutely regulate Na^+ transport in other epithelia, e.g., renal epithelia, including ANF, also appear to be ineffective in human airway epithelia. Therefore, it is difficult to compare the regulation of elements of the Na^+ transport path in CF versus normals.

A second problem with regard to the study of Na^+ transport in airway epithelia relates to the difficulty in identifying at the single channel level the channel that participates in transepithelial Na^+ transport. At least two cation channels have been recognized utilizing the patch clamp technique in human airway epithelia. The first channel is a small conductance (3-8 pS) channel that appears to be highly selective for sodium over potassium (Fig. 3). This channel exhibits very slow kinetics and in a small number of studies is sensitive to external amiloride. Because of these characteristics, this channel would be an attractive candidate as a participant in net transepithelial Na^+ transport. Major difficulties in studying this channel, however, have been the difficulty in identifying it in patches with any frequency, and the rapid run down of channel activity in the excised patch mode.

A second cation channel with a larger conductance (20-30 pS) is observed with much greater frequency. The selectivity of this channel for Na^+ over K^+ is rather poor (1.5:1). Unlike "nonselective" cation channels in some other epithelia, the channel is not inhibited by ATP. The amiloride sensitivity of this channel in the outside-out configuration has not been fully determined, but it appears rather insensitive to amiloride in the inside-out mode in concentrations to 10^{-4}M. The key question with regard to this channel is whether it is expressed normally in the intact airways epithelium. Theoretically, given the selectivity of this channel, and the driving force for K^+ across the apical membrane, a high degree of K^+ secretion should be exhibited by human airway epithelia if this channel is expressed and maintains the characteristics described in the excised patch. However,

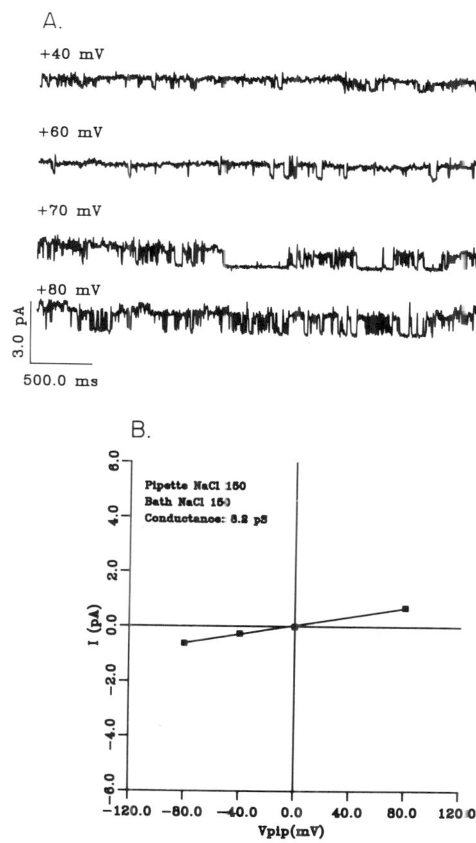

Fig. 3. The "small" cation channel. An excised, inside out patch derived from normal human airway epithelial cell. Panel A shows single channel record; Panel B shows the current voltage plot for this channel.

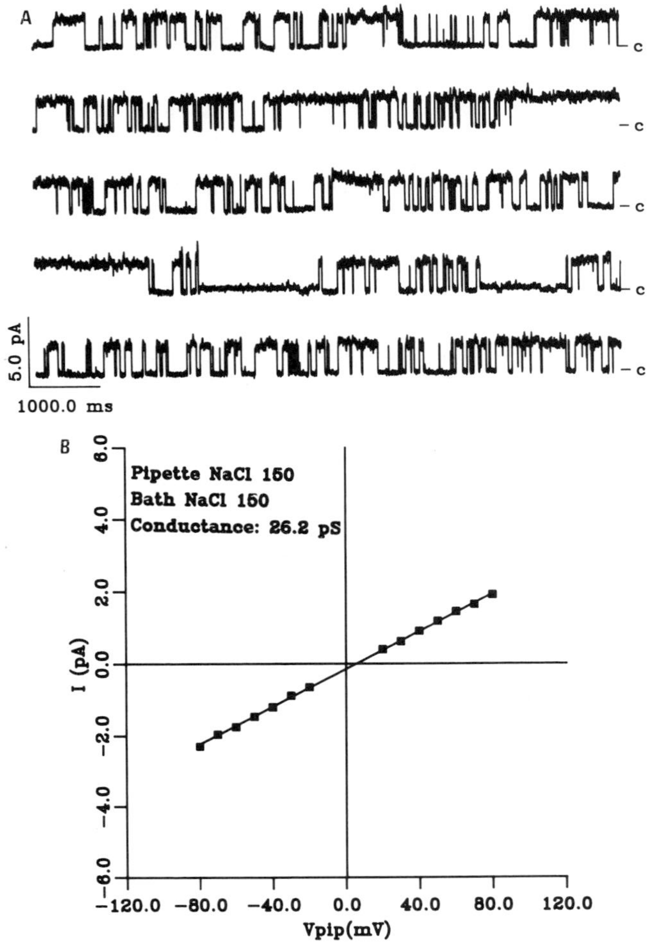

Fig. 4. The large, non-selective cation channel excised, inside out patch derived from normal human airway epithelial cell. Panel A shows single channel record; Panel B shows the current voltage plot for this channel.

recent studies measuring active K^+ secretion in human airways epithelia have detected only a very small rate (0.05 $mEq/cm^2/hr$) in human airway epithelia. Presently, the role of this channel in the translocation of ions across the apical membrane of the intact tissue is not known.

Given the general difficulty in identifying "the" Na^+ channel at the single channel level, it has been heretofore impossible to describe potential differences in kinetics, single channel conductance, and numbers of these channels in CF as compared to normal tissues. Studies that may be required in the future may include noise analysis and radio-ligand binding.

Finally, the importance of the accelerated rate of Na^+ transport in the pathogenesis of CF airways disease is not known. One approach for assessing the importance of this phenomenon in disease pathophysiology is to block this activity with the Na^+ channel blocker amiloride and test whether the course of CF lung disease is perturbed by this agent. This approach is dealt with in a subsequent chapter.

Chloride Transport Human Airway Epithelia

The identification of abnormalities in the regulation of the apical membrane Cl^- conductance in CF epithelia has focused much attention on this pathway in human airways epithelia. Despite this intensive investigation, several major questions with regard to the physiology, single channel correlate, and regulation of the apical membrane Cl^- permeability remain both in normal and CF airway cells.

One key question relates to the role of the apical membrane Cl^- conductance in the overall physiology of surface liquid metabolism in the human lung. For example, unlike the dog tracheal epithelium, there is no driving force for Cl^- secretion across the apical membrane of human nasal or bronchial airway epithelial cells under resting conditions. These data are consistent with previous evidence that indicated human airway epithelia do not tonically secrete Cl^-. Further, activation of the apical membrane Cl^- permeability, e.g., by isoproterenol, does not generate Cl^- secretion because, although apical membrane Cl^- permeability is stimulated, there is no net outward driving force favoring Cl^- exit. It had been speculated that the human proximal airway epithelium may convert to a Cl^- secretory epithelium in response to loss of water from airway surfaces. However, consequent to water loss the activity of Cl^- in the surface liquid increases so that a net inward driving force for Cl^- from surface liquid into the cell is generated. The cell response to the generation of an inwardly directed driving force (hypertonic airway surface liquid) is to decrease apical membrane Cl^- permeability, a response presumably triggered by cell volume loss. A clear understanding of the contribution of the abnormal regulation of the Cl^- channel in CF to disease pathogenesis awaits more complete understanding of the role of the Cl^- channel to normal airways mucosal physiology.

Second, at the single channel level, the "30 pS, outward rectifying" channel has been designated as the "secretory" Cl^- channel. An early question with respect to this channel was that it seemed to conduct ions rather poorly in the physiologic voltage range as compared to the non-physiologic, depolarized voltage range. Direct experimental data have subsequently been reported that also question the physiologic relevance of this outward rectifying channel. For example, current voltage analysis of the apical membrane of amiloride pre-treated airway epithelial cells employing double-barrelled Cl^- selective microelectrodes identified a rectifying Cl^- current. However, the rectification could be accounted for entirely by Goldman, Hodgkin, Katz relationships, i.e., the transapical membrane chemical Cl^- gradient. Further, other investigators have identified anion selective channels in airway epithelial cells that are non-rectifying and typically exhibit a somewhat smaller (20 pS) conductance (8). Finally, in studies of secretory cell types from colonic epithelia, which mirror some features of airway epithelial Cl^- secretory patterns, whole cell clamping studies have identified at least 3 Cl^- currents (9). One current may be a volume activated current that reflects basolateral Cl^- channels whereas a non-rectifying (cyclic AMP sensitive) and partially rectifying (Ca^{2+} sensitive) current have been detected. Thus, because of the complexity of the various regulatory paths in human airway epithelia, a more complete characterization of the types of anion channels in the apical membrane of human airway epithelia appears to be warranted.

Third, the number of regulatory pathways that impact on Cl^- channels in human airway epithelia is large and their interactions complex. Recent studies have indicated that the cAMP dependent and Ca^{2+} dependent paths appear to be at least partially separate in activating Cl^- conductances in human airway epithelia (10). Unresolved issues involve the mechanism(s) by which raising Ca_i^{2+} activates the apical membrane Cl^- conductance and which anion channels are involved (see above). More recently, protein kinase C has been shown to be involved in the regulation of human airway epithelial Cl^-channels. Studies in intact tissues indicate that the effects of protein kinase C are at least initially stimulatory with regard to the induction of Cl^- secretion and interactions occur between this path and the cAMP dependent path, suggesting potentially final common mechanisms. In the intact epithelium, there appears to be additivity between the Ca^{2+} (ionophore) dependent and PKC dependent paths, indicating distinct paths for regulating a common channel, or activation of different anion channel types. The data that relate to these issues at the single channel level are apparently more complex and controversial.

Receptor gated mechanisms that raise intracellular Ca^{2+} (Ca_i^{2+}) and induce Cl^- secretion in both CF and normal epithelial cells have been detected (10). It appears, for example, that both bradykinin and histamine raise Ca_i^{2+} and induce chloride secretion in both CF and normal airway epithelia. However, analysis of other receptor coupled mechanisms indicates that there may be a novel mechanism for coupling of receptor activation to channel activation, i.e. a direct coupling mechanism via G proteins. Recent double-barreled electrophysiologic data have indicated that external purinergic (ATP) receptors may be directly coupled to Cl^- channels. This is an area that will be of substantial interest because of the observation that ATP is effective when applied to the apical membrane in initiating Cl^- secretion in CF tissues.

Other Transport Mechanisms in Normal and CF Airway Epithelia

Relatively little work has been performed in normal human and CF airway epithelia with regard to transport of solutes other than Na^+ and Cl^-. Analysis of the sulfate ($SO_4^=$) concentration of high molecular weight glycoconjugates on the airway cell surfaces have indicated that $SO_4^=$ concentrations are increased in CF. Presumably this would indicate that abnormalities in the metabolism of $SO_4^=$ will be detected in CF cells, but neither the putative cellular defects nor their consequences for transepithelial $SO_4^=$ transport have been investigated. Similarly, the regulation of Ca_i^{2+} and protons (pH) have been studied in CF and normal airway epithelia. Again the metabolism of Ca^{2+} and protons appear to be relatively similar in CF and normal epithelia with regard to intracellular concentrations. However, the transepithelial fluxes, if any, of either of these ions have not been measured. Finally, although K+ concentrations are substantially raised in the airway surface liquids (20-40 mM), K+ transport has not been characterized in CF airways. As mentioned above, only a single study of K transport in cultured airway cells has been preliminarily reported. Depending on the putative function of the CFTR (see below), the importance of such studies may become emphasized.

CONCLUSION

The recent identification and cloning of the CF gene [cystic fibrosis transmembrane regulator, (CFTR)] permits two approaches to integrate functional data with concepts generated from the analysis of the CFTR cDNA [11 (see Fig. 5)]. One approach is to assign CFTR a function in accord with existing data. Perhaps the simplest analysis utilizing this approach is to conceptualize CFTR as: 1) a channel, either an anion or a cation channel, or both; or, 2) a regulator. With regard to CFTR function in airway epithelia, an attractive regulatory function would be to couple the activities of the apical membrane Cl^- and Na^+ channels. This function would serve to reciprocally coordinate the activities of the Na+ and Cl- channel, determining whether a single airway epithelial cell exhibits net Na^+ or Cl^- transport. The mutation in CFTR would lead to an excessive up regulation of Na^+ channel activity and the inability to reverse this function, i.e., activate Cl^- channels, via A and C kinases.

Alternatively, the analysis of the CFTR cDNA may lead to entirely novel concepts that couple CFTR function/dysfunction to airway epithelial cell regulation. For example, based on the homology to the multi-drug resistance gene, one must entertain the notion that CFTR is a member of a "super family" of active solute transporters. Dysfunctions of a solute pump must be integrated into the known abnormalities of ion channel regulation but may produce unique predictions with regard to other func-

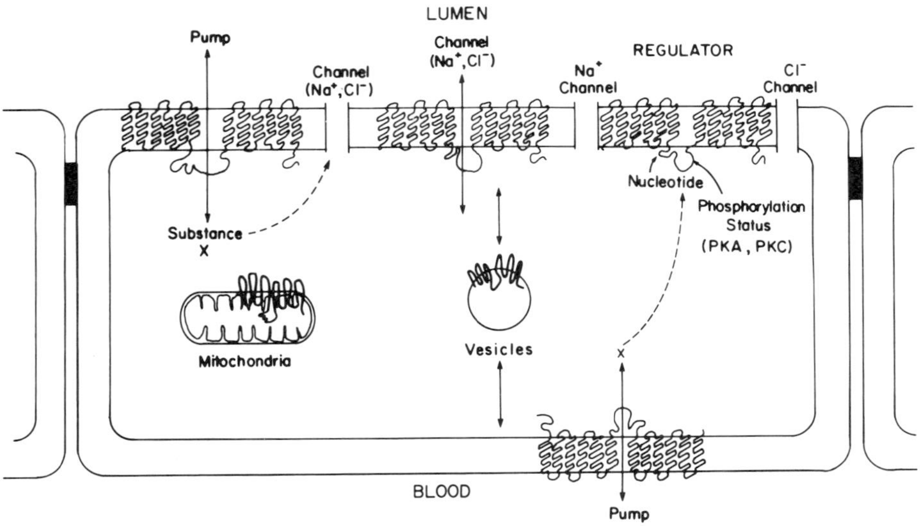

Fig 5. Schema describing possible cellular locations and functions of CFTR. CFTR is depicted as projected by Riordan et al (11). On the apical membrane, roles as a solute pump, channel, and regulator are depicted. Possible intracellular localizations are also shown.

tions/dysfunctions that may be more closely related to the CF phenotype in the airway, i.e., chronic airways infections.

Identification of the function of CFTR represents the challenge for the next several years in CF airways structure/function research. Key data that will be required will be the localization of CFTR to individual cell membranes and cell compartments utilizing antibody techniques and the determination of the level of CFTR expression in different cell types within the airway wall. Expression studies will be extraordinarily useful in determining whether or not the CFTR can function as a conductive ion channel. Studies that focus on potential regulatory functions of CFTR may be difficult and will require a variety of biochemical and physiologic techniques that will likely early focus on the nucleotide binding specificity of the individual nucleotide binding folds of CFTR. Finally, if CFTR is a solute pump, the identification of a substrate, most importantly a native substrate, and the level of activity of CF CFTR versus normal CFTR, will be key variables.

References

1. Cheng, P.,T. Boat, K. Cranfill, J. Yankaskas, and R. Boucher, 1989, Increased sulfation of glycoconjugates by cultured nasal epithelial cells from patients with cystic fibrosis, **J. Clin. Invest.**, 84:68.

2. Barker, P., C. Boyd, L. Strang, D. Walters, 1989, Pulmonary transport in the fetal sheep, **J. Physiol.**, 409:15.
3. Joris, L. and P. Quinton, 1989, Evidence for electrogenic Na- glucose cotransport in tracheal epithelium, **Pflugers Arch.**, 415:118.
4. Ussing, H., and K. Zerahn, 1951, Active transport of sodium as the course of electric current in short-circuited isolated frog skin, **Acta Physiol. Scand.**, 23:110.
5. Knowles, M., J. Gatzy and R. Boucher, 1981, Increased bioelectric potential difference across respiratory epithelia in cystic fibrosis, **N. Engl. J. Med.**, 305:1489.
6. Boucher, R., M. Stutts, M. Knowles, L. Cantley and J. Gatzy, 1986, Na+ transport in cystic fibrosis respiratory epithelia: Abnormal basal rate and response to adenylate cyclase activation, **J. Clin. Invest.**, 78:1245.
7. Willumsen, N., R. Boucher, 1989, Shunt resistance and ion permeabilities in normal and cystic fibrosis airway epithelia, **Am. J. Physiol.**, 256(Cell Physiol 25):C1054.
8. Duszyk, M., A. French and S. Man, 1990, The 20-pS chloride channel of the human airway epithelium, **Biophys. J.**, 57:223.
9. Frizzell, R., W. Cliff, R. Worrell and A. Morris, 1990, Different chloride conductances in Cl-secreting epithelia, **Biophys. J.**, 57:18a (Abstr).
10. Boucher, R., E. Cheng, A. Paradiso, M. Stutts, M. Knowles, and H. Earp, 1989, Chloride secretory response of cystic fibrosis human airway epithelia: Preservation of calcium but not protein kinase C- and A-dependent mechanisms, **J. Clin. Invest.**, 84:1424.
11. Riordan, J., J.M. Rommens, B.-S. Kerem, N. Alon, R. Rozmahel, Z. Grzelczak, J. Zielenski, S. Lok, N. Plavsic, J.-L. Chou, M.L. Drumm, M.C. Iannuzzi, F.S. Collins and L.-C. Tsui, 1989, Identification of the CF gene: Cloning and characterization of complementary DNA, **Science**, 245:1066.

DISCUSSION

FRIZZELL: Rick, I was wondering about the presence of the apical adenosine receptor in cultured airway cells.

BOUCHER: We don't see it. We don't see an adenosine receptor. Everything we do see seems to be ATP. And again, we have done the experiments in both freshly excized and cultured cells and we don't see an adenosine receptor media defect on short circuit current. We haven't definitively made membranes, looked at adenocyclase activity and seen whether there are A1 or A2 defects on cyclase, but the physiological response we are measuring is quite different from Daw.

GUGGINO: I was wondering about whether in the imaging experiments you had found any difference in the apical or the basolateral ATP response, or secondarily, whether you had looked at PI turnover. I'm thinking about translocation of PKC to the apical membrane versus the basolateral, since we do know there are different pools of calcium. Whether that is a part of the differential response.

BOUCHER: One of our mutual best friends is, I hope, doing those experiments this week. Bill Davis has a system where he can unilaterially perfuse tissues and pulse agonists on one or the other side and look to see whether or not a calcium response is triggered. I'm fairly neutral. I don't really know what the answer is going to be, and almost don't care. On the one hand, the bradykinin responses are exactly symmetric, one barrier to the other barrier. As far as the electrophysiology goes, activation of K-conductance is first, followed by an activation of the chloride permeability, which argues a bit against your notion, I think. On the other hand, the ionophores, when we use A23187 and expose them to the apical membrane, the pattern looks very similar to what we'd see with apical ATP, in that there is a major activation of the apical membrane chloride permeability, rather little of the base lateral least potencing conductance as we measure it. So I think there are two critical experiments that are ongoing, and one is the calcium experiments we talked about, the other is that we are obviously performing outside-out patch experiments, exposing the patches to external ATP in defined conditions so see whether or not we get direct activation.

FROMTER: In the talk you concluded that under physiological conditions Cl^- secretion was negligible and you extended this statement to cases of forced ventilation, eg. exercise, when water loss from the bronchial surface is increased. When making this statement, have you also considered the possibility of reflex stimulation of Cl^- secretion?

BOUCHER: I think we have to be very careful, and I hope I said "under these conditions" with regard to the way we measured. In vivo, it's very hard to know. We worried about this a lot and we've measured in vivo potentials a lot and we've tried to mimic in vivo the response to the various agonists that we use in vitro, culture, freshly excized tissue. So I don't think we can be absolute. Somewhere in the lung there has got to be chloride secretion and this fluid is coming from somewhere, and again, by dint of a little imagination I usually place that in the alveolus. I think the issue, though, is that, at least when we added surgical exogenous agonists in vitro, for example isoprome or isoprotenerenol or something that raises cyclic AMP, we see an activation of the apical membrane chloride permeability, we see changes in the fractional resistances, there is just no current, there is no driving force. The big caveat is, I think, maybe in cystic fibrosis. And that is, and I think it is sobering to us and makes us, hopefully, look at our data evenhandedly, that when we place amiloride on CF preparations, either freshly excised directly from the patient, or in cultured preparations, the PD and the current always go to zero, ± zero. It's a very routine phenomenon in the lab. When Mike Knowles performs that maneuver in vivo, and superfuses amiloride on nasal surfaces, the PD goes from -70 to -15: very finite. Now, is that residual PD carried by chloride secretion? In my heart of hearts I think it is. So that does imply that there may be some tonic stimulus in vivo that at least allows the expression of chloride secretion. So I think we have to temper all that, but at least by our best estimates and even in vivo with amiloride, the major potential is removed by amiloride.

GREGER: Rick, you just briefly mentioned - are you in the process of doing patch clamp experiment tests? This is the question, but as I understand your results thus far are negative in outside-out patches. Can you say this or it is premature?

BOUCHER: No, it's premature.

RIORDAN: You mentioned it is not an adenosine receptor, but in terms of substrates, do non hydrolyzable analogues or other nucleotides have effects?

BOUCHER: Yes, I skipped through that second-to-last slide. It does. Basically, the rank order potency of this specific receptor is UTP, and is indeed the most effective agonist. UTP is a pyrimidine, so the whole field may be filled with misnomers. But the rank order potency is UTP more than ATP, more than ADP. Adenosine is ineffective. Specifically, ATP gamma-S is about equally effective with ATP.

SATO: I'm doing studies on the effects of ATP on secretion, and of course it does. Is the effect of ATP repetitive or long-lasting or is it just a one-shot event?

BOUCHER: The generic wisdom for most of the cell surface receptors we have looked at is that they exhibit rather exquisite homologous desensitization. They quickly desensitize, but only to themselves. ATP gives us somewhat less desensitization than the other receptors, but clearly the effects that I am showing are lasting a time span of minutes, not hours.

SATO: What do you mean by desensitization? Is ATP simply reacting by disrupting the membrane that takes some time to recover from it?

BOUCHER: No, we're very confident that this is not, if you like, an ATP-induced permeabilization effect. Clearly, the electrophysiology suggests a cell type that is able to react with selective changes in permeability. It is non-specific. We very early looked to see whether or not the cell was non-specifically permeabilized to other external solutes, trypan blue, whatever. But also, the cell is perfectly able to regulate its calcium, as I showed you there, back to a plateau level in the presence of external calcium. There is a huge gradient, suggesting that it is not non-specifically permeabilized.

GARDNER: If this is a direct receptor ionophore-type mechanism, what would be the natural ligand on the luminal side for a purinergic receptor, do you know?

BOUCHER: Well, that has always been one of the issues about purinergic receptors. Did God really make ATP to get outside cells and serve as a messenger? I'm not convinced that it isn't a good thesis. We think that there may actually be copackaging of ATP with mucine granules, and this is an area that has certainly been well-known to you in the neurosecretory field. So that would be one potential source. There clearly is also ATP in afferent nerves that are just underneath the epithelial cell. And finally, there is probably a lot of work, I would suggest, going on in Sam Silverstein's lab at Columbia, looking for ATP release as a physiologic response in a variety of inflammatory cells. So any inflammatory cell in the region could likely release ATP either by lidic mechanisms or, one would hope, by a more controlled and regulated system. We think there are actually going to be multiple sources of ATP. Actually, my own private notion is I wonder if it doesn't coordinate a lot of mucosal physiology. If indeed you get ATP release with mucine granules, we have found that not only does it trigger, if you like, activation of chloride secretion, but I think a number of groups have shown that actually it accelerates ciliar beat frequency quite effectively. So the whole system may be primed so that if you get mucine release, you get a little bit of fluid out, and you get cilia beating more rapidly. So we think it may be a coordinator in that context.

FRIZZELL: Is there an ATP effect on sodium?

BOUCHER: Thank you for asking. To keep it simple (I focused on chloride), there is. It's harder to tease out, because of analogue techniques, in normal tissues. It's very clear in CF tissues that ATP inhibits sodium transport. It does appear to inhibit sodium permeability, at the same time activating chloride. We have, I think, pretty good inferential data that the same mechanism occurs in normal tissue, but with a large chloride permeability it's a little harder for us to resolve. But I think that this may be the coordinator of these two permeabilities that we've been looking at for awhile.

AEROSOLIZED AMILORIDE AS TREATMENT OF CYSTIC FIBROSIS LUNG DISEASE: A PILOT STUDY

Michael R. Knowles, Nina L. Church, William E. Waltner, James R. Yankaskas, Peter Gilligan, Malcolm King, Lloyd J. Edwards, Ronald W. Helms, and Richard C. Boucher

The University of North Carolina and
The University of Alberta

INTRODUCTION

Excessive Na^+ absorption[1,2] and defective regulation of apical membrane Cl^- secretion[3-6] likely leads to dehydration and abnormal rheology of CF airway secretions[7]. These abnormalities contribute to poor clearance of airway secretions, and chronic bacterial infection in CF. Because amiloride inhibits the absorption of Na^+ (and liquid) across airway mucosa[1,8,9] when applied to the apical surface of CF airway epithelia, we hypo- thesized that inhalation of amiloride may improve rheology and clearance of airway secretions[9,10] This hypothesis was tested in a double-blind crossover study to explore the safety and efficacy of long-term amiloride aerosol in the treatment of cystic fibrosis lung disease[11].

METHODS

Study Subjects: Eighteen patients (9F; age range 18-37 years) were recruited. Guidelines for enrollment included: clinical and sweat Cl^- criteria for the diagnosis of CF; no hypoxemia or chronic CO_2 retention; no *Pseudomonas cepacia*; and normal renal function.

Study Design: The protocol involved a randomized, double-blind, crossover study with baseline measurements before each treatment period and multiple visits scheduled during aerosol (Figure 1, Table 1). Twenty-five weeks was chosen as the nominal length of each treatment period because it was considered likely

The Identification of the CF (Cystic Fibrosis) Gene
Edited by L.-C. Tsui *et al.*, Plenum Press, New York, 1991

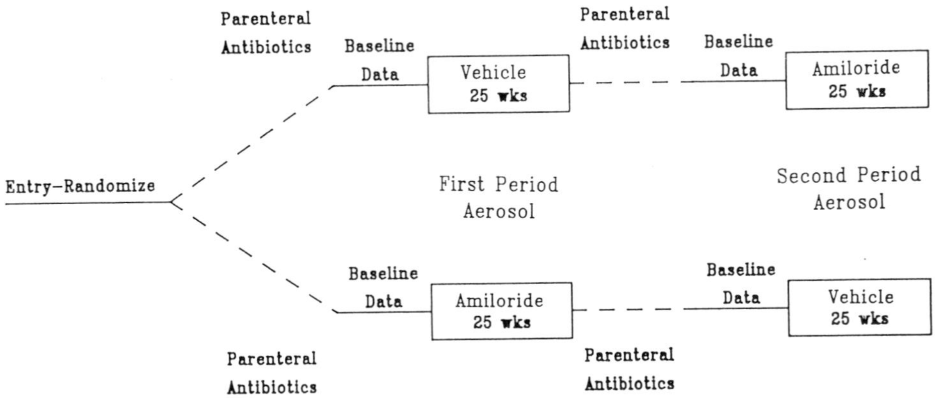

Fig. 1. Schema of the crossover study design (see text).

that during this interval most patients would suffer a measurable decline in pulmonary function that might be modified by a therapeutic agent. Each period was preceded by 10-14 days of parenteral tobramycin and ceftazidime. All respiratory medicines (oral and aerosol antibiotics and bronchodilators) were withdrawn and withheld during the study period unless specific guidelines for reinstitution were met[11].

The design for collection of outcome data is summarized in Table 1. Two baseline measurements of selected parameters were obtained after completion of parenteral antibiotics. Single baseline measurements of a smaller number of parameters were also obtained. Aerosol treatment was initiated after baseline measurements, and a small set of single measurements was obtained at intervals ranging from 3-25 weeks (Table 1). At the end of each phase (~25 weeks), duplicate or single measurements were obtained as at baseline. Sputum was obtained from 6 pa-

Table 1

Frequency and Interval of Data Collection During Each Phase of the Study

Test*	Baseline Data	During Aerosol (wks) 3	6	12	18	25
Spirometry, Phys. Exam, Diary, Chem 6, Sputum quantitative culture	++	+	+	+	+	++
ABG, Lung volume, DLCO	++					++
CXR, CBC, U/A, Aldosterone, Chem 18	+					+
Sputum Rheology						+

*Chem 6; electrolytes, BUN, cr; sputum quantitative bacterial culture; ABG, arterial blood gas, DLCO, diffusing capacity of lung-carbon monoxide; CXR, chest X-ray; CBC, complete blood count; U/A, urinalysis; aldosterone, 24 hr urinary excretion.

tients for biorheologic studies at the end of each phase.[12] Aerosol was discontinued after the end-of-period measurements. Parenteral antibiotics preceded the initiation of the second period of aerosol.

Therapeutic interventions with oral antibiotics (except quinolones) and bronchodilators were allowed for mild disease exacerbations. Criteria were also established for illness episodes that required "extraordinary" intervention (usually parenteral antibiotics; tobramycin and ceftazidime), e.g., for major hemoptysis or marked reduction in pulmonary function with systemic toxicity. Safety was monitored by a non-participating observer who reviewed coded clinical and laboratory data.

Aerosol Solution and Drug Delivery: Amiloride hydrochloride was dissolved in 1/3 normal saline (5×10^{-3} M; pH 7.0). Vials of unused aerosol solutions were returned at clinic visits to estimate compliance. The Devilbiss 646 nebulizer and a Pulmoaide compressed-air generator were used for aerosol delivery four times daily. Patients were taught a specific inhalational technique: exhalation to residual volume, a slow inhalation (4-6 seconds) to total lung capacity (TLC), and a 4-6 second breath-hold at TLC. Previous studies established that aerosolization of 3.5 ml of solution (requiring ~10 minutes) deposits an effective dose of amiloride ($\sim 10^{-4}$ M) on surfaces of proximal conducting airways[13,14]

Techniques: Pulmonary function tests were performed at least 3 hours after arising and at least 2 hours after aerosol. Quantitative sputum cultures were performed by a published technique[15]. Sputum for biorheology was obtained at the end of each period by a technique designed to minimize salivary contamination[12]. Rheological properties were determined using the magnetic oscillating sphere technique[16]. Chest radiographs were graded using the Brasfield method[17]. Twelve pulmonary and systemic symptoms were graded on an ordinal scale (1-5 or 1-4) weekly by patients using a diary designed for this study[11]: chest tightness; shortness of breath; wheezing; cough during the first 2 hours of day, the rest of day, and during sleep; hemoptysis; sputum color, consistency, and volume; appetite; and general well-being.

Data Analysis: Because of irregularities in evaluation schedules and unequal periods of follow-up, the use of complete, balanced ANOVA or MANOVA methods was precluded. The irregular schedules were anticipated and analyses were performed by two statistical methods, selected prior to the study: (a) analysis of mean change since baseline, i.e., difference between baseline values and mean of values during aerosol; (b) analysis of within-subject, within-treatment slopes of simple linear regressions, i.e., the difference in the slopes of the regression lines as a measure of the rate of decline from baseline during vehicle and amiloride. Methods a and b treatment comparisons were based on these differences in unweighted, two-tail, paired t-tests and signed rank tests. To test for differences in the biorheologic properties and bacterial counts of sputum obtained at the end of each period, paired t-tests were employed using $\log_{10}$ transformed data. A per comparison Type I error rate of 0.05 was used to judge statistical significance. All investigators except one remained blinded to the identity of the treatments during the course of the analyses. Results are reported as means ± SEM.

RESULTS

Fourteen of the 18 patients completed the study. Three subjects withdrew from the study when they were unable to follow the aerosol dosing schedule and/or complete required visits. One other patient was withdrawn after developing a hemolytic anemia associated with Epstein-Barr virus seroconversion. That patient subsequently completed a toxicity study of amiloride aerosol with no adverse consequences.

An equal number of the 14 patients who completed the study were randomized to vehicle and amiloride for the first period of aerosol. All patients demonstrated adequate inhalation technique as assessed by lung scan, but the percentage of lung regions with poor aerosol distribution varied (range 8-83%; mean 45.2±6.7%). The mean duration of vehicle and amiloride aerosol periods was well-matched (24.7±0.7 and 24.7±0.8 weeks), and there was no seasonal bias in either treatment arm. There was equal compliance for aerosol doses in the vehicle (79.0±6.0%) and amiloride (80.3±4.7%) periods. After the study was completed, six patients correctly identified the aerosol in each period, three patients guessed incorrectly, and 5 patients were uncertain.

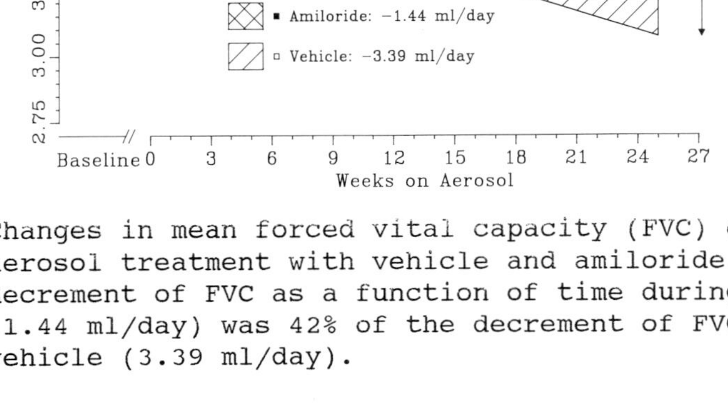

Fig. 2. Changes in mean forced vital capacity (FVC) during aerosol treatment with vehicle and amiloride. The mean decrement of FVC as a function of time during amiloride (1.44 ml/day) was 42% of the decrement of FVC during vehicle (3.39 ml/day).

Efficacy: Baseline spirometric data were the same for the subjects before vehicle (FVC=3.66±0.28 L; FEV_1=2.25±0.24 L) and amiloride (FVC=3.60±0.29 L; FEV_1=2.22±0.25 L). As analyzed by Method a, the mean decrease of FVC from baseline during vehicle (296±79 ml) was about twice that during amiloride [(160±53 ml), p<0.04]. The mean baseline to post-baseline decrease of FEV_1 during vehicle (291±95 ml) tended to be smaller by 33% during amiloride (202±51 ml) aerosol (p~0.20). A similar pattern was observed when post-bronchodilator values for FVC and FEV_1 were analyzed, i.e., there was a smaller loss of FVC (117±57 ml) during amiloride than vehicle (222±50 ml; p=0.03) and there was a tendency for a smaller decrease in FEV_1 from baseline during amiloride (170±54 ml) as compared to vehicle (225±50 ml; p~0.30). The mean duration of expiration for FVC was the same for vehicle and amiloride treatment (10.0±0.7 sec).

The summary of FVC slope analysis (Method b) is shown in Figure 2. The rate of the decrement in FVC during vehicle (3.39 ml/day) is more than halved during amiloride aerosol [(1.44 ml/day), p<0.04]. The trend for amiloride to spare the rate of decrement of FEV_1 was noted (3.21 ml/day during vehicle; 2.09 ml/day during amiloride) but the difference in these rates did not reach statistical significance (p~0.09).

Matched pairs of sputum samples for biorheology studies were obtained from 6 patients. The results of the sputum analyses are

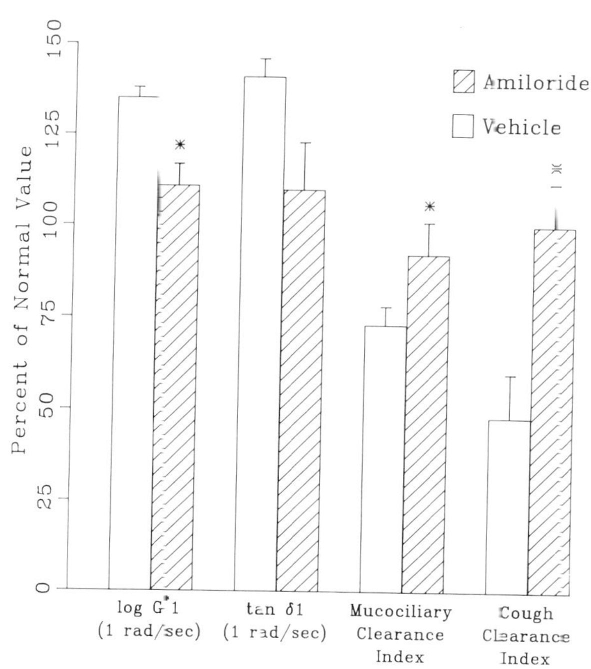

Fig. 3. Effect of vehicle and amiloride on sputum rheology (n=6 patients). Results are plotted as a percentage of normal values (reference 16). See text for details. *p<0.05- amiloride different from vehicle

illustrated in Figure 3[18]. The rheologic parameters of CF sputum obtained during chronic amiloride aerosol approximated values obtained at bronchoscopy from normal subjects (depicted as 100% in Figure 3). Log G*1 and 100, measures of mechanical impedance of mucus, were significantly improved during amiloride. The loss tangents (tan §1 and §100), indices of the ratio of viscous to elastic strain, tended to improve during amiloride but did not reach statistical significance. Calculated indices of mucociliary clearance[18] and cough clearance[19] were also significantly improved during amiloride. The percent solid content of sputum did not vary between samples obtained during vehicle (7.5±3.2%) as compared to amiloride (8.3±4.1%).

Several monitored indices were not significantly altered by amiloride. Measures of non-spirometric pulmonary function showed no difference between the vehicle and amiloride periods for changes in arterial blood gases (n=11), diffusing capacity (n=11), or lung volume (TLC, FRC, and RV) (n=9). The chest exam score worsened during vehicle (0.65±0.23; $p<0.01$) and was unchanged during amiloride (0.18±0.56; p~0.56), but the differences between vehicle and amiloride were not significant. There was no difference in respiratory rate at the beginning and during vehicle (18.3±0.9 and 18.7±0.7) and amiloride (17.3±0.9 and 18.0±0.7). There was no difference in the Brasfield score (maximum=25 points for normal) of the chest radiograph between vehicle and amiloride: the score was the same at the beginning of both periods (~19) and declined by ~1.0-1.6 points in each phase. No difference in bacterial density (log_{10}/ml sputum) was noted between vehicle and amiloride for all organisms (8.16±0.18 and 8.13±0.20) or Pseudomonas aeruginosa (7.96±0.21 and 7.79±0.35).

The changes in the symptom scores over the 25 weeks were small (~00.1 unit/week). During vehicle, significant worsening (t-test and signed rank; $p<0.05$) was observed for shortness of breath and morning cough, and statistically borderline worsening ($p<.05$ for t-test or signed rank) were observed for chest tightness, cough during the day, and general well-being. During amiloride, only chest tightness worsened significantly, but the rate of worsening was 30% of that during vehicle. A significant difference in symptom scores between vehicle and amiloride was seen only in increased shortness of breath during vehicle ($p<0.01$).

Therapeutic interventions: The number and duration of therapeutic interventions with oral antibiotics, bronchodilators, and prednisone did not differ between the vehicle and amiloride periods when analyzed as individual or grouped data, although paired analysis of the sum of usage weeks for all interventions for each patient tended to be greater during vehicle ($p=0.11$).

Safety: No evidence of pulmonary or systemic toxicity emerged from analyses of pulmonary function, symptoms, or lab work. Nor was there evidence for diuresis, hypovolemia, or a clinically significant change in serum electrolytes or renal function for the whole group[11].

Statistical power: The interpretation of the negative results, both for safety and efficacy, should be tempered by a recognition that the small number of subjects led to relatively low statistical power. For example, the probability of obtaining significant ($p<0.05$) FEV_1 results was ~0.24.

DISCUSSION

Three general features were important in this pilot study of amiloride aerosol in the treatment of cystic fibrosis. First, effective drug delivery to airway surfaces was essential. To mantain an effective drug concentration ($ED_{50}=10^{-6}$M) on airway surfaces, a peak concentration of amiloride of $5x10^{-4}$ M delivered four times daily is required.[13,14] Although a suboptimal concentration of ~$8x10^{-5}$ M amiloride in bronchi of CF patients was achieved using nebulized delivery,[14] this dosage was accepted because respirable amiloride was available only in this formulation. Second, because outcome variables that may be particularly pertinent to aerosolized amiloride are not defined, we tested outcome variables based on the ability of amiloride to inhibit excessive Na^+ absorption, hydrate secretions, and improve mucociliary clearance[20]. Spirometry was the main criterion of efficacy because it measures airway function as well as retention of airway secretions[21]. Third, this pilot study of aerosol diuretic in humans involved drug safety. Thus, we selected a cohort of reliable adult patients and monitored for any effect of amiloride as a K^+-sparing diuretic and for deleterious effects on pulmonary function.

Amiloride as administered in this study produced no evident pulmonary or systemic toxicity. No patient noted drug-induced pulmonary symptoms, or reductions in pulmonary function tests to suggest alveolar toxicity. Likewise, there were no adverse effects of amiloride on vascular volume status, renal function, serum electrolytes, or hematologic function.

The key finding of this study is that amiloride slows the disease-induced decline in FVC. This rate of loss of FVC was inhibited by more than 50% during amiloride as compared to vehicle. A similar trend but smaller difference was observed in the analysis of the FEV_1 data. Retrospective power analyses indicated that patient populations of 24 and 56 patients would be required ($p=0.05$, power=0.95) to detect significant changes in FVC and FEV_1, respectively, if the magnitude of the slope differences between amiloride and vehicle persisted.

Because the effects of vehicle and amiloride were compared following parenteral antibiotics, it is important to recognize that the decline in FVC may be more rapid than under usual circumstances, and the absolute reduction in the rate of loss of

FVC during amiloride cannot be extrapolated to prognostic predictions. However, the spirometric data indicates that amiloride may extend the duration of benefit of antibiotics.

Interventions with bronchodilators and antibiotics do not explain the beneficial effects of amiloride. There was a trend towards greater bronchodilator use and longer but not more frequent usage of oral antibiotics in the vehicle phase.

The mechanism of action of amiloride on spirometric parameters is not established. Amiloride probably acts via its demonstrated effect on Na^+ transport in CF airways, but alternative mechanisms are possible. These effects do not appear to reflect any direct bronchodilating activity because the differences in FVC between treatment arms persist after aministration of β-agonist bronchodilators, and no acute bronchodilating effect of amiloride has been seen in CF patients[13,20]. Also, the effect does not reflect antimicrobial activity, because bacterial densities were not different. Whereas it is likely that amiloride exerts its effects at least in part via increasing clearance of secretions as has been reported in acute studies[20], direct measurements of secretion clearance during chronic therapy will be required to rigorously test this notion.

In an attempt to test the feasibility of directly measuring sputum biorheologic properties as an index of therapeutic efficacy[22], we collected sputum specimens at the end of each period from 6 patients. The data illustrated in Figure 3 indicate that biorheologic properties of sputa obtained at the end of the amiloride period were different from those of the vehicle period and were similar to values for secretions obtained bronchoscopically from normal subjects.[18] These measured biorheologic changes are projected to improve mucociliary transport rates and cough clearance, as have been observed in short-term studies of amiloride aerosols in CF subjects.[20]

No change was detected in percent solids of microwave dried sputum. This finding may reflect the insensitivity of the techniques employed to detect small but functionally important changes in sputum water content. Other potential effects of amiloride that would tend to normalize sputum biorheology include a reduction in sputum purulence, changes in divalent cation concentrations and concentrations of serum-derived macromolecules, e.g., albumin, and/or alterations in the distribution of mucin types secreted onto the airway surface. Further studies will be required to resolve these issues.

Diary data during aerosolized vehicle indicated that the patients perceived their clinical course to be reasonably stable. This observation might suggest that a 500 ml mean decline in FVC over 6 months is not associated with significant changes in symptoms, or that the diary questionnaire was insensitive. However, of the five symptom categories that exhibited borderline or significant worsenings with time during vehicle, only one symptom showed statistically significant worsening during the amiloride phase.

In summary, this pilot study suggests that amiloride aerosol may be useful in the treatment of adult CF patients with established lung disease. These observations raise at least two major questions. First, can these effects be increased in adult patients? It is possible that improvements in aerosol delivery, e.g., perhaps via metered dose inhaler, may improve pharmacokinetics, compliance, and clinical efficacy. Combination therapies with other drugs, e.g., ones that may induce Cl^- secretion via intracellular Ca^{++}-dependent mechanisms,[23] may further enhance efficacy. Second, might greater benefit, e.g., a "protective' effect, be achieved by administering the drug to younger subjects with less severe lung disease? This latter question will require clinical trials in younger patients, preceded by toxicity testing and direct measurements of drug delivery.

Acknowledgements

We thank Lisa Brown for editorial assistance and preparation of the manuscript. Supported by grants from the Cystic Fibrosis Foundation (RRDP, R015, and A014), the National Institutes of Health (HL34322 and MOl RR00046), the Canadian Cystic Fibrosis Foundation, and the Mr. and Mrs. James Snyder Fund.

References

1. R.C. Boucher, M.J. Stutts, M.R. Knowles, L. Cantley, J.T. Gatzy, 1986, Na^+ transport in cystic fibrosis respiratory epithelia. Abnormal basal rate and response to adenylate cyclase activation, **J. Clin. Invest.**, 78:1245.
2. R.C. Boucher, C.U. Cotton, J.T. Gatzy, M.R. Knowles, J.R. Yankaskas, 1988, Evidence for reduced Cl^- and increased Na^+ permeability in cystic fibrosis human primary cell cultures, **J. Physiol. (London)**, 405:77.
3. M. Knowles, J. Gatzy, R. Boucher, 1983, Relative ion permeability of normal and cystic fibrosis nasal epithelium, **J. Clin. Invest.**, 71:1410.
4. M. Knowles, J. Gatzy, R. Boucher, Modulation of nasal epithelium ion permeability in normal and cystic fibrosis subjects in vivo, **Clin. Res.**, 31:858A (Abstract).
5. R.A. Frizzell, G. Rechkemmer, R.L. Shoemaker, 1986, Altered regulation of airway epithelial cell chloride channels in cystic fibrosis, **Science**, 233:558.
6. R.A. Schoumacher, R.L. Shoemaker, D.R. Halm, E.A. Tallant, R.W. Wallace, R.A. Frizzell, 1987, Phosphorylation fails to activate chloride channels from cystic fibrosis airway cells, **Nature**, 330:752.
7. J.L. Potter, L.W. Matthews, S. Spector, J. Lemm, 1967, Studies of pulmonary secretions. II. Osmolarity and the ionic environment of pulmonary secretions from patients with cystic fibrosis, bronchiectasis and laryngectomy, **Am. Rev. Respir. Dis.**, 96:83.

8. M. Knowles, G. Murray, J. Shallal, F. Askin, V. Ranga, J. Gatzy, R. Boucher, 1984, Bioelectric properties and ion flow across excised human bronchi, **J. Appl. Physiol. :Respirat. Environ. Exercise Physiol.**, 56(4):868.
9. M. Knowles, J. Gatzy, R. Boucher, 1981, Increased bioelectric potential difference across respiratory epithelia in cystic fibrosis, N. **Engl. J. Med.**, 305:1489.
10. W.E. Waltner, R.C. Boucher, J.T. Gatzy, M.R. Knowles, 1987, Pharmacotherapy of airway disease in cystic fibrosis, **TIPS**, 8:316.
11. M.R. Knowles, N.L. Church, W.E. Waltner, 1990, A pilot study of aerosolized amiloride for the treatment of cystic fibrosis lung disease, **N. Engl. J. Med.**, 322:1189.
12. E. Puchelle, J.M. Tournier, J.M. Zahm, P. Sadoul, 1984, Rheology of sputum collected by a simple technique limiting salivary contamination, **J. Lab. Clin. Med.**, 103:347.
13. W.E. Waltner, N.L. Church, J.T. Gatzy, R.C. Boucher, M.R. Knowles, 1986, Toxicity and pharmacokinetics of acute amiloride aerosol in normal and cystic fibrosis subjects, **Cystic Fibrosis Club Abstracts**, 27:121 (Abstract).
14. W.E. Waltner, N.L. Church, J.T. Gatzy, R.C. Boucher, M.R. Knowles, 1987, Deposition, pharmacokinetics and toxicity of amiloride aerosol in normal and cystic fibrosis (CF) subjects, **Am. Rev. Respir. Dis.**, 135:288A (Abstract).
15. R.L. Hodinka, P. Gilligan, P. Gage, K. Wait, M. Knowles, 1988, Quantitative bacteriology of sputum from cystic fibrosis patients, **American Society for Microbiology**, Miami Beach 334 (Abstract).
16. M. King, P.T. Macklem, 1977, The rheological properties of microliter quantities of normal mucus, **J. Appl. Physiol.**, 42:797.
17. D. Brasfield, G. Hicks, S. Soong, R. Tiller, 1979, The chest roentgenogram in cystic fibrosis: A new scoring system, **Pediatrics**, 63:24.
18. A. Jeanneret-Grosjean, M. King, M.C. Michoud, H.Liote, R. Amyot, 1988, Sampling technique and rheology of human tracheobronchial mucus, **Am. Rev. Respir. Dis.**, 137:707.
19. M. King, G. Brock, C. Lundell, 1985, Clearance of mucus by simulated cough, **J. Appl. Physiol.**, 58:1776.
20. D. Kohler, E. App, M. Schmitz-Schumann, G. Wuertemberger, H. Matthys, 1986, Inhalation of amiloride improves the mucociliary and the cough clearance in patients with cystic fibrosis, **Eur. J. Respir. Dis.**, 69(Suppl.146):319.
21. M. Demedts, 1987, Assessment of airway secretions by pulmonary function tests, **Eur. J. Respir. Dis.**, 71:310.
22. M. King, 1989, Mucus, mucociliary clearance and coughing, **in:** "Respiratory Function in Disease," D.V. Bates, ed., Saunders, Philadelphia.
23. R.C. Boucher, E.H.C. Cheng, M.J. Stutts, M.R. Knowles, H.S. Earp, 1989, The chloride secretory response of cystic fibrosis (CF) human airway epithelia: Preservation of calcium ionophore but not protein kinase C and A-dependent mechanisms, **J. Clin. Invest.**, 84:1424.

DISCUSSION

IANNUZZI: Is it known if amiloride orally has any effect in the lung? And likewise, do any diuretics affect PD in the lung when patients take them orally?

KNOWLES: Specifically regarding amiloride, the first study we did in 1981, when we considered the possibility that amiloride might be useful, was to administer a standard dose of oral amiloride to normal and CF patients, and we measured nasal and lower airway PD and collected samples of respiratory secretions. Oral amiloride does not have an effect on PD, and the concentration of drug achieved on airway surfaces is about 10^{-8} molar. One would not have anticipated achieving an effective concentration on airway surfaces, but we thought it was worth testing. We have studied many patients taking other oral diuretic agents, including furosemide, and we were unable to discern any effect on PD.

GEDDES: Do you have data on the rate of decline of lung function in the years before your study? Because, looking at the loss of vital capacity on active treatment, it comes out at about 300 milliliters a year, which would give the patients a prognosis of maximum 10 years, and yet they've achieved 25 years at 80% vital capacity.

KNOWLES: Yes, I attempted to emphasize that this study design is somewhat artificial compared to a "steady state circumstance" where patients are followed over several years. We gave i.v. antibiotics, and patients achieved their best pulmonary function in the past year, or several years, at the start of the study. We then followed the natural decline from that peak pulmonary function. So you cannot extrapolate the rate of decline during vehicle to a long-term steady-state kind of circumstance. We were attempting to set up a study design whereby, over a six-month period you would expect to see decrement in pulmonary function. Then you could perhaps see a therapeutic interventional effect.

GEDDES: You've shown very nicely that you have an effect by intervention, but still, it looks as if these patients have done worse during the trial period than in any other period of their lives. It would be reassuring to know that they hadn't.

KNOWLES: I can say that in general the pulmonary function in these patients was relatively stable. It depends on when you measure their pulmonary function. If one looked at their pulmonary function after i.v. antibiotics during the previous 3-4 years, it approximated what we achieved after i.v. antibiotics at the start of our study. And the rate of decline during vehicle was similar to that reported in several studies during follow-up of parenteral antibiotic efficacy.

QUINTON: Mike, does amiloride have any effect on normal lung function?

KNOWLES: Amiloride does not have any effect on normal lung function. We did extensive acute challenges with amiloride aerosol in normal volunteers measuring not only airway mechanics but also arterial blood gas and diffusing capacity. I did not see any deleterious effect acutely, nor did we see any deleterious effect on lung function after inhalation of this drug 4 times a day for a month, in normal volunteers.

GREGER: This is to you and also to Rick. I think the concept, at this point may have been a little distorted because Rick was saying, "We are looking to a hyperactively transporting tissue" and you are testing now one aspect, one side of this coin. I think this may well be the problem. In the normal, amiloride has fairly little effect on voltage, so the negative voltage is due to secretion. In your case, you are able now to block reabsorption - you are not able to turn on secretion. So I think, the outcome of this study, being somehow a little too little to be enormously exciting at this point, may well be due to the fact that you were unable to induce secretion while you were able to block reabsorption. I think neither maneuver alone may be really mimicking normal conditions. Would you agree with this point?

KNOWLES: I would respond to two parts of that. First, the relative magnitude of the protective effect in these patients is not complete. I would submit that this may well be because we are studying adult patients with established lung disease. As I indicated in my talk, if you do an aerosol lung scan in these patients, 45% of the lung doesn't receive aerosolized drug, so at best, whatever effect you're trying to see here, you're trying to pick out from good regions of the lung. Disease activity is unabated in almost half the lung in these patients, so I think the magnitude of the effect may be related to that sort of patient problem. Second, I agree that it is possible that if one were able to induce chloride secretion as well as block sodium transport, the response might be better. In fact, we are working to explore from that mechanisms whereby one might do that as well. But since under normal circumstances, under basal conditions without amiloride, we don't think there is chloride secretion, it may be that just blocking sodium absorption would return things toward normal. Having said both of those things, one must recognize that the dosage regimen we had in the study is sub-optimal. Even if patients took the drug 4 times a day they probably didn't have a fully effective dose to block sodium absorption around the clock. On the average our patients really took the drug about 3 times a day. So I think there are several factors which could contribute to the less-than-optimal protective effect, and that may be related in part to the question of chloride secretion.

QUINTON: The emphasis here has been clearly on the balance between secretion and absorption. In view of the fact that you see no change in lung function for normals, I wonder if perhaps the effect of the drug might be on the composition of the airway fluid as opposed to the volumes moved, because if you are having an effect with the drug, especially on sodium transport, you should be decreasing the airway electrical potential, and that decrease would be in the direction toward normal, so that we might expect that all ions that are displaced or distributed across the epithelium on the basis of electrical driving force would return more to a normal distribution and a more normal composition. Do you suspect that compositional changes may have had some effect on the study, or do you have evidence to the contrary?

KNOWLES: We don't have any direct evidence that the composition is changed or might be changed.

RECHKEMMER: I have a question for Dr. Knowles and Dr. Boucher. Where is the origin of the fluid in the lungs when there is only, or primarily, absorption going on?

KNOWLES: Well, I think Rick alluded to that this morning. We end up pointing down to the alveolar region and saying "Fluid originates there." We don't think fluid originates from airway surface cells. Certainly some fluid originates from submucosal gland secretion, but that extends only down to the small airways. If you look at cultured cell studies of alveolar epithelium, there is certainly no absorption there. Is that a culture artifact or does that reflect the native function of the tissue? Is it possible that liquid could sneak in due to biophysical forces at the corners of alveoli? The bottom line is that we don't know clearly where airway surface liquid originates.

FROMTER: I have a short question regarding the patients. Some of them must have known by the bitter taste that they were having amiloride. Did these patients, or even the others, ask for continuation of the amiloride treatment, or did they not care for it. What was the patients' subjective idea about the treatment?

KNOWLES: Four of the 14 patients had been exposed to amiloride as a part of previous studies, and they did know that amiloride had a taste. We attempted to blind the remainder (and them as well) of the patients by indicating that the vehicle had a taste, the other drug had a taste. We suggested that some people could taste vehicle, some could taste drug, and if they tasted something, they wouldn't know. At the end of the study we actually asked. The four patients who had been exposed to the drug before guessed correctly. I think 6 of the 14 patients guessed correctly, 3 guessed incorrectly, and 5 were uncertain. I would say that about half the patients in a final summary discussion thought they would have liked to continue amiloride, and perhaps a few more if drug administration were easier.

AUSIELLO: You're focusing on the effect of amiloride on alterations in reabsorption. Is there any evidence that there is any change in cell volume associated with the diuretic? To ask the question a little bit differently, you said there is no effect of amiloride on normals, but has amiloride or any diuretic been found to have an effect on patients with other kinds of lung disease where cell swelling might be a problem, such as bronchospasm or asthma?

KNOWLES: I think I'll defer the question on cell swelling to Rick Boucher. There have been some studies done on cultured cell systems, and Rick has a more informed answer to that than I do.

BOUCHER: We don't know directly. The indirect measurements would indicate that amiloride probably leads to very modest reductions in cell volume in this particular cell type. The other question, I think, that may be relevant is: are there any other amiloride-sensitive transporters on the apical membrane - sodium/calcium exchange, sodium/proton exchange? We're looking at that and we certainly have those exchangers on our cell somewhere. The answer to the second part of your question is that aerosolized diuretics to some extent are getting to be popular, at least in the experimental approach to asthma, largely fueled by work here in Italy by Di Bianco and colleagues, that aerosolized furosemide may indeed block antigen-induced and exercise-induced bronchoconstriction. There may be similar modes of action imputed to amiloride, but I haven't seen them as definitively written up as the Italian one.

AL-AWQATI: First, Mike, I think this is a really nice study. My question really was about the taste of amiloride, which you already answered. But I have a second question: When you did rheological studies with your collaborator, did you have a control to find out that amiloride itself had no effect on the rheological properties?

KNOWLES: We had a control only to the extent that the rheology was measured from samples that had been collected from patients at the end of both phases, one in which the patients had been inhaling amiloride and one in which they had been inhaling vehicle. It's very difficult to "add" exogenous amiloride to airway secretions. So we don't have any direct measurements in that regard.

SOME PROPERTIES OF SODIUM AND CHLORIDE CHANNELS IN RESPIRATORY EPITHELIA OF CF- AND NON-CF-PATIENTS

J. Disser, A. Hazama and E. Frömter

Zentrum der Physiologie
Johann-Wolfgang-Goethe-Universität
Frankfurt/Main, Germany

INTRODUCTION

Thus far patch clamp studies on single channel properties of respiratory epithelia have mainly focussed on Cl^- channels (9, 10, 13, 15, 16, 21, 23). This has been so for two reasons: 1. The disturbance in Cl^- conductance was the first transport defect to be discovered in CF epithelia (19, 22) and 2. In single channel studies on respiratory epithelia Cl^- channels are found more frequently and identified and studied more easily than the low conductance, slowly switching Na^+ channels. On the other hand, however, it has been known for almost ten years that sodium transport is increased in CF epithelia (12) and this alteration of sodium transport may even be more relevant as a pathogenetic factor since it may affect the transepithelial fluid balance in the bronchi more than the inability to secrete chloride (see contribution by Boucher and collaborators to this symposium). In the present paper, therefore, we shall mainly discuss our recent studies on properties of Na^+ channels and shall, at the end, only briefly mention some of our recent experiments concerning the regulation of Cl^- channels.

METHODS

All experiments were performed on the "apical" cell membrane of respiratory epithelia in cell culture which were derived from human nasal polyps of both control and CF patients. To isolate epithelial cells, the polyps were digested in 1 -2.5 mg/ml protease XIV (Sigma, Taufkirchen, FRG) either for 1 hour at 37°C or for up to 24 hours at 4°C. In the first case contamination by fibroblasts was reduced by repeated platings of the cells on fibronectin or laminin coated culture dishes. The cells were then seeded on collagen coated cover slips and cultured for up to two weeks in a B5060 EK-CO_2 incubator (Heraeus, Hanau, FRG) at 37°C. The patch clamp technique as introduced by Hamill et al (8) was used to determine single channel currents with the help of an EPC 7 electrometer (List, Darmstadt, FRG) and the data were analysed with the software from Instrutech (Elmont, N.Y., USA). Seals were obtained preferentially at the edge of small cell clusters during day 1 - 6 after seeding. In the present study only single channel currents were analysed from excised inside-out, or occasionally outside-out patches. Individual channel types were identified from the current-voltage (I.V.) relation in symmetrical NaCl solutions and in asymmetrical solutions after substitution of individual ions in the bathing fluids.

RESULTS AND DISCUSSION

A) Studies on Sodium Channels

It is generally held that the single channel equivalent of the amiloride inhibitable epithelial Na^+ conductance are the channels which have been described by Palmer and Frindt in patch clamp studies on the apical cell membrane of rat renal collecting duct epithelia (17, 18). These authors observed highly selective, low-conductance channels (of ≈ 5 pS) with a linear I.V. relationship which were not inhibited by amiloride from the cytoplasmic surface, but exhibited a flicker block, if 0.5 μmol/l amiloride was applied from the extracellular surface. The same type of highly selective Na^+ channel was also observed in our experiments. Fig 1 shows the I.V. relation of such a channel in symmetrical isotonic NaCl solutions, and after substitution of the greater part of bath sodium by N-methyl-D-glucamine. Note the shift in the reversal potential in direction towards the reversal potential of Na^+ (E_{Na}). Similar observations were also made after substitution of Na^+ by K^+ indicating the high selectivity of this channel for Na^+. The incidence of this type of channel, however, was rather low (ca. 5 % of all seals) and when trying to summarize the data we ran into the problem that the slope of the I.V. relation in some experiments was nearly immeasurable (values near 1 to 2 pS) while in other experiments it ranged up to 15 pS. We have also observed non-selective cation channels but their conductivity was distinctly larger (see e.g. Fig 2).

This wide range of single channel conductances contrasts with the reports by Palmer and Frindt on rat collecting duct (17, 18), but a comparison with other recent reports on Na^+ channels suggests that either Na^+ channels have a wide range of single channel conductances or that a variety of channels exists with different conductances. This view is supported by the results of our previous Na^+ channel studies on monolayer cultures of collecting duct principal cells (14). In these epithelia - which generated substantial mascroscopic amiloride-inhibitable

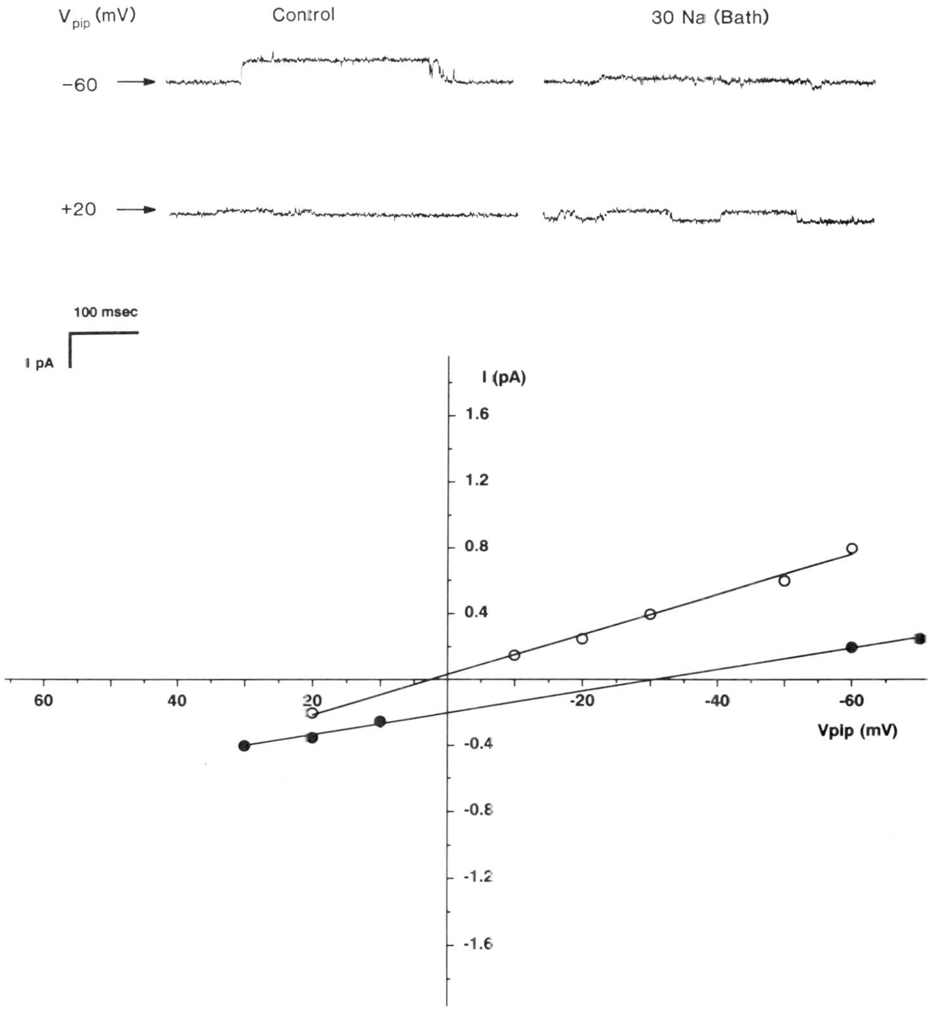

Fig. 1. Single channel currents and current voltage relation of respiratory cell Na^+ channel excized inside-out patch. Left traces and open circles: recordings with identical isotonic NaCl-Ringer solution present in bath and pipette. Right traces and filled circles: recordings after equimolar replacement of 120 mmol/l of bath Na^+ by N-methyl-D-glucamine. Arrows indicate channel closed state. Note that after bath exchange the I.V. relation intersects with the x-axis close to E_{Na} . This value (-39mV) is nearly reached if the liquid junction potential developing at the reference electrode is taken into account.

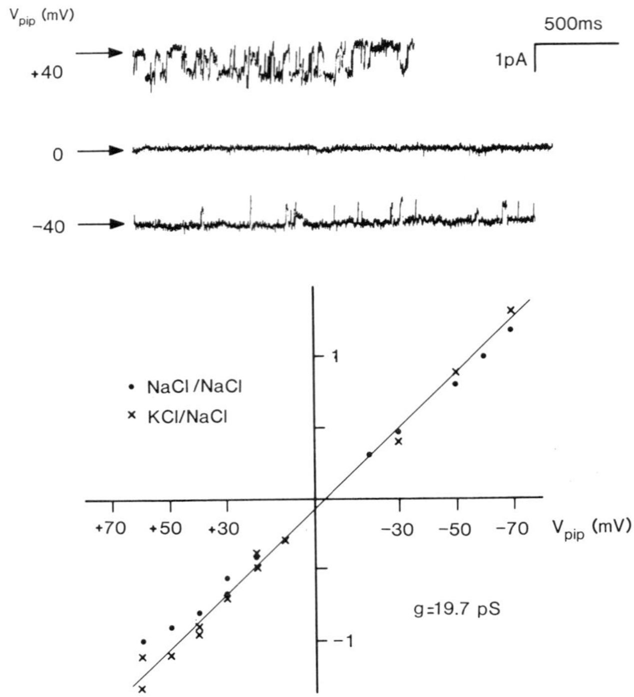

Fig. 2. Single channel currents and current voltage relation of non-selective cation channel of respiratory epithelium. Details as in Fig 1. Note steeper I.V. relation and missing response to Na^+ substitution.

Na^+ currents (7) - we were unable to identify single Na^+ channels, but noticed an asymmetry of the total background current when amiloride was present in the pipette (see Fig 3). This suggested that Na^+ channnels were present and that their blockage was interrupted when amiloride was pulled out from the channel mouth by strong pipette negative voltages. The fact that we have been unable to identify these channels from single channel events directly, must have meant that the single channel conductance was below the limit of resolution (below 2-3 pS).

Our contention that Na^+ channels do not necessarily exhibit a narrow range of single channel conductance values is also supported by the fact that all publications on Na^+ channels of 5 pS conductance are based on rather small numbers of observations (see Table 1), while quite on the contrary, due to the low single channel conductance, the incidence of channels should have been rather high. This suggests that also in previously published studies a discrepancy may have existed between the Na^+ channels that have been identified and those that may have been present but were not recognized. The reason why single channel conductances of Na^+ channels vary so greatly (different functional states, different genetic isozymes) is presently not known.

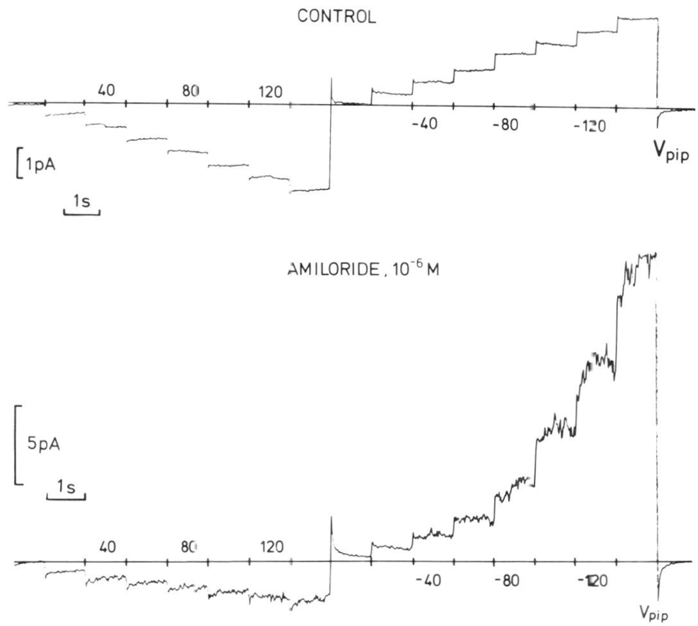

Fig. 3. Current recordings from two excized, inside-out patches of the apical cell membrane of collecting duct principal cells in culture. In the lower trace amiloride 10 mol/l was present in the pipette solution (isotonic NaCl as in bath). Note asymmetric current response in presence of amiloride suggesting release of amiloride block at pipette negative potentials inspite of absence of discernable single current events (from ref 14).

Table 1

Some properties of amiloride blockable Na⁻ channels

tissue	respir. epith.	cell (16) duct	toad (4) bladder	prox. (6) tubule	A6 (9 cells
g (pS)	1 - 15	4.1±0.2	5.0±1.6	12.0±2.1	8.4+1.3
n	35	7	21	13	5
P_{Na}/P_K	10:1	10:1	>10:1 oΘ 1÷1	>10:1	>3:1
t_{open} (ms)	25 220	3400	310		≈30
t_{closed} (ms)	40 1600	3900	2 1400		≈90
flicker block	yes	yes	yes	yes	(yes)

g = single channel conductance in pS,
n = number of observations,
P_{Na}/P_K =permeability ratio for sodium over potassium,
t_{open} and t_{closed} are open and closed time constants in ms.
Numbers in brackets are literature references.

A second typical feature of epithelial Na^+ channels, reported previously in the literature, is their extremely slow switching kinetics with time constant in the range of up to several seconds (4, 17, 18). These slow switching kinetics were not only observed in patch clamp studies but have previously been predicted from quasi macroscopic current fluctuation analysis experiments on frog skin (3). In our experiments on cultured human nasal polyp cells it appeared that the smaller the single channel conductance the slower the switching kinetics were, but in contrast to the observations by Palmer and Frindt (17, 18) we found that each two time constants were needed to fit the open and the closed time histograms (see Table 1).

Another typical feature of the epithelial Na^+ channel is the independence of its open probability (P_o) of the applied voltage and of cytoplasmic calcium activity (in isolated membrane patches) (17, 18). Both properties were confirmed in our experiments. In addition we confirmed the observation of Palmer and Frindt (17, 18) that amiloride (10^{-5}mol/l), if applied to the cytoplasmic surface, did not affect channel open probability or the single channel current. However, when applied in the concentration of 0.5 μmol/l in the patch pipette amiloride caused a typical flicker block.

Regarding a comparison of Na^+ channels from CF and non-CF tissue we have found that all single channel properties discussed above were equal in CF and non-CF tissue except for a difference in the mean open probability. By comparing P_o of 3 control and 4 CF patches in the on-cell configuration (total observation time 11.5 minutes each) we noticed that the open probability of CF channels was 2.50 times higher than the open probability of non-CF channels, while in the off-cell configuration the difference was less clear. In 3 off-cell patches of CF tissue (total observation time 9 minutes) P_o was 1.40 times greater than in 6 off-cell patches of non-CF tissue (total observation time 30 minutes). This finding agrees with the observations by Knowles and collaborators (1) who found a 2-3 times higher Na^+ permeability in the apical membrane of cultured nasal polyp epithelia from CF patients compared to controls. It suggests that the density of active Na^+ channels may be identical in patients and controls, but that Na^+ channel regulation may be altered in CF.

B) Studies on Chloride Channels

Starting from the idea that the voltage and/or heat induced activation of Cl^- channels in excised cell membrane patches might reflect the loss of a cytoplasmic inhibitor we have recently begun to search for such a substance. Initially we tested whether cytosolic ATP might have an inhibitory effect. This hypothesis had several attractive features: 1. If the CF-gene product CFTR (20) was identical with the Cl^- channel, the hypothetical inhibition by ATP could have explained why CFTR possesses

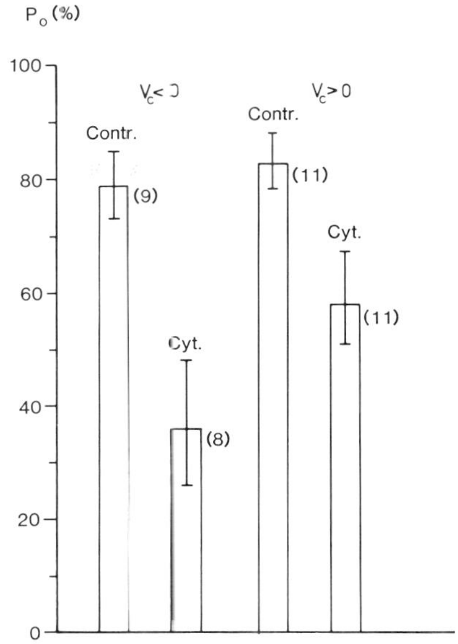

Fig. 4. Effect of cytoplasmic extract on open probability of CF Cl^- channel. The columns represent mean values + SE, number of observations are given in brackets. In the two columns on left the command voltage was held at 60 mV, cytoplasm negative, in the two columns on right polarity was reversed.

nucleotide binding folds (19). 2. If point 1 was correct, one could have postulated that in non-CF tissue phosphorylation by protein kinase A changed the binding affinity to ATP, so that ATP was released and the channel unblocked, while in CF tissue ATP was bound more firmly (because of the 508 mutation) and protein phosphorylation was unable to release the nucleotide. 3. Since ATP bears 3 negative charges per molecule, it would have been conceivable that a high cytoplasm positive voltage was able to pull off the ATP from its binding site, both in CF and non-CF thus explaining voltage induced activation of excised channels (2, 5, 23).

However, despite the conceptional beauty of the hypothesis it turned out to be wrong. When testing the effect of ATP on CF and non-CF channels in excised patches which had been activated by depolarization, we found that ATP (5 mmol/l) did not significantly change P_o. It increased P_o from 76 ± 12% to 78 ± 18% (n = 3).

In spite of this negative result we have continued to search for the putative cytosolic inhibitor by preparing in collaboration with Dr. W. Krick and Dr.G. Burckhardt cytosolic extracts from HT29 cells, a cell-line which is known to exhibit similar Cl^- channels as respiratory cells (11), and testing these on the cytoplasmic surface of excised patches. In addition cytosolic extracts of pig kidneys were prepared. Both extracts were found to reduce the open probability of outwardly rectifying Cl^- channels. In 4 non-CF channels the cytosolic extract reduced P_o from 79 ± 10% to 42 ± 17% and similar observations were made on Cl^- channels activated from CF cell membranes after excision (see Fig 4). These data suggest that an inhibitory substance may be present in the cytosol of epithelial cells which down-regulates the open probability of outwardly rectifying Cl^- channels. However, the chemical nature and physiological role of this substance remain to be established, and it is completely unknown at present whether the putative inhibitor has anything to do with the defective Cl^- channel regulation in CF.

References

1. Boucher, R.C., Stutts, M.J., Knowles, M.R., Cantley, L., Gatzy, J.T., 1986, Na^+ transport in cystic fibrosis respiratory epithelia, **J. Clin. Invest.**, 78:1245.
2. Disser, J., and Frömter, E., 1989, Single channel patch clamp-studies on sweat duct epithelium and respiratory epithelium of CF and non CF origin, **in:** "Mukoviszidose 1989: Ergebnisse aus Grundlagenforschung und Klinik," D.Kaiser, ed., Kali-Chemie Pharma, Hannover.
3. van Driessche, W. Zeiske, W., 1985, Ionic channels in epithelial cell membranes, **Physiol. Rev.**, 65:833.
4. Frings, S., Purves, R.D., Macknight, A.D.C., 1988, Single-channel recordings from the apical membrane of the toad urinary bladder epithelial cell, **J. Membrane Biol.**, 106:157.
5. Frizzell, R.A., Rechkemmer, G., and Shoemaker, R.L., 1986, Altered regulation of airway epithelial cell chloride channels in cystic fibrosis, **Science** 233:558.
6. Goegelein, H., and Greger, R., 1986, Na^+ selective channels in the apical membrane of rabbit late proximal tubules (pars recta), **Pflügers Arch.**, 406:198.
7. Gross, P., Minuth, W.W., Kriz, W., Frömter, E., 1986, Electrical properties of renal collecting duct principal cell epithelium in tissue culture, **Pflügers Arch.**, 406:380.
8. Hamill O.P., Marty, A., Neher, E., Sakmann, B., Sigworth, F.J., 1981, Improved patch-clamp techniques for high-resolution current recording from cells and cell-free membrane patches, **Pflügers Arch.**, 391:85.
9. Hamilton, K.L., and Eaton, D.C., 1985, Single-channel recordings from amiloride-sensitive epithelial sodium channel, **Am. J. Physiol.**, 24:C200.
10. Hwang, T.C., Lu, L., Zeitlin, P.L., Gruenert, D.C., Huganir, R., and Guggino, W.B., 1989, Cl^- channels in CF: lack of activation by protein kinase C and cAMP-dependent protein kinase, **Science**, 244:1351.
11. Hayslett, J.P., Goegelein, H., Kunzelmann, K., Greger, R., 1987, Characteristics of chloride channels in human colon cells (HT29), **Pflügers Arch.**, 410:487.
12. Knowles, M.R., Stutts, M.J., Spock, A., Fischer, N., Gatzy, J.T., and Boucher, R.C., 1983, Abnormal ion permeation

through cystic fibrosis respiratory epithelium, **Science**, 221:1067.
13. Kunzelmann, K., Pavenstädt, H., Greger, R., 1989, Properties and regulation of chloride channels in cystic fibrosis and normal airway cells, **Pflügers Arch.**, 415:172.
14. Laskowski, F.H., Christine, C.W., Gitter A.H, Beyenbach, K.W., Gross, P., Frömter, E., 1990, Cation channels in the apical membrane of collecting duct principal cell epithelium in culture, **Renal Physiol. Biochem.**, 13:70.
15. Li, M., McCann, J.D., Liedtke, C.M., Nairn, A.C., Greengard, P., and Welsh, M.J., 1988, Cyclic AMP dependent protein kinase opens chloride channels in normal but not cystic fibrosis airway epithelium, **Nature**, 331:358.
16. Li, M., McCann, J.D., Anderson, M.P., Clancy, J.P., Liedtke, C.M., Nairn, A.C., Greengard, P., and Welsh, M.J., 1989, Regulation of chloride channels by protein kinase C in normal and cystic fibrosis airway epithelia, **Science**, 244:1353.
17. Palmer, L.G., Frindt, G., 1986, Amiloride sensitive Na^+ channels from the apical membrane of the rat cortical collecting tubule, **Proc. Natl. Acad. Sci. USA**, 83:2767.
18. Palmer, L.G., Frindt, G., 1986, Epithelial sodium chanels: characterization by using the patch clamp technique, **Federation Proc.**, 45:2708.
19. Quinton, P.M., 1983, Chlcride impermeability in cystic fibrosis, **Nature**, 301:421.
20. Riordan, J.R., Rommens, J.M., Kerem, B.S., Alon, N., Rozmahel, R., Grzelczak, Z., Zielenski, J., Lok, S., Plavsic, N., Chou, J.L., Drumm, M.L., Iannuzzi, M.C., Collins, F.S., and Tsui, L.C., 1989, Identification of the cystic fibrosis gene: cloning and characterization of complementary DNA, **Science**, 244:1353.
21. Schoumacher, R.A., Shoemaker, R.L., Halm, D.R., Tallant, E.A., Wallace, R.W., and Frizzell, R.A., 1987, Phosphorylation fails to activate chloride channels from cystic fibrosis airway cells, **Nature**, 330:752.
22. Schulz, T.J., and Frömter, E., 1986, Mikropunktionsuntersuchun gen an Schweißdrüsen von Mucoviscidosepatienten und gesunden Versuchspersonen, **in:** "Mucoviscidose," 2. Deutsches Symposium, hrsg. von An A. Windorfer, U. Stephan, G. Thieme Verlag, Stuttgart.
23. Welsh, M.J.; and Liedtke, C.M., 1986, Chloride and potassium channels in cystic fibrosis epithelia, **Nature**, 322:467.

DISCUSSION

AUSIELLO: Do you think that there is a possibility that the highly selective and non-selective channels are various states of the same channel? If there are two different channels, under physiological conditions I would think that the cation-selective channel is basically working as a sodium channel. Therefore what do you think is the physiological relevance of the cation-selective channel?

FROMTER: This question cannot be answered with certainty as long as the channels have not been cloned. However, from the differences in regulatory properties of selective and non-selective channels I would tend to conclude that we are dealing with different entities here. On the other hand, it is as uncertain at present whether all selective and all non-selective Na^+ channels represent each one specific channel type or rather groups of different channels.

GREGER: May I just come back to this point of the cation channel? If I noted correctly, your own data on the collecting duct non-selective cation channel are in discrepancy with the data by Dr. Ausiello, Dr. Stanton and so forth with respect to amiloride. Is that correct? How do you explain it? They report a non-flickering amiloride block, if I'm not mistaken.

FROMTER: In the present experiments on Na^+ selective channels we have observed a flicker block in the presence of 0.5 μmol/l amiloride. However, in previous experiments on non-selective cation channels from collecting duct cell cultures we have also seen a reduction of single channel conductance during 10 μmol/l amiloride applied from the cytoplasmic surface without any discernable high frequency flicker effects.

AUSIELLO: In the two cell types that we have studied, one with Dr. Stanton that Dr. Greger alluded to, which is the intermedullary collecting duct cell type, and the A6 cell which I will discuss tomorrow, both of those have classical amiloride sensitivity when the drug is added to the extracellular surface. So at least in our hands, and in Doug Eaton's hands, the A6 cell clearly has amiloride sensitivity. I can't speak for Stanton with regard to his actual patching, but I believe he feels confident that the papillary collecting duct cells also are amiloride sensitive in a classical manner, despite the fact that these are cation-selective but not sodium-selective channels.

FROMTER: I have very carefully looked into the data from Stanton's laboratory. They show a non-selective cation channel of 27 pS conductance which is inhibited by 0.5 μmol/l amiloride from the extracellular surface in a voltage dependent manner. The difference from our data is the high channel conductance, the non-selectivity with respect to Na^+ over K^+, and the absence of a flicker block. We conclude that we are dealing with two different types of channels here.

AUSIELLO: Yes, I agree with that. Even in the A6 cells, that is not uncommon, but as you pointed out, the opening and closing time of these channels is so rapid, that distinguishing the flickering is sometimes a very difficult phenomenon. In addition, there is a clustering effect very often, as you point out, with regard to these channels. So whether or not one can discern that effect of amiloride, I'm not quite certain.

KUNZELMANN: I have a question concerning the cytosol effects, of course. My question is, did you use different doses, or dilution of cytosol and another question is, did you use cytosoles from different cell types - epithelial cells and non epithelial cells, for example?

FROMTER: No, thus far we have used only epithelial cell fraction. The cytoplasm was obtained from HT-29 cells or from kidney cells and both types of preparations were found to inhibit. Drs. W. Krick and G. Burckhardt from the Max Planck Institut für Biophysik in Frankfurt have also studied the Cl^- flux inhibition by cytosolic preparations on renal cell membrane vesicles and have also demonstrated gradual loss of inhibition after progressive dilution.

KUNZELMANN: One more question please, concerning these sodium channel, did you also observe the rundown phenomenon comparable to this description by Rick Boucher?

FROMTER: Yes, there is of course a rundown of channel activity. It varies from patch to patch. However, switching off a channel appears to be a sudden all-or-none event rather than a gradually developing process.

GREGER: What is the concentration of cytoplasm? Did you dilute it?

FROMTER: No, it may be too diluted during the preparation. It's just crude cytoplasm.

GREGER: You don't even centrifuge?

FROMTER: Sure, the cells are homogenized by sonication or better by pottering, and membrane material is then removed by centrifugation at 150,000 g for 1 hour.

GREGER: Second question, was this inhibition by cytoplasm overcome or reversed by a higher concentration of calcium.

FROMTER: I cannot answer this question, since we have not tested the effect of increasing Ca^{++} concentration. We have fixed the Ca^{++} concentration to 10^{-7} mol/l.

WINE: The distribution that you showed of channel conductances, are you certain that all of those channels in that conductance range were the highly sodium-selective channels?

FROMTER: We have not been able to analyze the selectivity of every channel, however, besides the single channel conductance there are other characteristic properties such as the slow switching kinetics, and the voltage independence of open probability which are readily recognized and may help to identify a channel as belonging to a certain channel type.

WINE: But did you see channels in the same range cell-attached that appeared not to be sodium channels?

FROMTER: No, as far as can be deduced from the current voltage regulation and the above given criteria: but I admit that channel characterization may not be conclusive in every case.

ROLES OF Ca AND cAMP ON Cl CHANNEL ACTIVITY IN CYSTIC FIBROSIS SWEAT CLEAR CELLS AS STUDIED BY MICROSUPERFUSION AND CELL VOLUME ANALYSIS

K. Sato, M. Ohtsuyama, Y. Suzuki, G. Samman,
K.T. Sato, and F. Sato

Marshall Dermatology Research Laboratories
Department of Dermatology
University of Iowa College of Medicine
Iowa City, Iowa

SUMMARY

In an attempt to study regulation of Cl channels in intact sweat secretory coils in cystic fibrosis and controls in the least invasive manner, isolated secretory coils were superfused with various drugs and K-efflux was determined as an indirect measure of Cl movement. Cl channel activity was also determined from the drug-induced cell volume increase in gramicidin (GC)-treated dissociated eccrine clear cells. We observed that while MCh-induced K-efflux from the CF secretory coils was entirely normal, K-efflux in the presence of isoproterenol (ISO), forskolin (FK), or IBMX was absent in CF, suggesting that these agents failed to stimulate Cl movement. Cl channel activity of dissociated CF clear cells, as studied by cell volume analysis, was entirely normal when stimulated by Ca-elevating agents but was defective when stimulated by cAMP-elevating agents. TPA (phorbol ester) does not appear to stimulate Cl channel activity nor does it modify the effect of other agents. The following observations from the present and previous studies are not necessarily consistent with the traditional thesis that the observed Cl movement is due to cAMP: CT-cAMP had no effect on cell swelling or on K-efflux; ISO is more potent in accumulating tissue cAMP than IBMX yet the latter is more potent in stimulating K-efflux; IBMX increases cytoplasmic [Ca] yet is unable to stimulate K-efflux in CF; K-efflux stimulated by cAMP-elevating agents was inhibited by removal of Ca from the bath; and, cell swelling of GC-treated cells in response to cAMP elevating agents was inhibited by removal of Ca. The inability of IBMX to stimulate Cl channels in the face of elevated cytoplasmic [Ca] and cAMP in CF cells deserves further scrutiny.

The Identification of the CF (Cystic Fibrosis) Gene
Edited by L.-C. Tsui *et al.*, Plenum Press, New York, 1991

INTRODUCTION

The eccrine secretory coil elaborates nearly isotonic fluid secretion in response to Ca-elevating agents such as cholinergic agonists and Ca-ionophores and cAMP-elevating agents such as isoproterenol (ISO), forskolin (FK), and vasoactive intestinal peptide (VIP) (1). Evidence accumulated in the authors' laboratories suggests that the Na-K-Cl cotransport system (2) may be the major transport mechanism for cholinergic sweat secretion (1, 3-7). The presence of both K and Cl channels has also been confirmed in freshly dissociated secretory cells (8). In contrast, the ionic mechanism of β-adrenergic sweat secretion, the role cAMP plays in regulation of sweat gland function, and the nature of the defective β-adrenergic function in cystic fibrosis sweat gland remain largely unknown. Recent patch clamp studies by others suggest that the cAMP-mediated regulation of Cl channels in cystic fibrosis (CF) is abnormal not only in transporting epithelia (9, 10, 11) but in lymphocytes (12). Although the patch clamp technique is a powerful method of studying Cl channels, the availability of an alternative method will further facilitate understanding of the abnormal regulation of Cl channels in cystic fibrosis. We have recently reported that the cell volume increase (swelling) induced by a given agonist in the presence of gramicidin (GC, a nonselective ionophore of Na and K) most likely reflects Cl channel activation by the agonist (13). In this review, we will briefly summarize our recent studies on Cl-channel activity as studied by GC-cell image analysis, stimulation-induced KCl efflux from secretory coils as studied by the superfusion method, and intracellular Ca concentration ([Ca]). Although our results are consistent with the abnormal regulation Cl channel activity in CF clear cells during stimulation with cAMP elevating agents, the exact roles of cAMP and Ca in β-adrenergic regulation of sweat secretory function remain largely unsolved.

MATERIALS AND METHODS

Isolation of Sweat Glands

Procedures for skin biopsy and isolation of rhesus monkey palm and human eccrine sweat glands have been repeatedly described previously (4,14,15). The excised tissue was blotted of blood, sliced into several pieces, and immediately washed in several changes of cold (about 10°C) modified Krebs-Ringer bicarbonate solution (KRB) containing (in mM): 125 NaCl, 5 KCl, 1.2 $MgCl_2$, 1.0 $CaCl_2$, 25 $NaHCO_3$, 1.2 NaH_2PO_3, 5.5 glucose, and 20 mg/100ml bovine serum albumin (BSA).

Superfusion of Isolated Secretory Coils

The method of constructing the superfusion chamber has been described in detail (16) and is shown in the insert to Fig. 1. It is constructed from a 0.8 mm o.d. glass capillary on a micro-

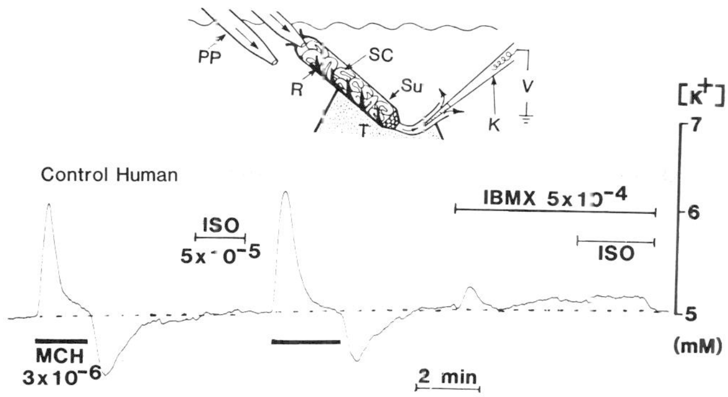

Fig. 1. Effect of isoproterenol (ISO) and isobutylmethylxanthine (IBMX) on K-efflux from superfused control human secretory coils. Methacholine (MCh) stimulation was inserted between stimulation with cAMP-elevating agents to minimize the decline of responsiveness to these agents (27). Insert is a schematic illustration of a superfusion chamber. In this and subsequent figures, unless otherwise noted concentrations of drugs are given in moles/l (M). pp, perfusion pipet made of polystylene tube; R, rod with thorns; SC, secretory coils; Su, superfusion chamber; T, epoxy to secure Su to the floor; K, potassium sensitive electrode; V, amplifier.

forge, is about two mm long and 0.4 mm internal diameter with a constriction at its orifice (to achieve a tight seal with a superfusion tubing) and a flared distal portion (to accommodate ion sensitive electrodes). Five to 10 secretory coils were placed inside the chamber and secured on the thorns of a polyethylene rod (about 0.15 mm o.d.). Each tube is connected to a perfusion syringe on a Harvard syringe pump. Perfusion speed was 0.5 to 1 fl per min. The superfusion chamber is submerged in the bath (Ringer's solution) kept at 37°C. The [K] in the effluent perfusate was determined with a K-sensitive electrode. Before each experiment, the absence of baseline shift by the perfusates alone was confirmed. Relative values of K-efflux were quantified from the chart paper using the commercial computer programs Data Scan (Brainpower, Calabasas, CA) and Microsoft Excel (Redmond, WA) (16).

Preparation of Dissociated Secretory Coil Cells and Cell Volume Analysis

Dissociated cells have been prepared by serial collagenase digestion essentially as described previously (5, 17) with modifications (18). The mixed dispersed cells thus prepared were used for cell volume analysis because clear cells can be readily identified by the presence of autofluorescent lipofuscin granules (5). Construction of the perfusion chamber has been

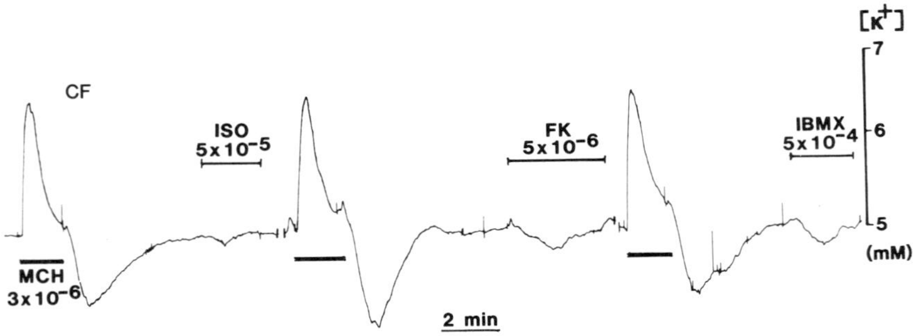

Fig. 2. Absence of K-efflux during stimulation with (ISO), forskolin (FK), or IBMX in CF secretory coils. The downward deflection of the K-trace indicates net K-uptake by the glands. All three traces are from the same glands.

described in detail (5, 18). Polylysine was used to immobilize dissociated cells on the cover glass mounted on the chamber. Temperature of the Olympus BH microscope was adjusted so that the center of the chamber was set at 37°C. The optical system consists of a 100x oil-immersion objective, a DIC condenser system (1.25x), and a 15x photomicrograph eyepiece (a total magnification of 1875x). The DIC image of the cells was captured with a CCTV camera connected to a VCR recorder. Digital paint brush system (Jandel Corp, CA) was used to determined the maximal cross section of the cell on the still frame of a high-resolution CCTV monitor.

Determination of Intracellular Ca Concentration

The fura 2 method (19) was used with minor modifications. Percol-purified clear cells were first loaded with 0.5-1.0 μM fura 2 AM (Molecular Probes, Eugene, OR) at 37°C for 60 min. A special chamber was constructed on a thermostated stage of a Leitz inverted microscope kept at 30°C. Cells were attached to the bottom of the chamber (made of a cover glass) coated with poly-D-lysin (Mr. 30,000).

A 100x fluorescence objective was used. Only the emission from the targeted cell reached the Leitz MPV photometer, with the help of a field diaphragm . A chopper filter (revolution 0.5 Hz, 340 and 380 nm) was used to apply two excitation wavelengths. $[Ca^{2+}]$ of the cell was thus calculated from the equation, $[Ca^{2+}]=K_d x \beta (R-R_{min})/(R_{max}-R)$ where K_d is 200 nM (at 30°C), β is the ratio of fura 2 fluorescence (at 380 nm excitation and 500 nm emission) when $[Ca^{2+}]$ is 0 and when $[Ca^{2+}]$ is 2 mM (the value is 5.24 in our particular equipment), R is the 340/380 nm ratio, R_{min} is the R value when $[Ca^{2+}]$ is zero, and R_{max} is the value of R when $[Ca^{2+}]$ is 2 mM.

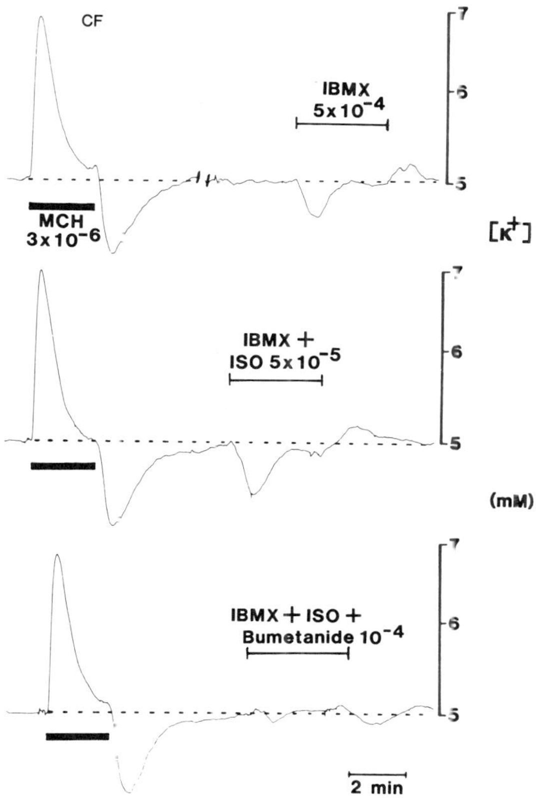

Fig. 3. Absence of K-efflux by IBMX (upper panel), augmentation of IBMX-induced K-uptake by ISO (middle panel), and inhibition of IBMX + ISO-induced K-uptake by bumetanide (lower panel). All traces from the same experiment.

RESULTS

Superfusion of Isolated Secretory Coils

K-efflux during stimulation of isolated human secretory coils with 5 x 10^{-5} M MCh (Fig. 1) was much less than that of rhesus glands (16). More prominent K-efflux was observed with IBMX alone or IBMX + ISO. After the initial K-efflux, the K-trace deflected downward indicating the net K-reuptake. However, in none of the CF secretory coils (n=14) did the K-efflux occur during perfusion of cAMP-elevating agents such as ISO, forskolin, IBMX (a phosphodiesterease inhibitor), or the combination of these agents. Interestingly, as shown in Figs. 2 and 3, varying degrees of K-uptake were observed. As shown in Fig. 3, K-uptake was completely inhibited by bumetanide, suggesting that the K-uptake in response to cAMP-elevating agents is due in large part to activation of Na-K-Cl cotransporters. In control human glands (not shown) and rhesus glands (16), bumetanide

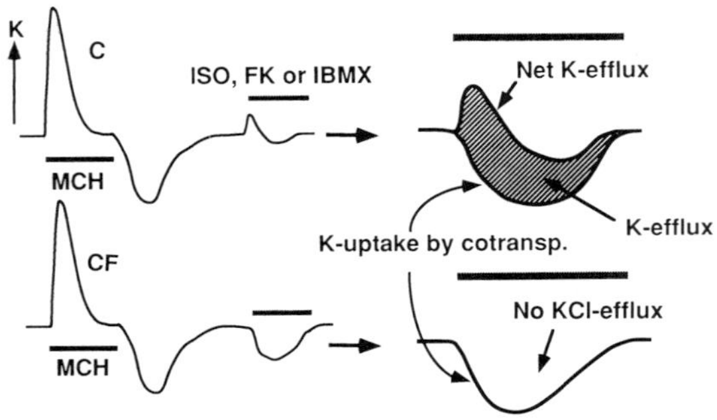

Fig. 4. Schematic illustration of K-efflux from control (C) and CF secretory coils indicating that the net K-efflux observed in the present study corresponds to the difference between K-efflux and K-uptake.

potentiated such K-efflux, further supporting the thesis that the net K-efflux we observed in control glands is the difference between K-efflux and the cotransporter-mediated K-uptake (16) (Fig. 4). That K-efflux during ISO stimulation is largely accompanied by Cl-efflux has been demonstrated in our previous X-ray microanalysis study (16). Hence, the absence of K-efflux in CF glands by cAMP-elevating drugs may be interpreted as the inability for Cl to move in response to elevation of cellular cAMP. It is puzzling, however, that the seemingly cAMP-induced initial transient K-efflux was completely inhibited in a Ca-free Ringer also containing 1 mM EGTA (16).

Cytoplasmic Ca Concentration during Stimulation with ISO, FK and IBMX

As shown in Fig. 5, 5 x 10^{-4} mM IBMX (with or without DMSO), normally used as phosphodiesterase, raised cytoplasmic [Ca] by 90 nM in a dissociate CF clear cell. Control clear cells behaved in the same manner. IBMX-induced increase in [Ca] was 94 ± 6.4 nM for CF (n=9) and 86 ± 5.9 nM for control (n=14). Neither FK nor ISO increased cytoplasmic [Ca] (not shown).

Cell Volume Increase of Gramicidin-Treated Cells as a Measure of Cl-Channel Activity

When an isolated rhesus clear cell was stimulated with MCh or A23187, the cell shrank by as much as 20 to 50% of the original volume (15) (Fig. 6). Some cells remained shrunken for as long as MCh was present, but the majority of cells partially restored the cell volume in the presence of MCh. MCh-induced cell shrinking is pharmacologically specific. It requires Ca in the bath for sustained shrinking, although cells shrink transiently in the Ca-free bath. MCh-induced cell shrinking was inhibited by quinidine (K-channel blocker) and in the Cl-free medium. Furthermore, the cell shrinkage was reversed in 120 mM KCl (replacing NaCl) in the bath. Thus we conclude that MCh-induced cell

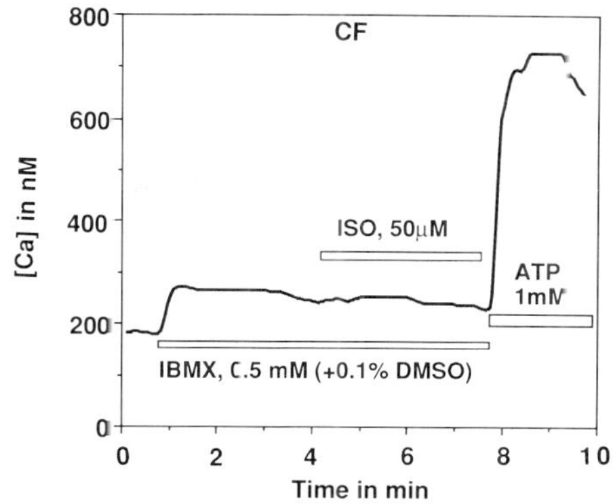

Fig. 5. Effect of IBMX or IBMX + ISO on cytoplasmic [Ca] in a dissociated CF sweat gland clear cell. At the end of the experiment ATP was added as positive control. Note the increase in [Ca] from 180 nM to 270 nM by IBMX with no further increase in [Ca] by ISO. An illustrative experiment. The effect of IBMX is independent of the presence of absence of DMSO (not shown).

shrinkage is due to stimulation of Ca-dependent K and Cl channels in the clear cell membrane, resulting in net flux of KCl (5, 20). However, as can be seen in Fig. 6, cell shrinkage does not occur when stimulated with cAMP-elevating agents, which could be due to the smallness of the net KCl efflux and/or the simultaneous activation of KCl uptake mechanisms such as cotransporters which minimizes the net K-loss. When gramicidin (GC, which forms artificial cation channels nonselective to Na or K) was added at 0.1-1.0 μM, cell volume remained unchanged for several minutes. At that point, addition of either Ca-elevating agents such as ionomycin (IO), ATP, or MCh or cAMP-elevating agents such as ISO, theophylline (TH), IBMX, or FK caused the cell to swell for about 30 sec (Fig. 6). At the peak of cell swelling the cell started blebbing, which was the end point of cell volume analysis. As expected, MCh- or IO + GC-induced swelling is dependent on extracellular Ca and Cl (20). Puzzlingly enough, the cell swelling due to cAMP-elevating agents (plus GC) is also inhibited by removal of extracellular Ca (Fig. 7). Thus, it is likely that the swelling of GC-treated cells may be due to activation of Cl-channels even though the effect of Cl channel blockers such as DPC or NPPB is only partial (not shown). Exactly what the intracellular mediator is, during ISO or FK stimulation, remains unknown. TPA (phorbol ester) at 10^{-6} M caused swelling in one cell but failed to convincingly swell 9 other cells (Fig. 8). Our experience with human clear cells is still in the preliminary stage (because preparation of cells is more difficult), however, as shown in Fig. 9, ISO + GC stimulated cell swelling in control human clear cells. Interestingly, none of 12 CF cells we have studied thus far failed to swell in response to cAMP-elevating agents in the presence of bumetanide, which was added to eliminate the contribution of cotransporters to cell swelling (Fig. 10).

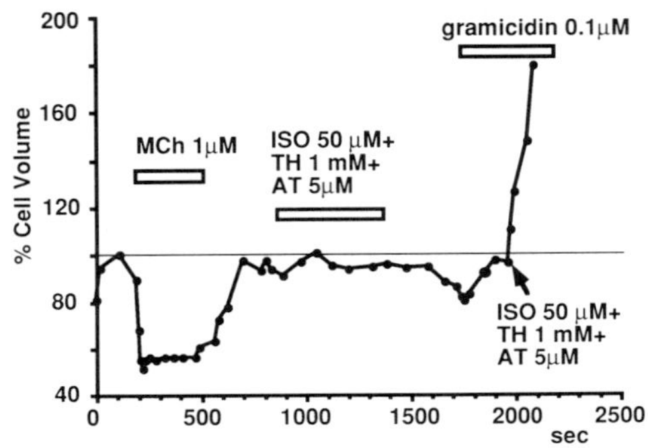

Fig. 6. ISO+GC-induced cell swelling
An illustrative example of cell volume response to MCh, ISO + TH (theophylline) + AT (atropine), and gramicidin + ISO + TH + AT in a dissociated rhesus clear cell. Note the marked increase in cell volume upon addition of the cAMP-elevating agents in the presence of gramicidin. The experiment was terminated at 2100 sec when the cell started blebbing.

DISCUSSION

The secretory coil of the eccrine sweat gland is uniquely qualified to serve as a model system for CF research because cholinergic sweating is entirely normal in CF in the face of abnormal β-adrenergic responsiveness. The entirely normal cholinergic responsiveness of CF sweat glands and thus the normal sweat secretory function may represent an interesting pharmacological model when new pharmacological approaches that bypass the basic CF defect are developed.

It appears that abnormal adrenergic responsiveness may be the underlying pathogenesis of CF (15, 21). Abnormal regulation of Cl channels by cAMP-cascades has been postulated by several investigators (9, 10, 11, 12). Although the gene responsible for the majority of CF patients has recently been cloned (22), it remains to be established whether the CF gene product is the putative regulatory protein responsible for regulation of Cl channels. Although investigators are unanimous relative to occurrence of abnormal regulation of Cl channels, they do not necessarily agree on the very nature of its abnormality (23, 24). It is not even settled whether seemingly cAMP-mediated cellular processes are mediated by intracellular Ca (25) or indeed by cAMP.

In this contribution, we have briefly reviewed our recent data on the effects of cAMP-elevating agents on K-efflux in both control and CF, intracellular [Ca] during stimulation with cAMP-elevating agents, and Cl channel activity of CF clear cells as determined by cell volume analysis.

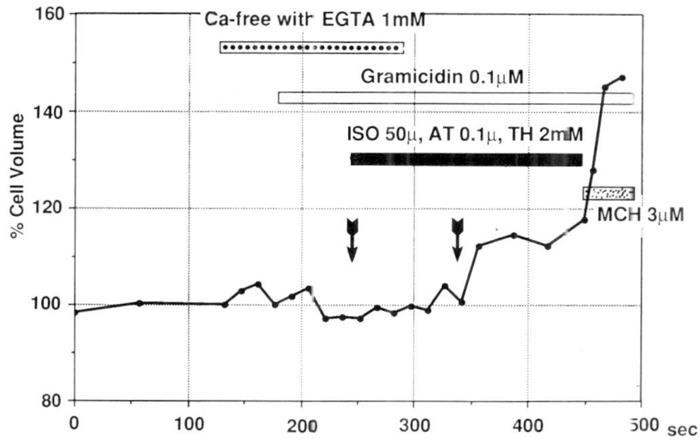

Fig. 7. Ca-free medium inhibits isoproterenol-induced cell swelling (monkey)
Effect of extracellular Ca on ISO-induced cell swelling in the presence of gramicidin in a rhesus clear cell. Note that after perfusion with Ca-free KRB (also containing 1 mM EGTA), the cell remained unresponsive to stimulation with ISO + TH + AT (as indicated by two arrows) until after readmission of Ca. Four experiments with the same results. At the end of the experiment, the cell responded normally to MCh.

We have recently observed (16) that stimulation of the rhesus secretory coils with ISO resulted in a dose-dependent, biphasic K-efflux consisting of the initial transient K-efflux followed by varying degrees of net K-reuptake. The initial ISO-induced K-efflux, which is less than 10% that of MCh_{max} [which

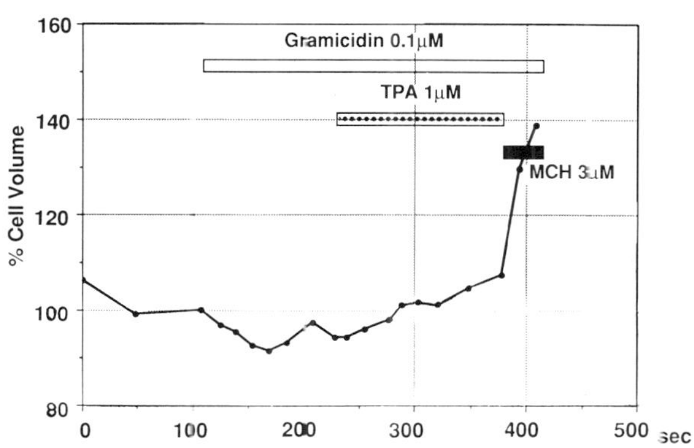

Fig. 8. Phorbolester (TPA) and Gramicidin fail to cause cell swelling in 90% of cells (monkey)
Effect of phorbol ester (TPA) on cell volume in a gramicidin-treated rhesus clear cell. The small (less than 10%) increase in cell volume after TPA may not be significant. An illustrative experiment.

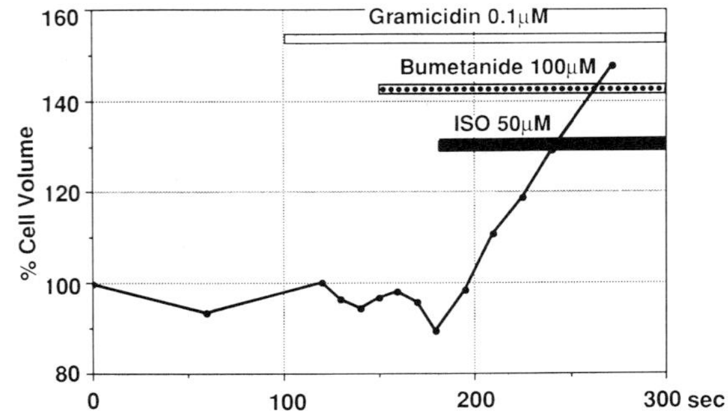

Fig. 9. Gramicidin and bumetanide and ISO-induced cell swelling in control human cell
Isoproterenol (ISO) stimulates cell swelling in the presence of gramicidin in a human control clear cell. Bumetanide was added to eliminate any contribution of the cotransporters to ISO + GC-induced cell swelling. Bumetanide alone had no effect on MCh-induced cell shrinkage in the absence of gramicidin (not shown). An illustrative experiment.

corresponds to about 45% of total cellular K (4,5)], is activated 2 1/2 fold by a phosphodiesterase inhibitor, IBMX, and is inhibited by propranolol but not by atropine or phentolamine, indicating that the ISO effect is mediated by β-adrenergic receptor activation. Furthermore, forskolin mimicked the effect of ISO, further suggesting that the ISO-induced K-efflux may be mediated by cAMP. The seemingly minute net K-efflux during ISO stimulation may be partly due to simultaneous activation of K-uptake by the cotransporters (16, see Fig. 4). We have also shown that ISO-induced K-efflux is accompanied by Cl-efflux (not just the exchange of K for Na) so that the observed ISO-induced K-efflux largely reflects the parallel movement of Cl (16). We are puzzled, however, that the ISO-induced K-efflux was completely inhibited in the absence of extracellular Ca (also containing 1 mM EGTA), although it was only partially inhibited in the nominally Ca-free medium (without EGTA).

With this background, the absence of K-efflux during stimulation with ISO, FK, IBMX, or the combination of these agents in CF secretory coils may be consistent with the thesis that Cl movement in CF clear cells is not activated by these agents. The marked bumetanide-sensitive K-uptake that was observed during stimulation by the cAMP-elevating agents in CF secretory coils may indicate that cAMP may regulate Na-K-Cl cotransporters normally in CF.

The cell volume increase of gramicidin (GC)-treated clear cells by both Ca- and cAMP-elevating agents most likely reflects the increase in Cl channel activity because when the cell membrane is made nonselectively permeable to Na and K, only the inward directing chemical gradient for Cl predominates the driving

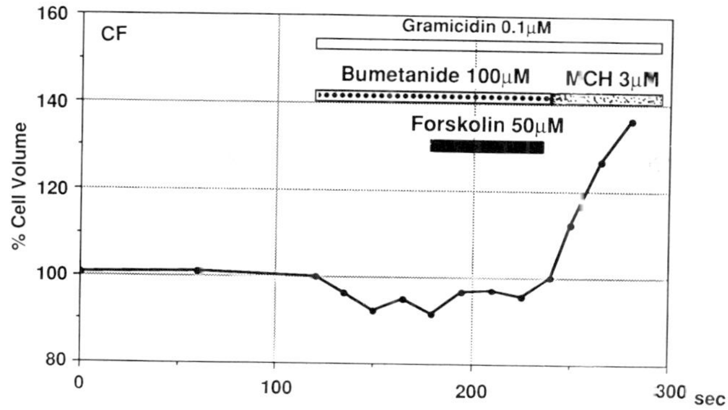

Fig.10. Forskolin fails to cause cell swelling in the presence of gramicidin and bumetanide

Absence of cell swelling after stimulation with forskolin in the presence of gramicidin in a dissociated cystic fibrosis clear cell. As in Fig. 9, bumetanide was also added.

force. In fact, when isolated secretory coils are stimulated with 1 μM GC alone, cell does not lose K, with or without MCh (unpublished). Thus if Cl channels are opened by drugs, Cl (accompanied by Na) enters the cell thus causing cell swelling. In fact, cell swelling by various stimulants does not occur in the absence of extracellular Cl and resumes upon readmission of Cl. Furthermore, GC + MCh-induced cell swelling is inhibited by atropine and ISO + GC-induced swelling is inhibited by propranolol, indicating pharmacological specificity. It is interesting to note that the CF clear cells studied to date fail to swell in response to various cAMP-elevating agents although they responded normally to Ca-elevating agents. This is consistent with the widely held postulate that regulation of Cl channels by cAMP may be abnormal in CF cells. As expected, both MCh- and ionomycin (IO)-induced cell swelling (in the presence of GC) are strictly dependent on the extracellular Ca. However, puzzlingly enough, cell swelling by ISO + GC is also inhibited by removal of extracellular Ca. The normal response of CF clear cells to MCh and IO suggests that Ca-mediated regulation of Cl channels may be entirely normal. However, IBMX, which elevates cytoplasmic [Ca] in addition to cAMP, also failed to induce K-efflux and cell swelling in CF cells. Furthermore, ISO, which accumulates tissue cAMP more than twice as much as does IBMX, has a lower effect on K-efflux in control glands. The results of these and other data are schematically presented in Fig. 11. It appears that the common denominator in the pathogenesis of Cl channel disfunction in CF appears to be the presence of cAMP. It is possible to speculate therefore that the inability for IBMX to open Cl channels in CF clear cells in the face of an elevated cytoplasmic [Ca] may be the presence of cAMP itself, which may directly or indirectly exert a negative influence on Cl channels which are normally activated by Ca but not by cAMP. In view of this

		ISO	FK	IBMX	MCH	TPA
Mediators (C & CF)		cAMP	cAMP	cAMP Ca	DAG+ Ca +X?	(DAG)
KCl-effl.	C	+	+	+	+++	-
	CF	-	-	-	+++	-
Cl-chann.	C	+	+	+	++	-?
	CF	-	-	-	++	-?

Fig.11. Simplified summary on the roles of Ca and cAMP on KCl-efflux and Cl channel functions in control and CF sweat gland. TPA does not appear to induce K-efflux or influence K-efflux induced by other agents. The preliminary data do not support the role of TPA on the Cl channel function as determined by the cell volume analysis. X, we speculate that the intracellular mediators of cholinergic stimulation may include more than just Ca and diacylglycerol (DAG).

connection, it is of interest to note that Kunzelmann et al. (23) recently reported that Cl channels in CF tracheal cells are abnormally regulated in the attached cell mode but become normal in the excised membranes. This observation has lead the authors to speculate that an endogenous inhibitor of Cl channels, if any, may be involved in abnormal regulation of Cl channels in CF cells.

Acknowledgement

This paper has been supported in part by NIH grants DK 27857, HL32731, AR 25339 and a Cystic Fibrosis Foundation Grant G124.

References

1. Sato, K., Kang, W.H., Saga, K., Sato, K.T., 1989 Biology of sweat glands and their disorders. 1. Normal sweat gland function, **J. Amer. Acad. Derm.**, 20:537.
2. O'Grady, S.M., Palfrey, H.C., Field, M., 1987, Characteristics and functions of Na-K-Cl cotransport in epithelial tissues, **Am. J. Physiol.**, 253:C177.
3. Saga, K., Sato, F., Sato, K., 1988, K^+ efflux from the monkey eccrine secretory coil during the transient of stimulation with agonists, **J. Physiol. (London)**, 405:205.
4. Saga, K., Sato, K., 1989, Electron probe X-ray microanalysis of cellular ions in the eccrine secretory coil cells during methacholine stimulation, **J. Membrane Biol.**, 107:13.
5. Takemura, T., Sato, F., Saga, K., Suzuki, Y., Sato, K., 1990, Intracellular ion concentrations and cell volume during cholinergic stimulation of eccrine secretory coil cells, Submitted for publication.

6. Takemura, T., Kang, W.H., Sato, F., Sato, K., 1989, Evidence that cholinergic stimulation is asscoiated with cell volume decrease in eccrine secretory cells as revealed by X-ray microanalysis and light microscopy, **Clin. Res.**, 37:640A.
7. Saga, K., Sato, K., 1988, Ultrastructural localization of Na, K-ATPase in the monkey palm eccrine sweat gland, **J. Histochem. Cytochem.**, 36:1023.
8. Sato, K., Sato, F., Characteristics of Na, K, and Cl channels in the clear cell membrane of the eccrine sweat gland as studied by the patch clamp techniques, **Clin. Res.**, 37:369A.
9. Welsh, M.J., Liedtke, C.M., 1986, Chloride and potassium channels in cystic fibrosis airway epithelia, **Nature**, 3222:467.
10. Frizzell, R.A., Rechkemmer, G., Shoemaker, R.L., 1986, Altered regulation of airway epithelial cell chloride channels in cystic fibrosis, **Science**, 233:558.
11. Li, M., McCann, J.D., Liedke, C.M., Nairn, A.C., et al., 1988, Cyclic AMP-dependent protein kinase opens chloride channels in normal but not cystic fibrosis airway epithelium, **Nature**, 331:358.
12. Chen, J.H., Schulman, H., Gardner, P., A cAMP-regulated chloride channel in lymphocytes that is affected in cystic fibrosis, **Science**, 243:657.
13. Ohtsuyama, M., Samman, G., Sato, F., Sato, K., 1990, Stimulation of Cl channel activity by Ca and cAMP elevating agents in dissociated eccrine clear cells as determined by cell volume analysis, **Clin. Res.** (in press).
14. Sato, K., Sato, F., 1981, Pharmacologic responsiveness of isolated single eccrine sweat glands, **Am. J. Physiol.**, 240:R44.
15. Sato, K., Sato, F., 1984, Defective beta adrenergic response of cyclic fibrosis sweat glands in vivo and in vitro, **J. Clin. Invest.**, 73:1763.
16. Sato, K.T., Saga, K., Ohtsuyama, M., et al., 1990, Movement of cellular ions during stimulation with isoproterenol in simian eccrine sweat secretory coils, Submitted for publication.
17. Sato, K., Sato, F., 1988, Relationship between Quin 2-determined cytosolic [Ca^{2+}] and sweat secretion, **Am. J. Physiol.**, 254:C310.
18. Sato, K., Sato, F., Methods for studying eccrine sweat gland function in vivo and in vitro, in: "Methods in Enzymology," S. Fleisher, 1990.
19. Grynkiewicz, G.M., Poenie, Tsien, R.Y., 1985, A new generation of Ca^{2+} indicators with greatly improved fluorescence properties, **J. Biol. Chem.**, 260:3440.
20. Suzuki, Y., Sato, F., Sato, K., 1989, Ionic basis for cellular volume regulation in dissociated sweat secretory cell during pharmacological stimulation, **Clin. Res.**, 37:618.
21. McPherson, M.A., Dormer, R., 1987, The molecular and biochemical basis of cystic fibrosis, **Bioscience Reports**, 7:167.
22. Riordan, J.R., Rommens, J.M., Kerem, B.S., Alon, N., Rozmahel, R., Grzelczak, Z., Zielenski, J., Lok, S., Plavsic, N.,

Chou, J.L., Drumm, M.L., Iannuzzi, M.C., Collins, F.S., Tsui, L.-C.,, 1989, Identification of the cystic fibrosis gene: cloning and characterization of complementary DNA, **Science**, 245:1066.
23. Kunzelmann, K., Pavenstädt, H., Greger, R., 1989, Properties and regulation of chloride channels in cystic fibrosis and normal airway cells, **Pflügers Arch.**, 415:172.
24. Wine, J.J., Solc, C.K., 1989, Chloride channels in cystic fibrosis patients, **Science**, 247:222.
25. Cook, D.I., Day, M.L., Champion, M.P., Young, J.A., 1988, Ca^{2+} not cyclic AMP mediates the fluid secretory response to isoproterenol in the rat mandibular salivary gland: Whole-cell patch clamp studies, **Pflügers Arch.**, 413:67.

DISCUSSION

GRUENERT: Kenzo, when you make your measurements on cell swelling it looks like they're two-dimensional measurements.

SATO: We can look at the top of the cell and the bottom of the cell, and in some experiments we did that, and we are satisfied that shrinking or swelling occurs equally to both dimensions: vertically as well as horizontally. It is very time consuming and technically tremendous to optically section and computerize everything, and I think that the horizontal measurement is adequate for this sort of analysis.

GREGER: Kenzo, I think this was a beautiful study. I have just one slight point on the gramicidin. When you use gramicidin in an attempt to take care of electron neutrality - I think that's your purpose. Then by doing so you still do not really clamp the voltage, so whenever the sodium goes in there will be a reequilibration of chloride, because there are chloride channels and therefore I think you'll only have a partial exchange of K vs. sodium and it will be also a chloride efflux or influx depending on the voltage.

SATO: Well, since the chloride in sodium chloride moves in as a neutral (or in effect neutral) particle, membrane potential has less effect on this, provided that potassium doesn't come out. Also, we are just looking at the initial 15-30 seconds. After that, we don't care, because the cell begins to swell anyway. So it's biophysically uncertain there, so we have to measure the potential in the cell, yes. No question.

REGULATION OF ABSORPTION IN THE HUMAN SWEAT DUCT

P.M. Quinton and M.M. Reddy

Biomedical Sciences, UC Riverside
Riverside, CA 92521, USA

It is now well established that cystic fibrosis is expressed physiologically as defects in certain fluid and electrolyte transport systems that are highly active in exocrine tissues (Quinton, 1983; Knowles et al., 1983; Kopelman et al., 1982; Orlando et al., 1989). More specifically, two basic physiological defects have been demonstrated directly in at least two of the organ systems characteristically affected in cystic fibrosis - the sweat glands and the airways. The two defects in these affected epithelia appear as 1) an inherently reduced permeability to Cl- (Quinton, 1983; Widdicombe et al., 1985; Knowles, 1983), and 2) a failure of fluid secretion to respond to β-adrenergic stimulation (Sato and Sato, 1984; Widdicombe et al., 1985). The refractory response to β-adrenergic stimulation is most clearly seen in secretory functions and is thought to be due to an inactivatable chloride channel (Welsh, 1986; Frizzell, 1986). While other defects have been associated with cystic fibrosis, e.g., increased Na^+ permeability (Knowles et al., 1983; Boucher et al., 1986; Willumsen and Boucher, 1989), abnormal sulfate metabolism (Cheng et al., 1985), and abnormal calcium levels (Sorcher and Beslow, 1982; Shapiro et al., 1980), the basic physiological abnormality in the disease seems to center upon chloride impermeability.

The sweat gland has played an important role in the discovery of both the abnormality in electrolyte absorption (Quinton, 1983) and the abnormality in stimulated fluid secretion (Sato and Sato, 1984). The clarity of the expression of abnormal chloride conductance in the absorptive duct of the sweat gland is probably due to the fact that the function of this tissue is purely absorptive (Slegers, 1967; Schultz, 1969). Likewise, the clear loss of β-adrenergic stimulation in the coil of the sweat gland is probably due to the fact that the function of this tissue is purely secretory. The fact that normal sweat secretion (Sato and Sato, 1984) can be achieved with cholinergic stimulation in addition to evidence from several patch-clamp studies

The Identification of the CF (Cystic Fibrosis) Gene
Edited by L.-C. Tsui *et al.*, Plenum Press, New York, 1991

(Schoumacher et al., 1987; Li et al., 1988), strongly indicates that the generalized defect in cystic fibrosis is limited to an inability to activate existing chloride channels via β-adrenergic stimulus coupling. Since cystic fibrosis is a single gene defect, we should expect the same basic biochemical abnormality to occur in both secretory and absorptive functions. That is, as in secretion, the defect in absorption should also be a defect in regulation of chloride conductance.

Adrenergic Regulation of Absorption

We sought to test this unifying hypothesis of the CF defect by demonstrating regulation of chloride permeability in the isolated reabsorptive sweat duct. We expected that absorption in the duct followed secretion in the coil and that absorption was activated only when secretory activity provided electrolytes in fluid to the duct lumen. Hence, we developed an initial notion that sodium was absorbed actively via sodium-potassium ATPase in the basal membrane of duct cell and chloride was moved passively through highly conductive pathways as a result of Na^+ driven electrochemical gradients (Quinton and Tormey, 1984; Reddy and

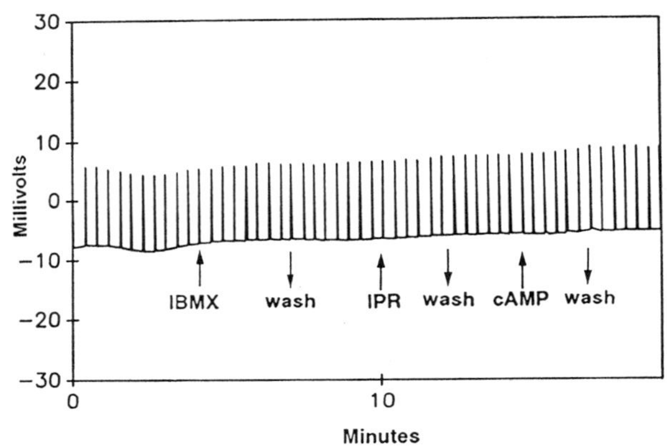

Figure 1. Lack of effects of β-adrenergic agents on transepithelial potential (V_t) and transepithelial resistance (R_t). A segment of normal reabsorptive duct was perifused and microperfused with isotonic Ringer's solutions with a double-barreled microperfusion pipette. A 100 nA current pulse of 500 msec duration was passed through one barrel of the pipette. The transepithelial potential was measured through the second barrel. The upward deflections in V_t (delta V_t) increase as a function of R_t. At the upward directed arrow, IBMX (isobutylmethylxanthine) was added at a concentration of 10^{-4} M to the perifusing solution and washed from the duct at the downward directed arrow. IPR (isoproterenol) at 10^{-4} M were also applied and cyclic AMP were applied in the perifusing fluid as indicated by the upward arrows, and washed, as indicated by the downward arrows, respectively. No effect of any of these maneuvers on either V_t or R_t were detectable.

Quinton, 1989). But since the defect in secretion is so closely coupled to Cl^- impermeability via defective β-adrenergic regulation (Widdicombe et al., 1985), we must now examine the high Cl^- conductance of the sweat duct also in terms of its regulation by agonists related to β-adrenergic stimulation. Consequently, we examined the effects of isoproterenol, cyclic AMP, forskolin, IBMX, and the putative inhibitor of protein kinases, H-8. We measured the transepithelial potential (V_t) and the resistance (R_t) with and without electrolyte gradients across the microperfused ductal epithelium with the prediction that if these agents changed the chloride conductance, a change would occur in both V_t and R_t. All experiments were carried out on tissue of normal subjects, since CF tissues were expected to be unresponsive.

In more than 2/3 of ducts examined, we found no detectable response in either V_t or G_t as a function of the application of any of the β-adrenergic agonists (Fig. 1). Similarly, epinephrine (10^{-5} M) also had no detectable effect. We further investigated a possible role for protein kinases by perfusing the microperfused duct with the putative permeable inhibitor H-8 of protein kinases. Again, no detectable effects on V^t or G^t were observed (Fig. 2). The lack of β-adrenergic stimulation in non-responding ducts is not due to chronic, spontaneous stimulation of the duct by endogenous adrenergic transmitter release such that subsequent exogenous stimulus is ineffective. Neither hyperpolarization of the pre-synaptic terminals by perfusing the preparation with low potassium Ringer's solution, nor depleting the bath of Ca^{2+} (required for transmitter release) altered the

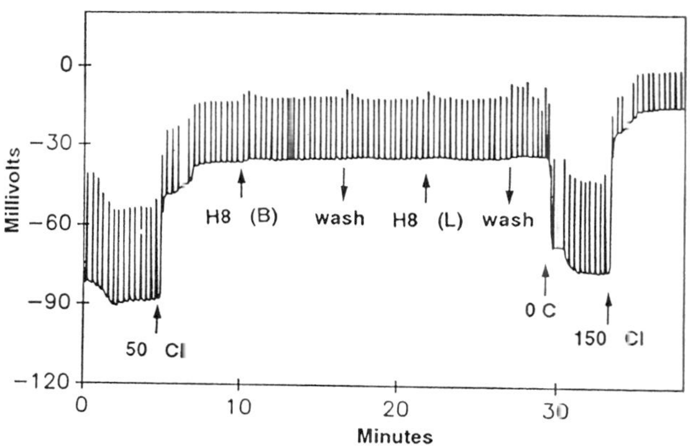

Figure 2. Effect of the kinase inhibitor, H-8, in the bath (B) and lumen (L). 50 mM NaCl replaced 150 mM Na gluconate in the perfusion solution at the first indicated upward arrow. H-8 was added first to the bath and washed and then to the lumen and washed, as indicated by the upward and downward directed arrows, respectively. The 50 mM NaCl solution was then replaced with 150 mM Na gluconate which was subsequently replaced by 150 mM NaCl. H-8 at 10^{-4} M had no detectable effect on either the serosal or luminal surface. Pulses = 50 nA at 1 sec.

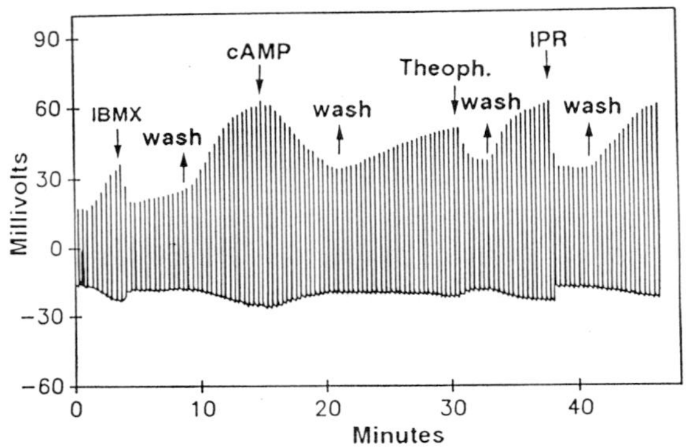

Figure 3. Positive effects of β-adrenergic agonists on sweat duct. In contrast to the preparation represented in Figure 1, the R_t began to spontaneously increase. The application of IBMX to the perifusing fluid, as indicated by the downward arrow, reversed the trend of increasing resistance, which continued to increase when the drug was washed (upward arrow). The further addition of cAMP (10^{-4} M) mimicked the effect of IBMX in initiating a decrease in R_t, which again reversed when cAMP was removed. Similarly, theophylline (10^{-4} M) and isoproterenol (IPR, 10^{-4} M) also reversed the trend of increasing R_t. Conditions as in Figure 1.

duct's insensitivity to β-adrenergic stimulation. Further, neither the transport status of the tissue (manipulated by inhibiting Na^+ conductance with amiloride), the concentration gradient (luminal Na^+ concentration was changed from 150 mM to 15 mM), nor the rate of perfusion (perfusion pressure varied up to 100 mmHg) had any effect on inducing β-adrenergic sensitivity.

However, in a few ducts, we did observe some decrease in R_t after stimulating with isoproterenol. The effect was sensitive to the β-blocker propanolol, was reversible, and could be induced with cAMP, IBMX, and forskolin as well, so that if the tissue responded to one of these agonists it was responsive to the others as well (Fig. 3). The decrease in R_t appears to be due to a decrease of Cl^- permeability because inhibition of Na^+ conductance by amiloride did not block the β-adrenergic effect. In addition, increases in conductance of more than 25 mS/cm^2 almost certainly are due to Cl^-, since 80% of transepithelial conductance (125 mS/cm^2) is due to Cl^- (Quinton, 1986). We conclude that in the normal function of the sweat duct, β-adrenergic control is not an integral component of normal Cl permeability regulation.

Non-Adrenergic Mechanisms of Regulation

Lacking consistent cAMP-related stimulatable effects, we then examined a few of the parameters of the cholinergic pathway

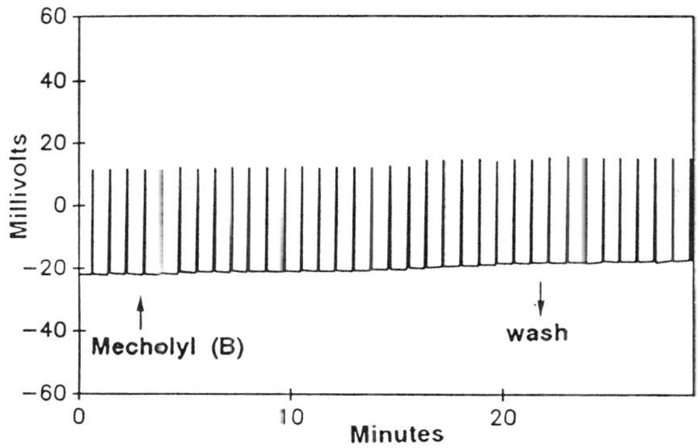

Figure 4. Lack of effect of mecholyl (10^{-5} M), a cholinergic agonist, on V_t and R_t. Conditions are as in Figure 1.

found in stimulus/secretion coupling. We found no detectable effects from applications of mecholyl (a cholincmimetic) at concentrations as high as 10 μM (Fig. 4). However, the Ca^{2+} ionophore ionomyocin (1 μM), when applied to the basal or apical membrane surface, induced slow, small oscillations of V_t and R_t (1 cycle/1-2 minutes) and a consistent, but transient hyperpolarization of the basal membrane by about 20 mV (Fig. 4). Ionomyocin had no effect in Ca^{2+}-free media. The data suggest the effect is more consistent with an increase in a Ca^{2+}-dependent K^+ conductance than with an increase in Cl^- conductance. Further, overnight incubation of the glands with indomethacin and perifusion of the duct with PGE_1 and PGE_2 had no detectable effects either (Fig. 5).

Physiological Reasons for Regulating Cl^- Permeability

Even though we found little evidence for pharmacologic control of Cl^- permeability, we expect from other data that the chloride conductance in the sweat duct is regulated in the course of carrying out normal absorptive functions. That is, it is well established that the concentration of NaCl in the sweat can decrease from essentially isotonic values at relatively high secretory rates to between 10 and 15 mEq/L at relatively slow secretory rates (Bijman and Quinton, 1984). Ion sensitive microelectrode studies show that when the luminal Cl^- concentration of microperfused ducts is set at 15 mEq/L, the intracellular Cl^- concentration drops to about 19.5 ± 9.0 (mean ± S.E.; n=5) and the apical membrane potential assumes a value of -21.4 ± 3.7 mV (mean ± S.E.; n=5), cytoplasm negative. These combined gradients indicate that a driving force of about 30 mV should force Cl^- out of the cell into the lumen, opposing rather than favoring absorption. Hence, at low, but physiological Cl^- concentrations, we anticipate that a regulated Cl^- conductance in the apical membrane must operate in concert with some other mechanism (carrier or exchanger) for moving chloride uphill into

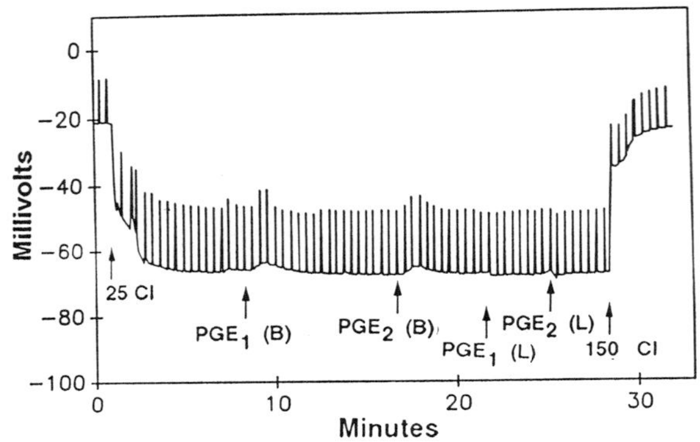

Figure 5. Lack of effect of prostaglandins (PGE) on the sweat ducts. A 25 mM NaCl, balance Na gluconate, solution replaced the isotonic Ringer's solution at the first upward arrow. PGE_1 was added to the perifusing solution at the second upward arrow and PGE_2 was added to the perifusing solution at the third arrow. PGE_1 was then added to the luminal perfusion and then replaced by PGE_2. The 150 mM NaCl Ringer's solution was then reinstated as the perfusion fluid. No effects of either PGE_1 or PGE_2 in either the bath or lumen were detectable. Conditions as in Figure 2.

the cell. That is, at low luminal Cl^- concentrations, the apical membrane Cl^- conductance should be significantly reduced (if not closed) in order to prevent back flux of Cl^- from the cell into the lumen (Quinton and Reddy, 1989).

Special Conditions indicating Regulation of G_{Cl}

In the course of searching for regulation of Cl^- permeability, we have encountered more direct evidence that regulation does, indeed, occur. In brief, we occasionally observed that in some preparations, the luminal diameter and the transepithelial conductance began to spontaneously decrease, indicative of cell swelling. We tested the hypothesis that this phenomena might be due to decreased metabolic energy by blocking ATP production with metabolic inhibitors. Addition of iodoacetate (Fig. 6), as well as cyanide, and DCCCP all caused a significant increase in R_t. Additionally, we found that depleting the duct cells of cytoplasmic constituents of solutes of molecular weights of about 1,000 or less by exposing the basal membrane to the pore-forming agent, staph aureus α-toxin (Fussle et al., 1981) caused first a decrease and then a subsequent increase in R_t (Fig. 7). In contrast, the application of the ionophore, amphotericin D, which opens the cell membrane to small ions (< 100 mol. wt.), resulted in a decrease in R_t only (Fig. 8).

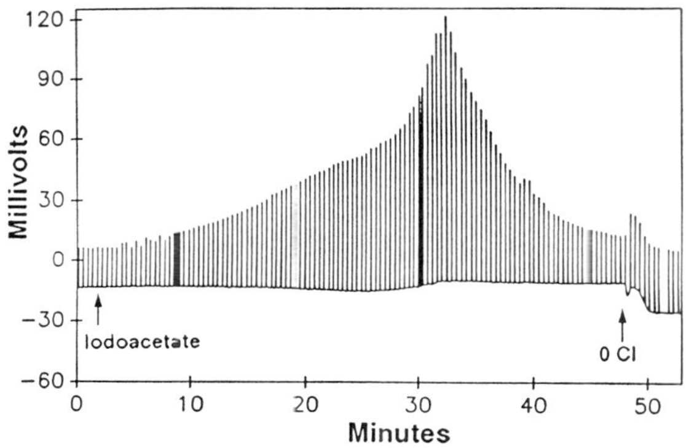

Figure 6. Effect of iodoacetate (10^{-3} M) in the perifusing fluid of a sweat duct. The metabolic inhibitor of oxidative phosphorylation causes a gradual but increasingly larger R_t resistance, as indicated by an increasing magnitude of the delta V_t. After reaching a peak value, the resistance then began to fall, which is probably due to deterioration of the epithelial barrier. This conclusion is supported by the fact that substitution of the isotonic Ringer's perfusion solution with Na gluconate elicits only a small increase in the negative luminal potential (second upward arrow). Normally, this maneuver would shift V_t by at least -50 mV, due to the prevalent Cl permeability of this tissue. Conditions as in Figure 2.

It is difficult to explain the increase in R_t after metabolic poisons and after exposing the duct to agents which create relatively larger pores. However, a common element may be ATP, since in both cases cytoplasmic levels of ATP should decrease. We attempted to test this possibility by adding ATP to microperfused ducts. Preliminary studies indicate that ATP (20 mM) in the bath slightly decreased R_t. Further, when the basolateral membrane was permeabilized to ATP with digitonin, the decrease in R_t due to addition of ATP in the bath seemed much more pronounced (Fig. 9a,b), suggesting that ATP might be involved in the regulation of transepithelial conductance. Prior permeabilization of the basal membrane with α-toxin and addition of ATP to the bath caused rapid and reversible changes in R_t and V_t consistent with ATP-dependent changes in Cl conductance (Fig. 10).

These results could suggest that a minimal concentration of ATP in the cell may be required to open ion permeabilities in the cell membrane. Since transport of ions out of the cell is ATP dependent, an ATP-dependent permeability would protect the cell from overloading with extracellular ions when energy sources are inadequate to support active transport.

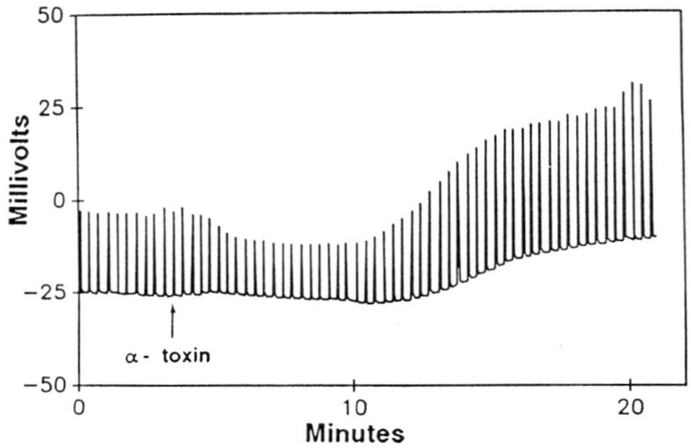

Figure 7. Effect of staph aureus α-toxin (200 units/ml) on R_t. This pore-forming agent, when added to the perifusing fluid (upward arrow) induced the anticipated decrease in R_t as reflected by the decreased amplitude of delta V_t, followed by the unexpected increase in R_t as seen by the increasing deltaV_t in response to 50 nA/1 sec current pulses.

The response of our tissue to β-adrenergic components during metabolic depression is also puzzling. But cells respond to β-stimulation with increased rates of metabolism. Exogenous stimulation of the sweat duct may increase metabolic activity at least temporarily, with a concomitant rise in ATP levels. If membrane ionic permeability is dependent on ATP, a temporary rise in ATP after β-stimulation in energy starved duct cells could explain the observed concurrent increase in permeability in metabolically depressed cells. In normal, energy replete cells, β-stimulation may not cause a detectable increase because permeability is already maximal.

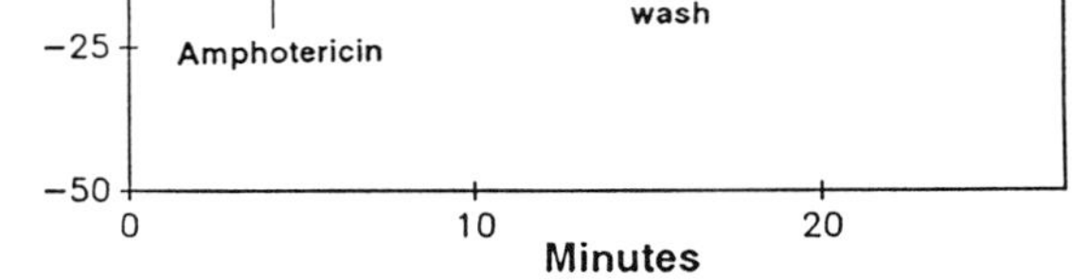

Figure 8. Effect of ionophore amphotericin (10^{-5} M) in the perifusing fluid of a sweat duct. The drug causes a slight, reversible increase in R_t.

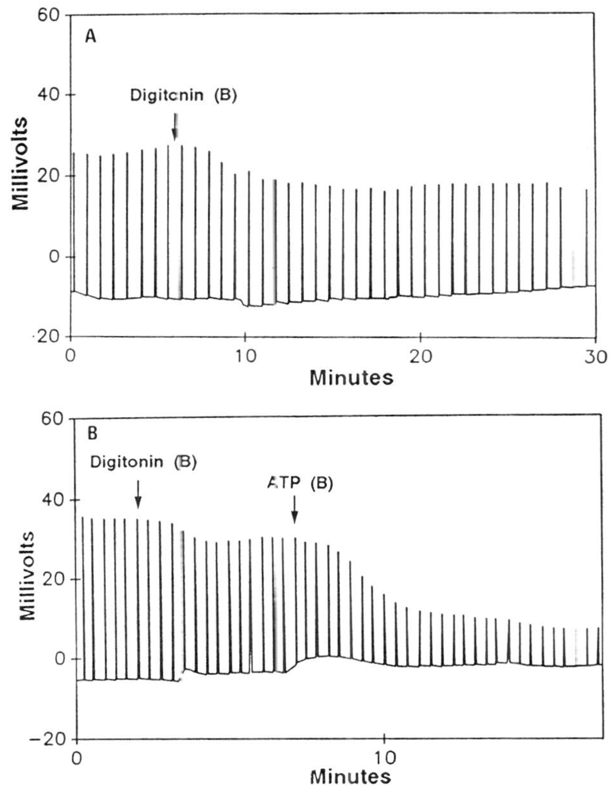

Figure 9. A. Effect of digitonin (10^{-6} M) added to the bath (B) at the downward arrow. Digitonin causes a decrease in R_t within a minute after application. R_t declines relatively rapidly at first, but then generally decreases relatively slowly over a period of half an hour or more.

B. Effect of ATP added to the bath (B). After treating the sweat duct with 10^{-6} M digitonin in the bath at the first downward arrow, disodium ATP (20 mM) was substituted for 40 mM Na gluconate at the downward arrow. Digitonin effects a relatively rapid decrease in R_t which presumably reflects permeabilization of the basal membrane of the duct cells. Subsequent addition of ATP elicits a further decrease in R_t.

Speculation on the Role of Nucleotides and CFTR

In view of the recent finding of a genetic defect for cystic fibrosis (Riordan et al., 1989) in a putative nucleotide-binding fold (NBF) of the protein produced by the cystic fibrosis gene (CFTR), our data tempt us to speculate that ATP may be an important, permissive modulator of chloride permeability. That is, ATP may be required in the NBF of CFTR before alosteric changes can occur which allow activation of chloride conductance.

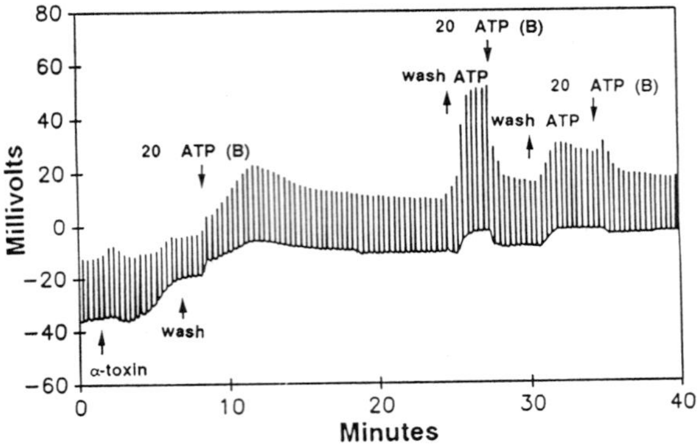

Figure 10. Reversible effects of ATP (20 mM) in the bath (B) on a sweat duct previously treated with the pore-forming agent, α-toxin, in the perifusing fluid. As shown by delta V_t after the first upward arrow, α-toxin induces an initial decrease and a secondary increase in R_t (compare with Figure 7). The addition of 20 mM ATP to the bath (first downward arrow) appeared to have reversed the secondary phase of increasing R_t due to α-toxin. However, after washing ATP from the bath, R_t increased, but decreased by the re-addition of ATP to the bath (third upward arrow). Subsequent application and replacement of ATP had a smaller effect (last two arrows).

In this scheme, the defect in the NBF would prevent CFTR from recognizing ATP and thereby block other events from activating the Cl^- channel and increasing Cl^- permeability; that is, the defect in CF may be in permissive control, rather than in direct stimulatory regulation.

Whatever the role of CFTR and its NBF, a unifying hypothesis must attend to the question of how both secretion and absorption are linked through the defect in the CFTR to Cl^- impermeability in CF.

References

Boucher, R.C., M.J. Stutts, M.R. Knowles, L. Cantley and J.T. Gatzy, 1986, Na^+ transport in cystic fibrosis respiratory epithelia. Abnormal basal rate and response to adenylate cyclase activation, **J. Clin. Invest.**, 78:1245.

Cheng, P., T. Boat, K. Cranfill, J. Yankaskas and R. Boucher, 1989, Increased sulfation of glycoconjugates by cultured CF nasal epithelial cells, **Ped. Pulmon. Suppl.**, 4:130a.

Feigel, R.J. and B.L. Shapiro, 1979, Mitochondrial calcium uptake and oxygen consumption in cystic fibrosis, **Nature**, 278:276.

Frizzell, R.A., G. Rechkemmer and R.L. Shoemaker, 1986, Altered regulation of airway epithelial cell chloride channels in cystic fibrosis, **Science**, 233:558.

Fussle, R., S. Bhakdi, A. Sziegoleit, J. Tramm-Jensen, T. Kranz, and H.J. Wellensiek, 1981, On the mechanism of membrane damage by Staphylococcus aureus α-toxin, **J. Cell. Biol.**, 91:83.

Knowles, M., J. Gatzy and R. Boucher, 1981, Increased bioelectric potential difference across respiratory epithelia in cystic fibrosis, **New Engl. J. Med.**, 305: 1489.

Kopelman, H., P. Durie, K. Gaskin, Z. Weizman, and G. Forstner, 1985, Pancreatic fluid secretion and protein hyperconcentration in cystic fibrosis, **New Engl. J. Med.**, 12: 329.

Li, M., J.D. McCann, C.M., Liedtke, A.C., Nairn, A.C., P. Greengard and M.J. Welsh, 1988, Cyclic AMP dependent protein kinase opens chloride channels in normal but not cystic fibrosis airway epithelium, **Nature**, 331:358.

Li, M., J.D. McCann, J.P. Anderson, C.M. Clancy, C.M. Liedtke, A.C. Nairn, P. Greengard and M.J. Welsh, 1989, Regulation of chloride channels by protein kinase C in normal and cystic fibrosis airway epithelia, **Science**, 244:1353.

Orlando, R.C., D.W. Powell, R.D. Croom, H.M. Berschneider, R.C. Boucher and M.R. Knowles. Colonic and esophageal trans epithelial potential difference in cystic fibrosis, 1989, **Gastroenterology**, 96:1041.

Quinton, P.M., 1983, Chloride impermeability in cystic fibrosis, **Nature**, 301:421.

Quinton, P.M., 1986, Missing Cl conductance in cystic fibrosis, **Am. J. Physiol.**, 251:C649.

Quinton, P.M. and J.Mc.D. Tormey, 1976, Localization of Na/K-ATPase sites in the secretory and reabsorptive epithelia of perfused eccrine sweat glands: A question to the role of the enzyme in secretion, **J. Membrane Biol.**, 29:383.

Reddy, M.M. and P.M. Quinton, 1989, Localization of Cl conductance in normal and Cl impermeability in cystic fibrosis sweat epithelium, **Am. J. Physiol.**, 257:C727.

Reddy, M.M. and P.M. Quinton, 1989, Altered electrical potential profile of human reabsorptive sweat duct cells in cystic fibrosis, **Am. J. Physiol.**, 257:C722.

Riordan, J.R., J.M. Rommens, B. Kerem, N. Alon, R. Rozmahel, Z. Grzelczak, J. Zelenski, S. Lok, N. Plavsic, J.L. Chou, M.L. Drumm, M.C. Iannuzzi, F.S. Collins and L.C. Tsui, 1989, Identification of the cystic fibrosis gene: cloning and characterization of complementary DNA, **Science**, 245: 1066.

Sato, K. and F. Sato, 1984, Defective beta adrenergic response of cystic fibrosis sweat gland in vivo and in vitro, **J. Clin. Invest.**, 73:1763.

Schoumacher, R.A., R.L. Shoemaker, D.R. Halm, E.A. Tallant, R.W. Wallace and R.A. Frizzell, 1987, Phosphorylation fails to activate chloride channels from cystic fibrosis airway cells, **Nature**, 330:752.

Schultz, I.J., 1969, Micropuncture studies of the sweat formation in cystic fibrosis, **J. Clin. Ivest.**, 48:1470.

Slegers, J.F.G., 1964, The mechanism of sweat secretion, **Pflugers Arch.**, 279:265.

Sorscher, E.J. and J.L. Breslow, 1982, Cystic fibrosis: A disorder of calcium-stimulated secretion and transepithelial sodium transport, **The Lancet** 1:368.

Welsh, M.J. and C.M. Liedtke, 1986, Chloride and potassium channels in cystic fibrosis airway epithelia, **Nature**, 322: 467.

Widdicombe, J.H., M.J. Welsh and W.E. Finkbeiner, 1985, Cystic fibrosis decreases the apical membrane chloride permeability of monolayers cultured from cells of tracheal epithelium, **Proc. Natl. Acad. Sci. USA**, 82:6167.

Willumsen, N.J. and R.C. Boucher, 1989, Shunt resistance and ion permeabilities in normal and cystic fibrosis airway epithelia, **Am. J. Physiol**, 256:C1054.

DISCUSSION

GREGER: With respect to the ATP effects, I think there are two issues. One, if it is receptor mediated you would run into things like we have been hearing from Rick this morning, testing which kind of purinergic receptor am I looking at. And the second issue is, you can use the non-metabolisable ATP, beta-methyl-ATP for instance, and then ask the question whether you still see the effect because of the K-channels, for instance in the heart muscle and in the pancreas and many other cells. These non-metabolisables act as well because they don't need to be phosphorylated. Have you done anything of that kind?

QUINTON: No, I haven't. It has occurred to us and we started to do it but then we found out how expensive it was, so we went ahead with just the plain old ATP. We'll probably do that in the future, but at this point I don't have any additional data.

WINE: Paul, I have two questions. One is, when you have a permeabilized basolateral membrane, you have a perfect situation to test the ion selectivity of the apical membrane of the duct, and have you done that?

QUINTON: No, we have not run the selectivity. I'm hoping to do that, but at this point, no. These are preliminary data.

WINE: The second question is, because there is such a very high chloride conductance in the normal sweat duct, can you rule out that there might be some change in a CF duct in the sodium permeability that is just masked because the chloride change is just so massive there.

QUINTON: No, I can't rule that possibility out. I've long wanted to do it, and so has Dr. Boucher, but still to this point the only data we have on that is the data that we have with respect to amiloride changes and induced changes in conductance with and without chloride. When we get rid of the chloride conductance by removing it, then it looks as though the conductance for sodium is the same in normal and in CF, but the situation gets complicated because if you do the same experiment in the presence of chloride, then you get bigger changes. So I don't know that Rick isn't right, that even in the sweat ducts there may be some feedback between sodium and chloride conductance as a function of chloride concentration.

WINE: Is there any significance to the CFTR floating around inside the cell? Do you know something that I don't? Or is that just schematic?

QUINTON: Just schematic. I don't know anything you don't know, for sure.

MEHTA: Could I ask you, when we do permeabilized cell preparations we use calcium buffers. What was the calcium concentration in your external media?

QUINTON: 1 mM.

MEHTA: And we found, certainly at those sort of levels, a sort of direct toxic effect in the absence of calcium buffers. Have you tried lower concentration?

QUINTON: Not as yet. We've been playing a little bit with different concentrations trying to match the cytoplasmic concentration better, and we thought that might perhaps give us a longer window with which to work with the permeability properties of the apical membrane, because in these initial studies we just sort of dumped it in with sodium present in the bath, so it is essentially an extracellular solution that we're dumping into the cytoplasm, which I think is a significant criticism, yes.

AUSIELLO: Have you tried adding ATP without permeabilizing the cell? Or any of its analogues?

QUINTON: Yes, we have and at very high concentrations we see a slight effect. It will decrease the conductance at a 20 Mmol concentration.

RIORDAN: The other comment I would make is that the permeabilization on one side of the cell is interesting, but I don't really see the need of it. I mean, one can normally use cyanide and 5 or 10 Mmol glucose and go back and forth and manipulate ATP levels pretty much at any level you want, and if it's a good tissue.

QUINTON: That's true, you can use reversible inhibitors to do that. We thought this would be a direct way to elevate it for sure, and quite frankly I was just kind of jealous of the patch clampers, who have a single membrane there, and I wanted to try to catch up with them.

DEARBORNE: Now, just to GTP binding proteins, have you tried ADP? That is, the ratio of ADP to ATP may be very crucial.

QUINTON: We've considered it but we haven't done it yet.

BOUCHER: Paul, a couple of years ago Jack Stutts did similar studies in airway cells. We were at that time looking for inhibition on the rate of sodium transport and we found the expected. The unexpected, as we mentioned, was that the chloride conductance also deactivated in airway cells, and most predominantly the apical membrane chloride conductance. This happened when we metabolically inhibited cells with either cyanide or nitrogen (glucose, at least in our hands, does not work in airway cells). What was different about it from CF is that the calcium-activated chloride conductance response was blocked by that response, and the ISO, or the cAMP dependent system was also blocked by that response. So we lost both limbs of the activation sequence. Have you looked at that at all yet?

QUINTON: No. Obviously we have lots of things to look at.

ALTERED BIOCHEMICAL REGULATION OF SECRETION IN CYSTIC FIBROSIS EPITHELIAL CELLS

M.A. McPherson, D.K. Shori, C. Lloyd Mills,
M.C. Goodchild* and R.L. Dormer

Dept. of Medical Biochemistry
University of Wales College of Medicine and
*Dept.of Child Health, U.H.W., Heath Park
Cardiff, U.K.

INTRODUCTION

A link between alteration in the protein coded for by the CF gene (Riordan et al., 1989) and the abnormal epithelial cell physiology in CF must be established in order to recognise potential site(s) of intervention to correct the secretory abnormality. Defective B-adrenergic responses are seen in all affected epithelial cells in CF (McPherson and Dormer, 1987, 1988). Thus, sweating was stimulated by isoproterenol in control but not in CF sweat glands (Sato and Sato, 1984). Secretion of mucins was increased in response to isoproterenol in control but not in CF submandibular acinar cells (McPherson et al., 1985, 1986). CF submandibular acinar cells also showed a much reduced secretion of amylase in response to isoproterenol and noradrenaline, both agonists acting via a B-adrenergic effect (McPherson et al., 1985, 1986). Cl^- transport in cultured CF sweat duct cells (Pedersen, 1986) and airways epithelial cells (Welsh and Liedtke, 1986; Frizzell et al., 1986; Widdicombe et al., 1986; Cotton et al., 1987) was also shown to be defective in response to isoproterenol. In all of these epithelial cells cyclic AMP rise was shown to be normal in response to isoproterenol (McPherson and Dormer, 1987, 1988), indicating either that the defect is beyond the site of cyclic AMP formation, or that this measurement is not relevant to the function of isoproterenol in opening Cl^- channels or stimulating protein secretion. Patch clamp studies in airway epithelial cells (Welsh and Liedtke, 1986; Frizzell et al., 1986) directly demonstrated that in a cell attached mode the B-agonist opens Cl^- channels in normal but not in CF cells. It is possible that data showing unresponsiveness of secretion of mucins and amylase to B-adrenergic stimulation (McPherson et al, 1986) is measuring the same end point. Thus, Cl^- channels have been shown to be present on secretory granule membranes and are opened when the cell is stimulated (DeLisle and Hopfer, 1986). If opening of Cl^- channels on secretory granules is necessary for protein secretion to occur, then inability to open Cl^- channels in CF cells could account for the defect in regulation of protein secretion observed in CF salivary cells (McPherson et al., 1986).

Defective B-adrenergic stimulation of secretion of mucins and amylase from CF submandibular acinar cells was partially restored by addition of the cyclic nucleotide phosphodiesterase inhibitor IBMX (3-isobutyl-1-methyl xanthine) (McPherson et al., 1986, 1988a). This led to the hypothesis that the autonomic disturbance in CF is due to an overactivation of this enzyme mediated by Ca^{2+}/calmodulin. (McPherson et al 1986; 1988a,b). Altered calmodulin-like activation of cyclic nucleotide phosphodiesterase was observed in CF extracts (McPherson et al., 1988a; Shori et al 1988). However, the structure of calmodulin was shown to be normal; calmodulin was not the site of the CF gene (Scambler et al., 1987). The data are consistent with alteration in a modulator protein which interacts with and modulates the biological actions of calmodulin in cells. Alteration in properties and phosphorylation of a 61,000 molecular weight (61kDa) calmodulin binding protein was seen in CF submandibular tissues (Shori et al., 1988, 1989). To investigate the links between cyclic AMP and Ca^{2+} pathways mediating secretion, the actions of protein kinase A and cyclic AMP on phosphorylation of calmodulin binding proteins have been examined in control and CF submandibular glands. In addition, the role of Ca^{2+} in B-adrenergic stimulation has been directly investigated using a method of manipulation of intracellular regulators in living epithelial cells.

MATERIALS AND METHODS

Human Submandibular tissues

Tissues were obtained from submandibular glands from five control subjects (aged 4-55 years) either during operation for ductal calculi (2 glands) or within 3h of death (3 glands) and from seven CF individuals (aged 15-44 years) within 3h of death.

Preparation of calmodulin binding protein fractions

Tissues (100 mg) were homogenised in 1.5 ml buffer (0.3M sucrose, 5mM dithiothreitol, 1mM benzamidine, 1mM phenylmethylsulphonyl fluoride, 40mM imidazole-HCl, pH 7 at 4°C, using a motor driven Potter homogeniser. Homogenates were centrifuged at 10,000g for 15 min at 4°C, supernatants were mixed with DEAE-Sephacel and eluted with 0.22M KCl as described (Shori et al., 1989). Eluates were then incubated for 90 min at 4°C with Calmodulin-Sepharose 4B, Pharmacia in the presence of 10mM $CaCl_2$.

After washing in buffer containing 0.1mM $CaCl_2$, calmodulin binding protein fractions were eluted in imidazole buffer pH 7 containing 2mM EGTA and 0.15M KCl.

Protein content of extracts was determined by the method of Lowry; where control and CF extracts were directly compared, the same amount of protein was used in the phosphorylation reaction and when applied to the gels.

Phosphorylation of calmodulin binding proteins

Calmodulin binding protein fractions were phosphorylated as described (Shori et al., 1989). Incubation of fraction (20μl) with ^{32}P -ATP (2uCi, 2μM) was carried out in imidazole buffer pH 7 containing 40mM LiCl, 110mM KCl, 0.8mM EDTA, 4mM $MgCl_2$ 0.4mM EGTA and 1mM benzamidine for 5 min at 30°C. Where extra additions were made, cyclic AMP was added at a concentration of 10μM; protein kinase A (Sigma, holoenzyme) at a concentration of 4ug/ml and protein kinase A inhibitor (Walsh inhibitor, Sigma) at a concentration of 10mg/ml. Proteins were precipitated using trichloroacetic acid (final concentration 10%), solubilised and fractionated on 9% SDS polyacrylamide gels. Gels were stained with Coomassie Blue, dried and autoradiographed to detect the ^{32}P-labelled proteins. The autoradiographs were scannned densitometrically using an LKB laser densitometer and area under the main phosphorylated peaks was measured to determine the amount of ^{32}P incorporated into the proteins.

Preparation of rat submandibular acini

Submandibular glands were obtained from two male Wistar rats (250-280g) which had been starved overnight. Acini were prepared as described (McPherson and Dormer, 1984). Briefly, glands were injected with collagenase solution (Ca^{2+} and Mg^{2+} free Hank's balanced salt solution (Gibco) containing collagenase (Worthington II, 64 units/ml) and hyaluronidase (500 units/ml), cut into fragments and incubated for 60 min at 37°C with shaking at 120 cycles/min. Tissue was mechanically dispersed by pipetting; acini were filtered through cheese cloth and purified by centrifugation through Krebs bicarbonate medium pH 7.4 containing 4% (w/v) bovine serum albumin.

Measurement of mucin secretion

This was carried out as described previously (McPherson and Dormer, 1984) with the following modifications. Acini were labelled with ^{3}H-glucosamine (5mCi/ml) for 90 min at 37°C, with shaking at 60 cycles per min, followed by washing (3 x 10 ml) in Krebs bicarbonate buffer containing glucosamine (1mM), glucose (12mM) and bovine serum albumin (20 mg/ml) and a cold-chase by incubation in similar medium for 60 min at 37°C. After short-term exposure to isotonic or hypotonic medium (see below : Incorporation of macromolecules) acini were incubated under the experimental conditions shown for 30 min at 37°C in Krebs bicarbonate buffer pH 7.4. Aliquots of acini suspensions were taken at zero time and 30 min, centrifuged (10s in an Eppendorf microfuge), the supernatant removed and an equal volume of ice-cold trichloroacetic acid/ phosphotungstic acid (20%/1%w/v) added to precipitate ^{3}H-labelled mucins as previously described (McPherson and Dormer, 1984). Radioactivity in the pellets was counted on a liquid scintillation counter (Packard). Results are expressed as a % of the basal secretion in the absence of agonist.

Incorporation of impermeant molecules by hypotonic swelling

This was carried out as described (Bradbury et al., 1989). Isolated acini which had been labelled with ^{3}H-glucosamine as described above, were washed three times in TES-buffered saline (10mM TES pH 7.4 containing 143mM NaCl, 4.7mM KCl, 1.1mM $MgCl_2$, 1 mg/ml bovine serum albumin). To acini suspended in TES-buffered saline was added an equal volume of either 10mM TES, pH 7.4 or TES-buffered saline each containing 5mM ATP and the appropriate impermeant molecule as described in the text and Figure legends.

After 1 min at room temperature, acini were centrifuged and washed three times in TES-buffered saline containing 1 mg/ml bovine serum albumin and finally resuspended in Krebs bicarbonate buffer containing 20 mg/ml bovine serum albumin. Following 15 min recovery at 37°C, mucin secretion was measured in the presence or absence of agonists for 30 min as described. Acini which had been incubated for 1 min at room temperature in hypotonic medium are referred to as swollen acini and those incubated in isotonic medium throughout as unswollen acini.

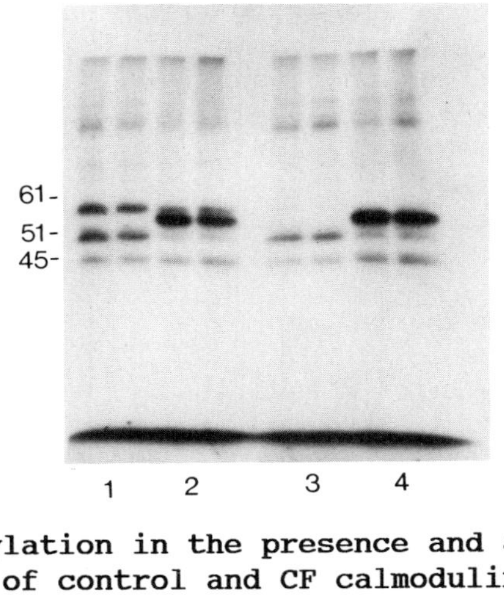

Figure 1. Phosphorylation in the presence and absence of protein kinase A of control and CF calmodulin binding proteins. Samples were run in duplicate. Lanes: - 1, endogenous, control; 2, + protein kinase A (Sigma, holoenzyme) + cyclic AMP, control; 3, endogenous, CF; 4, + protein kinase A (Sigma, holoenzyme) + cyclic AMP, CF.

RESULTS

1. Defective phosphorylation of a 61kDa calmodulin binding protein in CF submandibular tissues: actions of protein kinase A and C

Calmodulin binding protein fractions (prepared by EGTA elution of proteins from calmodulin-Sepharose) contain an endogenous protein kinase(s) distinct from protein kinase A. In these fractions (Fig. 1) a 61,000 molecular weight (61kDa) protein was phosphorylated to a much greater degree in control than CF glands. In the absence of any addition (endogenous phosphorylation), quantitation of data revealed that the peak area of the 61kDa band from densitometric scans of autoradiographs was 79.9 ± 10.9 mm^2 (n= 5 control glands) and 9.4 ± 1.1 mm^2 (n= 7 CF glands), p< 0.001 for difference. Although variation in calmodulin binding protein phosphorylatlon patterns was seen in individual glands (see also Fig. 2), the only consistent finding under these conditions was the markedly defective phophorylation of the 61kDa protein in CF glands compared to control.

When intact protein kinase A (holoenzyme) and cyclic AMP was added, an intense band of approximately 57-58kDa appeared on the autoradiographs, which is presumably the R subunit of protein kinase A. Under these conditions, although it was apparent that the 61kDa band was phosphorylated in the presence of protein kinase A, it was not possible to quantitate this effect. Using purified catalytic subunit of protein kinase A (kindly given by Dr. P. England, Smith Kline and French, U.K.) the data showed that the 61kDa protein was also less phosphorylated in CF glands than control. Thus, addition of exogenous catalytic subunit of protein kinase A (500nM) resulted in ^{32}P incorporation into the control 61kDa molecular weight protein of 206.8 ± 25.1 (n=4) mm^2 and into the CF protein of 36.3 ± 5.4 (n=4) mm^2. Similarly, addition of activators of protein kinase C (phosphatidyl serine, 25μg/ml; diacylglycerol, 25μg/ml and Ca^{2+}, 0.4mM) increased phosphorylation of the 61kDa protein band in control glands to 180.0 ± 24.7 (n=3) mm^2 and in CF glands to 18.4 ± 2.2 (n=4) mm^2. Thus although phosphorylation was increased by approximately the same extent in control and CF glands by protein kinase A or protein kinase C, the amount of ^{32}P incorporated into the 61kDa protein was much less in CF glands than control.

2. Actions of Cyclic AMP on the phosphorylation patterns

The effects of cyclic AMP and protein kinase A inhibitor on endogenous phosphorylation in the calmodulin binding protein preparations in the absence of added protein kinase were examined. Fig 2 shows representative data from control and CF glands. No significant effects of cyclic AMP or protein kinase A inhibitor were seen on phosphorylation of the 61kDa band in control or CF glands (Shori et al 1989), as is seen in Fig. 2; although the 61kDa band was hardly visible in the CF gland, as was often the case.

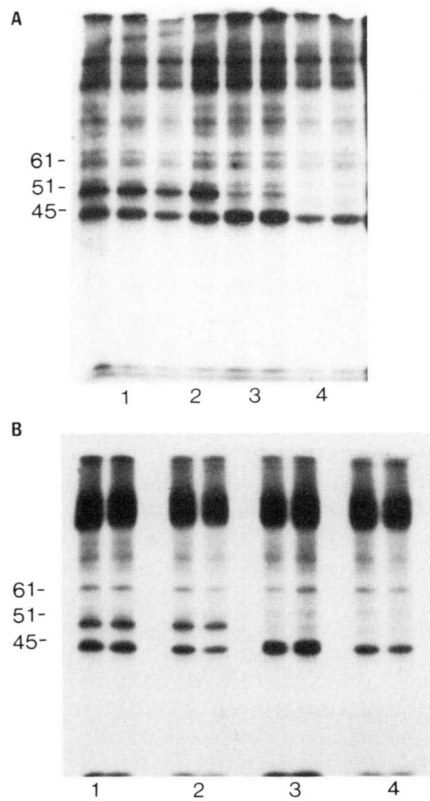

Figure 2. Actions of cyclic AMP and protein kinase A inhibitor on phosphorylation patterns in control and CF submandibular glands.
Phosphorylation of calmodulin binding proteins in control (A) and CF (B) submandibular glands. Samples were run in duplicate. Lanes : - 1, endogenous; 2, + cyclic AMP (10μM); 3, + protein kinase A inhibitor (10μg/ml); 4, + cyclic AMP + protein kinase A inhibitor.

However, cyclic AMP altered the protein patterns on the gels in both control and CF glands. Fig 2 shows that in control glands (Fig 2A) cyclic AMP decreased phosphorylation of the 51kDa band and increased phosphorylation of a 47kDa band. Protein kinase A inhibitor reversed the cyclic AMP-induced increase in phosphorylation of the 47kDa band but not the decrease in phosphorylation of the 51kDa band. In CF glands (Fig 2B) similar effects of cyclic AMP on the 51kDa and 47kDa bands were seen. In addition, an apparent increase in phosphorylation of a band of 57kDa molecular weight was noted. The explanation for the actions of cyclic AMP on the protein phosphorylation patterns is unknown. However, it does suggest that cyclic AMP dependent protein kinase is present in the calmodulin binding protein preparations, but in insufficient quantity to phosphorylate the 61kDa protein band, which as shown above (Section 1) is a substrate for protein kinase A.

3. Characteristics of the altered 61kDa phosphoprotein in CF

Two possibilities as to the identity of the 61kDa phosphoprotein have been suggested from the characteristics of this protein.

a) The 61kDa protein band in both control and CF submandibular glands was shown to cross-react with an antiserum raised against human platelet calcineurin which was kindly given by Dr. R.W. Wallace (Shori et al., 1989). However, the cross-reaction was seen on a one dimensional gel; so it is not certain that the altered phosphoprotein has a calcineurin-like function. However a defect in phosphorylation of calcineurin under the conditions of endogenous phosphorylation described might be compatible with increased activity of its calmodulin-dependent phosphatase actions (Hashimoto et al., 1988; McPherson et al.,1988b). Such an abnormally overactive Ca^{2+}/calmodulin-dependent phosphatase in CF would lead to neutralising the actions of protein kinase A and protein kinase C in opening Cl^- channels (McPherson et al., 1988b). It could also explain the apparent overactivity of cyclic AMP phosphodiesterase observed in CF submandibular acinar cells (McPherson et al., 1986) since this enzyme is regulated by phosphorylation and is more active in a dephosphorylated form (Sharma and Wang, 1986).

b) The altered 61kDa phosphoprotein has properties consistent with it being a cytosolic fragment of the CF gene protein, CFTR (Table 1) which is either naturally occurring or has arisen as a result of proteolytic activity during collection and homogenisation of the human material. The first cytoplasmic region of CFTR (Table 1), which comprises the first nucleotide binding domain and the regulatory domain has a similar molecular weight

Table 1

Characteristics of the altered 61kDa protein and CFTR

61kDa protein	1st cytoplasmic domain of CFTR
1. Molecular wt. 61k on on SDS PAGE	Approx. 510 amino acids
2. Substrate for PKA and PKC	Recognition sites for PKA and PKC
3. Calmodulin binding	Hydrophobic domain: P-glycoproteins interact with CaM antagonists
4. Defective phosphorylation in CF	Common phe deletion - might mask a phosphorylation site.

Abbreviations: PKA - protein kinase A; PKC - protein kinase C; CaM - calmodulin; PAGE - polyacrylamide gel electrophoresis.

to the 61kDa protein. Both have recognition sites for protein kinase A and protein kinase C. The calmodulin binding properties of the altered 61kDa phosphoprotein could indicate a hydrophobic domain, which might well be present in the `cytoplasmic domain' of CFTR, depending on the cleavage point. CFTR has homology with P-glycoproteins (Riordan et al., 1989), which are coded for by the multidrug resistance genes. The finding that multidrug resistance is reversed by calmodulin antagonists, presumably by their interaction with P-glycoprotein transporters suggests that CFTR-like proteins might possess calmodulin-binding domains. The defect in phosphorylation observed in the 61kDa protein in CF could explain many of the cellular physiological abnormalities in CF and it is possible that the common phenylalanine deletion present in the cytoplasmic domain of CFTR (Riordan et al., 1989) might mask a phosphorylation site.

4. Determining the function of the altered 61kDa phosphoprotein and CFTR

A method of intracellular manipulation, using hypotonic swelling has been developed. This will enable proteins, including antibodies, to be incorporated into the cytoplasm of intact living epithelial cells and their actions on a variety of cellular parameters such as mucin secretion and Cl^- transport determined.

Results have shown that fluorescently labelled immunoglobulins are taken up by cells during hypotonic swelling (Bradbury et al 1989). Examination of individual cells by fluorescence microscopy indicates that the majority of cells have incorporated antibodies in the cytoplasm by this technique. When antibodies to the CF gene protein and the altered 61kDa phosphoprotein become available, it will be possible to directly determine whether they are related proteins and to examine their actions in whole cells on a wide range of parameters.

5. Role of Ca^{2+} in B-adrenergic stimulation of mucin secretion

Incorporation of intracellular cyclic AMP phosphodiesterase into rat submandibular acini by hypotonic swelling abolished the cyclic AMP rise in response to isoproterenol as would have been predicted. However, surprisingly,in the absence of a rise in cyclic AMP, mucin secretion in response to isoproterenol was unaffected (Bradbury et al., 1989), suggesting that cyclic AMP is not the sole mediator of B-adrenergic function.

To directly investigate the role of intracellular Ca^{2+} in B-adrenergic stimulation of mucin secretion, the selective Ca chelator BAPTA(1,2-bis(2-aminophenoxy)ethane-N,N,N',N'-tetraacetic acid) was introduced into intact rat submandibular acini by hypotonic swelling. Fig 3 shows the results obtained. As has been shown previously for rat pancreatic (Dormer, 1984) and submandibular (Bradbury et al., 1989) acini, cells which have been swollen in hypotonic medium and allowed to recover show the same secretory response as cells which have not previously been sub-

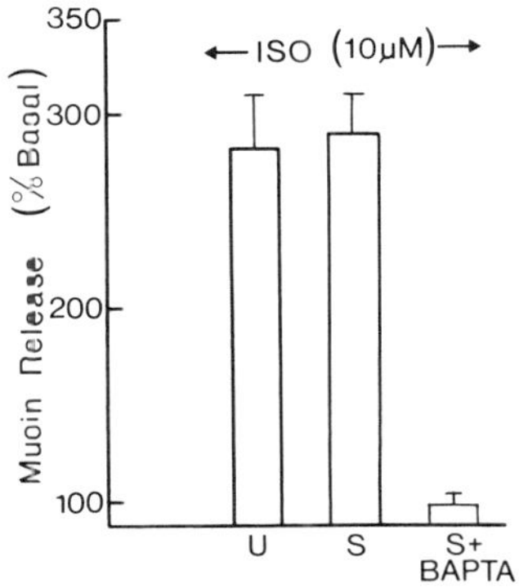

Figure 3. Actions of intracellular BAPTA on isoproterenol-stimulated mucin release.
Normal acini (unswollen, U) or acini subjected to hypotonic swelling (swollen, S) in the presence (S + BAPTA) or absence of BAPTA were allowed to recover in isotonic medium and then incubated +- isoproterenol (ISO) (10μM) for 30 min. Mucin secretion was measured as described in the Methods and expressed as a % of the basal secretion in the absence of agonist.

jected to hypotonic swelling. Fig 3 shows that submandibular acini swollen in the absence of BAPTA showed similar mucin secretion in response to isoproterenol as did unswollen acini, whereas acini swollen in the presence of BAPTA showed no response to the B-agonist isoproterenol. The data are the first direct demonstration that Ca^{2+} is necessary for mediating B-adrenergic stimulation of secretion.

DISCUSSION

Although current dogma indicates that B-adrenergic stimulation is mediated by a rise in cyclic AMP, the results show that Ca also plays a crucial role. This is in agreement with previous findings showing that B-adrenergic stimulation of mucin secretion mediated by either isoproterenol or noradrenaline is partially dependent on extracellular Ca^{2+} (McPherson and Dormer, 1984) and that isoproterenol mobilises Ca^{2+} without affecting inositol phosphate formation (McPherson and Dormer, 1984; Doughney et al., 1987). The present data suggest that an intracellular Ca rise induced by isoproterenol plays a direct role in triggering secretion.

The Ca^{2+} necessary for B-adrenergic stimulation of secretion could be to activate Ca^{2+}/calmodulin-mediated events or might be the Ca^{2+} necessary for protein kinase C activation by diacylglycerol (DAG). The finding of a rise in DAG formation in response to isoproterenol in airways epithelial cells (Li et al., 1989) would tend to support the latter hypothesis. However, the DAG increase in response to B-stimulation is likely to result from breakdown of phospholipids other than phosphatidylinositol bisphosphate (Doughney et al.,1987). In addition, the cytosolic free Ca^{2+} in submandibular acini following incorporation of BAPTA as described in Fig. 3 is unlikely to be below resting levels (Dormer, 1984; Ochs et al., 1985).

The finding of a defect in phosphorylation of a calmodulin binding protein in CF submandibular glands provides a link between alteration in events mediated by phosphorylation (McPherson and Dormer, 1988), including the direct demonstration that Cl^- channels can be opened in control but not CF airway epithelial cells by protein kinase A or protein kinase C (Hwang et al., 1989; Li et al.,1989) and altered Ca^{2+}-mediated events (Pedersen, 1987., Taylor et al., 1987; DeJonge et al., 1989), including alteration in calmodulin activity (McPherson et al 1988a,b; Shori et al., 1988) seen in CF cells. An overactive Ca^{2+}/calmodulin-dependent protein phosphatase in CF cells would be predicted to neutralise the actions of protein kinase A and protein kinase C in opening Cl^- channels. There is evidence for phosphatase activity in membrane patches (Li et al., 1989). In addition, the finding that Cl^- channels in excised membrane patches from both control and CF cells spontaneously activate at 37°C, (Kunzelmann et al., 1989) could be explained by removal of a cytoplasmic inhibitor or a protein phosphatase(s) from the cell membrane patches at this temperature. The latter hypothesis would require the presence of a tightly bound membrane protein kinase.

It is not clear how the altered regulation of secretion in CF epithelial cells is related to the alteration in the protein coded for by the CF gene; since although the computer model predicts a membrane transport protein, it is not known what type of molecule is transported by CFTR (Riordan et al., 1989). It is also not known whether the secretory events defective in CF are all related to Cl^- channel opening; this remains a possibility in the case of defective regulation of protein secretion from salivary cells (McPherson et al., 1986; DeLisle and Hopfer, 1986). It is intriguing that the altered 61kDa phosphoprotein in CF submandibular glands shows several characteristics in common with the major cytoplasmic domain of CFTR (Table 1), raising the question as to whether the 61kDa protein is a fragment of the CF gene protein itself. Nevertheless, the finding of a protein phosphorylation defect in CF which has the characteristics to explain many of the alterations in epithelial cell function will be crucial in unravelling events linking the CF gene mutation to physiological abnormalities. This will be vital for determining site(s) of intervention to correct the secretory abnormality in CF.

Acknowledgements

We are very grateful to Dr. P.K. Jeffery, Mr. R. Sharma and Dr.

M. Alfaham for obtaining submandibular material. This work was supported by the CF Research Trust, U.K.

References

Bradbury, N.A., Dormer, R.L., McPherson, M.A., 1986, Introduction of cyclic AMP phosphodiesterase into rat submandibular acini prevents isoproterenol-stimulated cyclic AMP rise without affecting mucin secretion, **Biochem. Biophys. Res. Commun.**, 161:661.

Cotton, C.U., Stutts, M.J., Knowles, M.R., Gatzy, J.T. and Boucher, R.C., 1987, Abnormal apical cell membrane in cystic fibrosis respiratory epithelium, **J. Clin. Invest.**, 79:80.

DeJonge, H.R., van den Berghe, N., Tilly, B.C., Kansen, M. and Bijman, J., 1989, (Dys)regulation of epithelial chloride channels, **Biochem. Soc. Trans.**, 17:816.

DeLisle, R. and Hopfer, U., 1986, Electrolyte permeabilities of pancreatic zymogen granules: implications for pancreatic secretion, **Am. J. Physiol.**, 250:G489.

Dormer, R.L., 1984, Introduction of calcium chelators into isolated rat pancreatic acini inhibits amylase release in response to carbamylcholine, **Biochem. Biophys. Res. Commun.**, 161:661.

Doughney, C., Dormer. R.L., McPherson, M.A., 1987, Adrenergic regulation of formation of inositol phosphates in rat submandibular acini, **Biochem. J.**, 241:705.

Frizzell, R.A., Rechkemmer, G. and Shoemaker, R.L., 1986, Altered regulation of airway epithelial cell chloride channels in cystic fibrosis, **Science**, 233:558.

Hashimoto, Y., King, M.M., Soderling, T.R., 1988, Regulatory interactions of calmodulin binding proteins:phosphorylation of calcineurin by autophosphorylated Ca^{2+}/calmodulin-dependent protein kinase II, **Proc. Natl. Acad. Sci. U.S.A.**, 85:7001.

Hwang, T.C., Lu, L., Zeitlin, P., Gruenert, D.C., Huganir, R. and Guggino, W.B., 1989, Cl^- channels in CF : Lack of activation by protein kinase C and cyclic AMP dependent protein kinase, **Science**, 244:1351.

Kunzelmann, K., Pavenstadt, H. and Greger, R., 1989, Properties and regulation of Cl^- channels in CF and normal airways cells, **Pflueg. Arch.**, 415:172.

Li, M., McCann, J.D., Anderson, M.P., Clancy, J.P., Liedtke, C.M., Nairn, A.C., Greengard, P. and Welsh, M.J., 1989, Regulation of Cl^- channels by protein kinase C in normal and cystic fibrosis airway epithelia, **Science**, 244:1353.

McPherson, M.A. and Dormer, R.L., 1984, Mucin release and calcium fluxes in isolated rat submandibular acini, **Biochem. J.**, 224:473.

McPherson, M.A. and Dormer, R.L., 1987, The Molecular and Biochemical Basis of Cystic Fibrosis, **Biosci. Rep.**, 7:167.

McPherson, M.A. and Dormer, R.L., 1988, Cystic fibrosis; a defect in stimulus response coupling, **TIBS**, 13:10.

McPherson, M.A., Dormer, R.L., Dodge, J.A. and Goodchild, M.C., 1985, Adrenergic secretory responses of submandibular tissues from control subjects and cystic fibrosis patients, **Clin. Chim. Acta**, 148:229.

McPherson, M.A., Dormer, R.L., Bradbury, N.A., Dodge, J.A., Goodchild, M.C., 1986, Defective B-adrenergic secretory responses in submandibular acinar cells from cystic fibrosis patients, **Lancet**, ii:1007.

McPherson, M.A., Dormer, R.L., Bradbury, N.A., Shori, D.K. and Goodchild, M.C., 1988a, Regulation of secretion of amylase and mucins in control and cystic fibrosis submandibular acinar cells: role of cyclic AMP and Ca^{2+}, **In:** "Cellular and Molecular Basis of Cystic Fibrosis," G. Mastella and P.M. Quinton, eds., San Francisco Press.

McPherson, M.A., Shori, D.K. and Dormer, R.L., 1988b, Defective regulation of apical membrane chloride transport and exocytosis in cystic fibrosis, **Biosci. Rep.**, 8:27.

Ochs, D.L., Korenbrot, J.I. and Williams, J.A., 1985, Relation between free cytosolic calcium and amylase release by pancreatic acini, **Am. J. Physiol.**, 249:G389.

Pedersen, P.S., 1986, Quantitatively abnormal beta-adrenergic response on cystic fibrosis sweat duct cell culture, **IRCS Medical Science**, 14:701.

Pedersen, P.S., 1987, Cholinergic influence on chloride permeability in sweat duct cell cultures from normals and patients with cystic fibrosis, **Med. Sci. Res.**, 15:769.

Riordan, J.R., Rommens, J.M., Kerem, B.S., Alon, N., Rozmahel, R., Grzelczak, Z., Zielenski, J., Lok, S., Plavsic, N., Chou, J.L., Drumm, M.L., Iannuzzi, M.C., Collins, F.S., Tsui, L.-C., 1989, Identification of the cystic fibrosis gene : cloning and characterization of complementary DNA, **Science**, 245:1066.

Sato, K. and Sato, F., 1984, Defective B-adrenergic response of cystic fibrosis sweat glands in vivo and in vitro, **J. Clin. Invest.**, 73:1763.

Scambler, P.J., McPherson, M.A., Bates, G., Bradbury, N.A., Dormer, R.L. and Williamson, R., 1987, Biochemical and genetic exclusion of calmodulin as the site of the basic defect in cystic fibrosis, **Hum. Genet.**, 76:278.

Sharma, R.K. and Wang, J.H., 1986, Calmodulin and Ca^{2+}-dependent phosphorylation and dephosphorylation of 63kDa subunit containing bovine brain calmodulin-stimulated cyclic nucleotide phosphodiesterase isozyme, **J. Biol. Chem.**, 261: 1322.

Shori, D.K., Dormer, R.L., Goodchild, M.C., McPherson, M.A., 1989, Defective phosphorylation of a calmodulin binding protein in cystic fibrosis submandibular glands, **Biochem. J.**, 263:613.

Shori, D.K., Bradbury, N.A., Goodchild, M.C., Dormer, R.L., McPherson, M.A., 1988, An altered calmodulin binding protein in cystic fibrosis - a clue to the biochemical defect, **Clin. Chim. Acta**, 174:283.

Taylor, C.J., Baxter, P.S., Hardcastle, J. and Hardcastle, P.T.,

1987, Absence of secretory response in jejunal biopsy samples from children with cystic fibrosis, **Lancet**, ii: 107.
Welsh, M.J., Liedtke, C.M., 1986, Chloride and potassium channels in cystic fibrosis airway epithelia, **Nature**, 322: 467.
Widdicombe, J.H., 1986, Cystic fibrosis and B-adrenergic response of airway epithelial cell cultures, **Am. J. Physiol.**, 251:R818.

DISCUSSION

(Unidentified)(inaudible)...KD protein might be the equivalent of the first cytoplasmic domain of CFTR. Are you trying to approach this also by the straight technique of sequencing, at least a little bit, the amino acids of this protein and trying to see if it corresponds?

McPHERSON: So far, we've got very little of this protein, because you don't actually see a staining pattern on the gels where you see the phosphorylation. So yes, we would like to find out its structure and function.

QUINTON: Maggie, is it possible that the increase in phosphorylation is due to an increase in protein?

McPHERSON: I think we would think that it probably isn't, at least the decrease in the CF cell we think is not due to a decrease in protein, but I can't absolutely rule that out. The reason we think it isn't, is because we are putting the same amount of protein on our gels and we are seeing the same amount of I125 calmodulin binding. But that's not to say there are not other calmodulin binding proteins not bound, and I think we could only resolve that with 2D or sequencing.

CORCELLI: I would like to have a comment about the presence of chloride transporter on secretory granules and the relationship between this chloride transporter and the channel on the luminal membrane.

McPHERSON: Yes, I don't know whether that's the same chloride channel. I don't know whether anyone else can provide that information.

CORCELLI: But do you think that under beta-adrenergic stimulation this transporter of the granules could, after fusion with the luminal membrane, significantly contribute to the chloride flux?

McPHERSON: It could, I don't know. It's a possibility. Certainly the channels have been shown to be opened when the cell is stimulated and it is thought that it may be necessary to wash out the granule contents, but that's as far as I know.

DEARBORNE: There are a lot of phosphorylated proteins around the size of 60 kilodaltons. Have you done two-dimensional electrophoresis and/or have you looked at other antibodies to cytokeratins, to band-3-like proteins, or other candidates for this identity?

McPHERSON: No, we haven't done that yet. We would like to. But these are calmodulin-binding proteins so I think that rules out a lot of other 61k proteins.

CHLORIDE ION TRANSPORT IN TRANSFORMED NORMAL AND CYSTIC FIBROSIS EPITHELIAL CELLS

A.L. Cozens (1), M.J. Yezzi (1), L. Chin (1), E.M. Simon (1), D.S. Friend (1,2) and D.C. Gruenert (1,3)

[1]Cardiovascular Research Institute
[2]Department of Pathology and
[3]Department of Laboratory Medicine
University of California San Francisco
San Francisco, CA

ABSTRACT

The inability of β-adrenergic agonists (eg. isoproterenol) to activate a specific apical membrane chloride channel in epithelial cells is characteristic of cystic fibrosis (CF). The study of these channels has been facilitated by the transformation of human airway epithelial cells. Clonal populations of SV40 large T-antigen transformed airway epithelial cells from both normal and cystic fibrosis individuals have been established. A limitation in the use of these cell lines has been the loss of the ability to form tight junctions after multiple subcultures. In particular, this loss appears to be associated with cell "crisis". A selection protocol that involves growing cells in medium that is high in Ca^{2+} and supplemented with fetal bovine serum (FBS) has been employed to facilitate progression through crisis.

One cell line (1HAEo-) that has a normal phenotype and retains tight junctions post-crisis has been produced using this protocol. Immunofluorescent staining with a monoclonal antibody to the E-cadherin adhesion molecule shows a characteristic pericellular localization, indicating the presence of the junctional complex. The presence of tight junctions has been confirmed by electron microscopy. These cells produce elevated (>30 fold) levels of cAMP in response to exposure to isoproterenol or forskolin. Chloride ion transport, as measured by $^{36}Cl^-$ efflux, is stimulated >2 fold by these agents.

Three post-crisis CF cell lines which grow in serum-containing medium have been established. These do not exhibit tight junctions. Elevated (>25 fold) levels of cAMP are detected in these cells after addition of isoproterenol or forskolin, but this increase in cAMP is not accompanied by an increase in $^{36}Cl^-$ efflux. Both normal (1HAEo-) and CF cells show increased $^{36}Cl^-$ efflux following addition of the calcium ionophore A23187. Activation of the apical membrane chloride channel by elevated intracellular calcium is not thought to be affected in CF.

When these transformed cells are grown in serum-free medium containing epinephrine the response to isoproterenol or forskolin is attenuated. This is consistent with down regulation of the β-adrenergic receptor upon chronic stimulation.

One of the post-crisis CF lines (2CFSMEo-) is homozygous for the deltaF508 deletion; the remainder do not have this mutation. This may make it possible to correlate specific deletions with the degree of severity of the CF phenotype.

INTRODUCTION

Cystic fibrosis (CF) is characterized by abnormalities of chloride transport and secretion in epithelia (1). In the airways this leads to the production of thick, underhydrated mucus (2). The underlying defect has been pinpointed to the inability of a specific apical membrane chloride channel in epithelial cells to be activated by β-adrenergic agonists (3-7). The gene that is defective in CF, the cystic fibrosis transmembrane conductance regulator (CFTR), has now been cloned (8-10), but it is still unclear what relationship CFTR bears to the demonstrated chloride transport defect.

Analysis of the pathways underlying the biochemical defect in CF has been hampered by the lack of suitable tissue. Cultured human airway epithelial cells have proven useful (11), but these have a limited life-span and senesce or terminally differentiate even under optimal culture conditions (12). To circumvent this problem, we have transformed normal (13) and CF (14) human airway epithelial cells with a plasmid bearing an origin-defective SV40 genome. Clonal populations of SV40 T-antigen transformed cells have been isolated and characterized. Clones that retain epithelial cell specific features relevant to CF research have been identified. These cell lines will facilitate considerably the study of both the regulation of chloride ion transport and the function of the CFTR gene product.

In addition to the provision of large amounts of experimental material, there are other important advantages to be gained by the use of transformed cell lines for studies of the CF defect. Experimental conditions are more reproducible, as repeated measurements can be made with cells of the same clone, at the same passage, avoiding the differences between individual human donors. The native epithelium contains a mixed population of cell types, while clonally derived transformants are relatively homogeneous. Finally, the transformed epithelial cell system is readily manipulable. Plasmids containing exogenous foreign or altered DNA can be introduced by transfection and stable clones containing the added sequences can then be isolated and studied.

Establishment of Human Airway Epithelial Cell Lines

Transformed cell lines have been established from human airway epithelial tissue, by transfection with the plasmid pSVori$^-$, which contains an origin-defective SV40 genome (15). All clones show greatly increased growth potential (13), and all exhibit characteristic nuclear staining with antibodies to the SV40 large T antigen (Table I). Pre-crisis, these clones maintain

Table I. Properties of transformed human airway epithelial cells. HAE: human airway epithelium, HTE: normal human tracheal epithelium, HBE: normal human bronchial epithelium, FHTE: fetal human tracheal epithelium, CFTE: cystic fibrosis tracheal epithelium, CFBE: cystic fibrosis bronchial epithelium, CFNPE: cystic fibrosis nasal polyp epithelium, CFSME: cystic fibrosis submucosal gland epithelium. The presence (+) or absence (-) of SV40 large T antigen or keratin have been demonstrated by immunocytochemistry. Chloride channels were demonstrated by $^{36}Cl^-$ efflux or by patch clamp analysis. Tight junctions or junctional complexes were demonstrated by staining with antibodies to the E-cadherin adhesion molecule and by electron microscopy. NT: not tested.

Transformed cells	Clones isolated	SV40 large T antigen	Keratin	Chloride channels	Tight junctions
HAE	1	+	+	+	+
HTE	4	+	+	+	-
HTE-2	9	+	+	NT	+
HTE-23	4	+	+	NT	+
HBE-6	1	+	+	NT	NT
HBE-7	4	+	+	NT	NT
HBE-14	13	+	+	NT	+
FHTE-8	47	+	+	+	+
FHTE-27	5	+	+	NT	+
CFTE-21	13	+	+	NT	NT
CFBE-21	13	+	+	NT	NT
CFNPE-9	1	+	+	+	-
CFNPE-14	5	+	+	NT	NT
CFSME	5	+	+	+	-
CFSME-2	1	+	+	+	-

epithelial cell "cobblestone" morphology, including tight junction formation and the presence of microvilli, and expression of epithelial cell specific proteins: cytokeratins and the junctional complex adhesion molecule E-cadherin (13, 16). Table I summarizes the characteristics of the human airway epithelial cell clones isolated in this laboratory to date. Individual clones have distinct growth properties, dependent upon the stage of neoplastic progression. Non-transformed cells and newly transformed cell clones rapidly undergo squamous differentiation in the presence of fetal bovine serum (FBS). However, after repeated subculturing, and particularly when approaching and subsequent to crisis, these transformed cell lines grow more rapidly in the presence of 10% FBS, and may eventually become serum-dependent. Indeed, successful recovery from crisis can be facilitated by replacing the serum-free culture medium (MLHC8e), prior to crisis, with MEM containing 10% FBS (17). The post-crisis transformants are no longer contact-inhibited and may also become capable of anchorage independent growth in soft agar, characteristic of a greater degree of cell transformation (18) (Table II).

Table II. Growth properties of transformed clones of human airway epithelial cells. +:good growth, +/-: poor growth, -: no growth, NT: not tested.

Clone	Pre/post crisis	Growth in 10% FBS	Growth in serum-free MLHC8e	Anchorage independent growth
1HAEo-	Pre	-	+	+
	Post	+	+/-	NT
3HTEo-	Post	+	+/-	NT
9HTEo-	Post	+	+/-	+
4AHTE2o-	Pre	-	+	NT
	Post	+	+	+
56FTHE8o-	Crisis	+	+/-	-
CFNPE9o-	Post	+	+/-	NT
2CFSMEo-	Crisis	+	+/-	-
	Post	+	NT	NT
3ACFSMEo-	Pre	-	+	-
6CFSMEo-	Pre	-	+	-
5ACFSME2o-	Pre	-	+	NT
	Post	+	+/-	NT

The 1HAEo- Cell Line

Many transformed epithelial cell lines tend to lose differentiated functions, including the ability to form tight junctions, with increasing time in culture. This loss seems to be exacerbated by cell crisis. However, there is considerable clonal variation, and we have produced one cell line, 1HAEo-, that does retain morphological and functional features characteristic of cells in the intact airway epithelium, including tight junctions, post-crisis. Immunofluorescent staining with a monoclonal antibody to the E-cadherin adhesion molecule shows a characteristic pericellular localization. The presence of tight junctions has been confirmed by electron microscopy. When grown in MEM containing 10% FBS, 1HAEo- cells produce elevated (>30 fold over controls) levels of cAMP in response to either the β-adrenergic agonist isoproterenol, or forskolin, which acts directly on adenylate cyclase. This is measured by a commercial competitive radioimmune assay. In addition, these cells retain proper β-adrenergic stimulation of chloride ion transport, as measured by a simple $^{36}Cl^-$ efflux assay (5, R. Frizzell, personal communication). $^{36}Cl^-$ efflux is stimulated >2 fold over background by isoproterenol or forskolin. Single channel patch clamp studies have confirmed the presence of the characteristic apical membrane voltage-gated outwardly-rectifying anion-selective channel in these cells (S. Garber, unpublished results).

This cell line should prove invaluable in the study of the function of the CFTR gene product and its relationship to control of Cl^- ion transport. Large quantities of relatively homogeneous cells are available for use in a variety of studies, with all of the advantages outlined earlier.

Cystic Fibrosis Airway Epithelial Cell Lines

A number of transformed cell lines derived from airway epithelial tissue of individuals with CF have also been established. Five of these cell lines have been tested for the most common CFTR mutation, deltaF508. Three CF cell lines are homozygous for the deltaF508 mutation; two do not have this deletion. These clones must therefore contain one (or more) other mutation(s) in the CFTR gene.

Comparative studies with cell lines containing different mutations may therefore make it possible to correlate specific mutations in CFTR with particular effects on cellular biochemistry or ion transport, and perhaps with the severity of the disease.

Three post-crisis CF cell lines have been established (2CFSMEo-, CFNPE9o- and 5ACFSME2o-) which grow in medium containing 10% FBS, but do not exhibit tight junctions. These cells produce elevated (>25 fold) levels of cAMP following addition of isoproterenol or forskolin, but, as predicted, this is not accompanied by an increase in $^{36}Cl^-$ efflux. However, it is possible that this loss of responsiveness is a result of neoplastic progression, rather than CF, or at least that the effect of the CF mutation(s) may be masked by other perturbations of cellular ion fluxes. Additional types of ion transport analyses will be necessary to characterize the CF defect in these cells. At least one of these cell lines has been shown to contain the apical membrane chloride channel by patch clamp analysis pre-crisis (20). Numerous pre-crisis CF cell lines have been generated and it should be possible to select additional post-crisis CF cell lines which retain tight junctions and other epithelial cell functions in the same way as the 1HAEo- cell line.

Both the normal 1HAEo- cells and two CF clones, CFNPE9o- and 5ACFSME2o-, show increased $^{36}Cl^-$ efflux following addition of the calcium ionophore A23187. Activation of the apical membrane chloride channel by elevated intracellular calcium is not thought to be affected in CF (1). One CF line, 2CFSMEo-, is homozygous for the deltaF508 mutation and does not respond to A23187 in the $^{36}Cl^-$ efflux assay. The reason for this is not yet known.

Down-Regulation of the β-Adrenergic Response upon Chronic Exposure to Epinephrine

Epinephrine is added to the serum-free medium (MLHC-8e)(12) in which both non-transformed and pre-crisis transformed airway epithelial cells are grown, since it is a potent mitogen for these cells (21). The presence of epinephrine is thought to cause chronic stimulation of the β-adrenergic receptors and a down-regulation of cAMP production, and therefore of cAMP regulated pathways within the cell. When transformed cells grown in MLHC-8e were subsequently stimulated with isoproterenol a greatly attenuated response, as measured by intracellular cAMP production, was observed. The response to forskolin was also diminished, suggesting that elements further along the regulatory pathway can also be down-regulated. The effect of chronic stimulation on Cl^- ion efflux has not been measured directly, but is predicted to be considerable.

CONCLUSION

Permanent post-crisis SV40 T-antigen transformed airway epithelial cell lines from both normal and CF individuals have been established. However, there is considerable variation in retention of specific morphological features between individual transformed cell clones. It is not surprising that many of these cell lines also appear to change their functional characteristics with increasing passage number. For example, several of the normal (non-CF) post-crisis cell lines have lost their ability to respond to isoproterenol or forskolin in the $^{36}Cl^-$ efflux assay, even though at least one of these lines was able to respond to forskolin in this assay at earlier passages. Two other normal post-crisis clones were previously shown to contain the apical membrane chloride channel by patch clamp analysis (13). The cause of this loss of ability to respond to stimulation is not yet known, but these lines do have a high background rate of $^{36}Cl^-$ efflux. There is considerable heterogeneity in the native airway

airway epithelium, and it is by no means clear which of the epithelial cell subtypes can be transfected with exogenous DNA or transformed by SV40 T-antigen. The extent to which this heterogeneity contributes, if at all, to the observed clonal variation, requires further study.

To avoid any question of misrepresentation, we would like to note that we believe that the 1HAE0- cell line, although originally mislabelled as CF, must have originated from normal (ie. non-CF) airway cells. It does not have the same genotype as other cells thought to have come from the same CF airway sample.

References

1. Frizzell, R.A., 1987, Cystic fibrosis: a disease of ion chan nels?, **Trends Neurol. Sci.**, 10:190.
2. Frizzell, R.A., 1988, Role of absorptive and secretory pro cesses, **Am. Rev. Respir. Dis.**, 138:S3.
3. Bartelson, R.A., Jacoby, D.B. and Widdicombe, J.H., 1987, Regulation of chloride secretion in dog tracheal epithelium by protein kinase C, **Am. J. Physiol.**, 253:C802.
4. Frizzell, R.A., Rechkemmer, G. and Shoemaker, R.L., 1986, Altered regulation of airway epithelial cell chloride channels in cystic fibrosis, **Science**, 223:553.
5. Li, M., McCann, J.D., Anderson, M.P., Clancy, J.P., Liedtke, C.M., Nairn, A.C., Greengard, P. and Welsh, M.J., 1989, Cyclic AMP-dependent protein kinase opens chloride channels in normal but not in cystic fibrosis airway epithelium, **Science**, 244:1353.
6. Schoumacher, R.A., Shoemaker, R.C., Halm, D.R., Tallant, E.A., Wallace, R.W. and Frizzell, R.A., 1987, Phosphorylation fails to activate chloride channels from cystic fibrosis airway cells, **Nature**, 330:752.
7. Welsh, M.J. and Liedtke, C.M., 1986, Chloride and potassium channels in cystic fibrosis airway epithelia, **Nature**, 322:467.
8. Rommens, J.M., Iannuzzi, M.C., Kerem, B.-S., Drumm, M.L., Melmer, G., Dean, M., Rozmahel, R., Cole, J.L., Kennedy, D., Hikada, N., Zsiga, M., Buchwald, M., Riordan, J.R., Tsui, L.-C. and Collins, F.S., 1989, Identification of the cystic fibrosis gene: chromosome walking and jumping, **Science**, 245:1059.
9. Riordan, J.R., Rommens, J.M., Kerem, B.-S., Alon, N., Rosmahel, R., Grzelczak, Z., Zielenski, J., Lok, S., Plasvic, N., Chou, J.-L., Drumm, M.L., Iannuzzi, M.C., Collins, F.S. and Tsui, L.-C., 1989, Identification of the CF gene: cloning and characterization of complementary DNA, **Science**, 245:1066.
10. Kerem, B.-S., Rommens, J.M., Buchanan, J.A., Markiewicz, D., Cox, T.K., Chakravarti, A., Buchwald, M. and Tsui, L.-C., 1989, Identification of the cystic fibrosis gene: genetic analysis, **Science**, 245:1073.

11. Jetten, A.M., Yankaskas, J.R., Stutts, M.J., Willumsen, N.J. and Boucher, R.C., 1989, Persistence of abnormal chloride conductance regulation in transformed cystic fibrosis epithelia, **Science**, 244:1472.
12. Gruenert, D.C., Basbaum, C.B. and Widdicombe, J.H., 1990, Long-term culture of normal and cystic fibrosis epithelial cells grown under serum-free conditions, **In Vitro Cell. Dev. Biol.**, 26:411.
13. Gruenert, D.C., Basbaum, C.B., Welsh, M.J., Li, M., Finkbeiner, W.E. and Nadel, J.A., 1988, Characterization of human tracheal epithelial cells transformed by an origin-defective simian virus 40, **Proc. Natl. Acad. Sci. USA**, 85:5951.
14. Gruenert, D.C., Basbaum, C.B., Welsh, M.J., Li, M., Finkbeiner, W.E. and Nadel, J.A., 1988, Transformation and characterization of normal and cystic fibrosis airway epithelial cells, **Pediatric Pulmonology Suppl.**, 2:100 (Abstract).
15. Small, M.B., Gluzman, Y. and Ozer, H.L., 1982, Enhanced transformation of human fibroblasts by origin-defective simian virus 40, **Nature**, 296:671.
16. Wheelock, M.J., Buck, C.A., ., Bechtol, K.B. and Damsky, C.H., 1987, soluble 80-kd fragment of cell-CAM 120/80 disrupts cell-cell adhesion, **J. Cell. Biochem.**, 34:187.
17. Cozens, A.L., Chin, L., Kattan, F.A.B., Reiss, T., Li, M., Welsh, M.J., Finkbeiner, W.E. and Gruenert, D.C., 1989, Transformed cystic fibrosis epithelial cells: Growth, biochemical characterization and ion transport, **Pediatric Pulmonology Suppl.**, 4:56 (Abstract).
18. Gruenert, D.C., 1987, Differentiated properties of human epithelial cells transformed in vitro, **BioTechniques**, 5:740.
20. Gruenert, D.C., Welsh, M.J., Li, M. and Finkbeiner, W.E., Transformation and characterization of cystic fibrosis submucosal gland epithelial cells, 1989, **J. Cell. Biol.**, 107:140a (Abstract).
21. Lechner, J.F. and LaVeck, M.A., 1985, A serum-free method for culturing normal human bronchial epithelial cells at clonal density, **J. Tissue Cult. Methods**, 9:43.

DISCUSSION

KUNZELMANN: Did you see any differences in forming tight junctions if you used glucose or glucose-free media?

GRUENERT: We haven't looked at that. What we've done, actually (I didn't mention this) with the line that does form tight junctions, we used a new selection protocol which involved growing the cells in the presence of the serum-containing media which usually induces terminal differentiation in the pre-crisis state. Right before crisis, and we're doing that now with some of our CF clones to see if we can't generate clones that have a more differentiated phenotype. But, in answer to your question, we haven't looked at glucose and its effect. We have looked at dexamethasone and other agents that will induce differentiation and at present they don't seem to have any effect on the formation of tight junctions. But that may be due to the fact that the cells have progressed beyond a point where they are affected by these differentiating agents.

FROMTER: Just a short question: at what age, at what generation does the crisis occur? Is it always the same?

GRUENERT: Well, there is a variation, in terms of passage and culture. We generally allow ourselves to come to about 80 to 90% confluence and then do a 1-2-3 split so you can make your calculation from that, but anywhere from passage 10 to passage 15, though we have had crises at late as passage 25 and as early as passage 5. But in general it falls within that particular range. It's a very crucial time and if you have a number of vials that you've frozen down, you may lose something but be able to go back and then nurse a particular line through crisis.

METHA: Have you looked at any phosphorylation patterns in this tissue? Any phosphoproteins in the membranes of the cytosol?

GRUENERT: I haven't, actually. Jonathan Widdicombe has something sitting in his freezer right now that's waiting to be run on a two-dimensional gel. We haven't done that ourselves, so I can't answer that.

METHA: The reason I'm asking is that in our group we've been comparing to see what happens to the phosphoprotein pattern between fresh cells taken from brushings of the nasal epithelium and those in primary tissue culture and those in culture for 18 days and 40 days, and we find certain proteins are characteristically preserved, which you can say are very much identical between the two, but there is a characteristic change. It doesn't matter whether you start with polyp tissue or cultured tissue, there is a change in a particular area of the membrane phosphorproteins with time, which is, you know, a very specific effect.

GRUENERT: That's been seen before. When you take cells you are obviously taking a very heterogeneous population of cells, putting them into culture and they maintain their original morphology for a certain period of time, and then once the cells start proliferating, they become more de-differentiated in terms of their morphology. And as a result of that de-differentiation, they may contain some original proteins and factors, but I would say that in general, there is some loss. We've been able to subculture non-transformed epithelial cells with very defined conditions for as many as 10 subcultures, and after multiple subcultures we still see the formation of transepithelial resistance and we still are able to stimulate with isoproterenol and bradykinin and forskolin. So at least those features are maintained after multiple subcultures.

GREGER: I'm very intrigued by your studies. But I stumble over the term "immortal", because you know we live 78 years and still are not immortal, so the question really would be how immortal are your immortal cells. And a very practical question for us as possible users of such cells, how rapidly do they grow?

GRUENERT: Okay. Immortality is a very relative term. We'll say that if they last over 2 years I would say that, under constant maintenance they would probably last indefinitely, unless you have some kind of accident. As far as their growth properties, they have pretty much the same cell cycle time as non-transformed cells, and that is anywhere between 24 and 30 hours as far as doubling time is concerned. You will find individual clones that will grow slower or that will grow faster. I should also mention that as far as kariotype is concerned, most of these cells are very abnormal in terms of their chromosome number. You can have anywhere from very close to diploid, a hypodiploid form, to a bimodal population of 100 and 200 chromosomes. There doesn't seem to be any correlation though, between the differentiated properties and the number of chromosomes, not specifically. One line that we have has over 100 chromosomes and it's very differentiated.

QUINTON: Have you verified that the 508 region is present in your "normals"?

GRUENERT: Yes, we've done normal PCR and we've used a normal oligomer and the CF lines are essentially the reverse of what you see from that particular one. That was just the probe, with the deltaF508 oligomer.

QUINTON: Do you know if you have RNA expression?

GRUENERT: Yes, our lab is doing it now and Giovanni informed me yesterday that one of the CF lines definitely has the CFTR expression.

CONDUCTANCE PATHWAYS INVOLVED IN CHLORIDE SECRETION AND THEIR REGULATION

William H. Cliff, Roger T. Worrell, Andrew P. Morris and Raymond A. Frizzell

Department of Physiology and Biophysics
Gregory Fleming James Cystic Fibrosis Research Center
University of Alabama at Birmingham - UAB Station
Birmingham, Alabama 35294 USA

Cystic fibrosis impairs the secretory activities of a variety of exocrine glands and other secretory epithelia in the intestines and airways. The secretion of salt and water across epithelia of this type is driven by a secondary active Cl transport mechanism (Frizzell et al., 1979). Chloride enters secretory cells due to the combined activities of three basolateral membrane transport events: Na/K/Cl co-transporters, Na/K pumps, and K channels. Chloride leaves secretory cells across the apical membranes by diffusion, and alterations in apical Cl conductance represent a pivotal control point that determines Cl secretion rate. A variety of hormones and neurotransmitters stimulate salt secretion via their intracellular mediators, cAMP and Ca. The overall Cl secretory process is electrogenic so that the co-ion, Na, accompanies Cl to the lumen via paracellular pathways, driven by the lumen-negative voltage arising from Cl secretion.

A variety of epithelial cell lines and primary cultures have been shown to possess this mechanism for cAMP- and Ca-mediated control of salt secretion. Single-channel patch-clamp studies performed using many of these systems have identified an agonist-activated Cl channel that appears to account for the increase in apical Cl conductance elicited by secretagogues. Activation of Cl channel activity by cAMP-mediated agonists is defective in cells derived from cystic fibrosis patients, but channel activity can be elicited by increased cell Ca in sweat gland secretory coil and airway cells (Frizzell, 1987). In this respect, the attention of many investigators has been focused on the properties of an outwardly-rectified Cl-selective channel which is easily detected because of its biophysical properties:

The Identification of the CF (Cystic Fibrosis) Gene
Edited by L.-C. Tsui *et al.*, Plenum Press, New York, 1991

current rectification and rapid open-closed kinetic activity. This channel, which we refer to as the outward rectifier, has been observed in a variety of cells involved in secondary active Cl secretion. Perhaps the most compelling evidence for the involvement of this channel in Cl secretion is derived from studies conducted in excised inside-out membrane patches from airway cells challenged with purified protein kinases (Schoumacher et al., 1987; Li et al., 1988). Under appropriate conditions, the activity of this channel can be elicited by protein kinases A and C, and this activation occurs in membranes excised from normal but not CF cells (Li et al., 1989; Hwang et al., 1989).

Whole-Cell Recordings from Secretory Epithelial Cells

In an effort to assess the properties of the macroscopic epithelial cell Cl conductance stimulated by secretory agonists, we have employed the tight seal whole-cell patch-clamp technique. Most of our experiments have been performed on the human colonic tumor cell line, T84, these studies have recently been extended to a variety of other cell types, both cell lines and primary cultures. Whole-cell current recording offers the opportunity to define the macroscopic Cl conductances of secretory cells in isolation from other ionic conductances. In this method the cell retains a more intact intracellular milieu than that offered by excised, inside-out membrane patches.

Since increases in Cl and K conductance are the major electrical events detected during secretory stimulation, we used bath-to-pipette ion gradients that fix the equilibrium potentials for Cl and K at different values. In this manner, the K and Cl currents evoked by secretagogues can be monitored independently at clamp voltages equal to the reversal potentials for these ions. To simulate physiological conditions, the pipette contained high KCl and the bath high NaCl. Under these conditions, the Cl equilibrium potential was 0 mV, while the K equilibrium potential was about -80 mV.

The Swelling-Induced Chloride Conductance: G^{swell}_{Cl}

Under conventional whole-cell recording conditions, with solutions of equal osmolarity in pipette and bath, T84 cells spontaneously swell after the membrane patch is ruptured and whole-cell recording begins (Worrell et al., 1989). Visible cell swelling takes place over several minutes and is accompanied by an increase in membrane Cl conductance, G_{Cl}. This increased conductance and cell swelling are reversible upon raising bath osmolarity and can be offset entirely by an appropriate bath-pipette osmotic pressure difference (30-50 mOsm/l depending on conditions). The swelling-induced current exhibits a characteristic time-dependence during depolarizing and hyperpolarizing voltage pulses, as shown in Fig. 1. Depolarizing voltages greater than +40 mV cause inactivation of currents during the voltage pulse, whereas hyperpolarizing voltage pulses greater than -40 mV cause activation of currents during the voltage pulse.

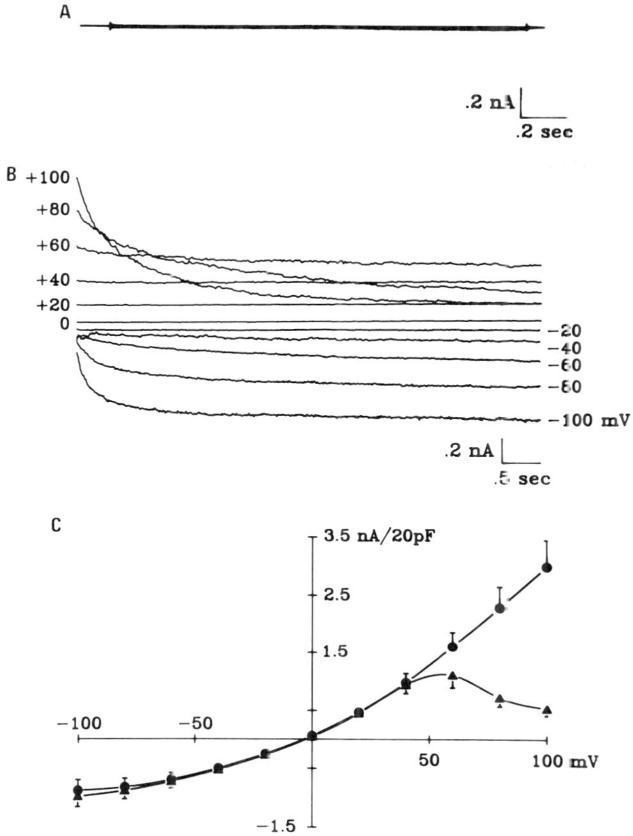

Figure 1. Swelling-induced Cl currents. Currents recorded before (A) and during (B) steady-state cell swelling at membrane voltages ranging between ±100 mV. Bath and pipette solutions contained NMDG-Cl as the principal salt. (C) Instantaneous (•) and steady-state (▲) current-voltage relation of the swelling-induced Cl conductance. The instantaneous currents were sampled 6 msec after the initiation of the voltage pulse and the steady-state currents were recorded after the time-dependent changes in current (B) were complete. In this experiment a holding potential of -100 mV was used to fully activate these currents.

The instantaneous currents (recorded 6 msec after the initiation of a voltage pulse) are outwardly-rectified, but the steady-state currents are markedly influenced by the voltage- and time-dependent properties of the swelling-induced conductance pathway. As shown in Fig. 1C, after activation of G^{swell}_{Cl} using a holding potential of -100 mV these currents remain maximally activated in the physiological voltage range; however, depolarizations to positive voltages progressively inactivate the steady-state currents. These swelling-induced currents are carried by

Cl because they can be evoked in solutions containing Cl as the major permeant ion. In addition, changes in extracellular Cl concentration evoke shifts in the I-V relation that are anticipated from the imposed Cl gradient.

The single-channel events responsible for G^{swell}_{Cl} have been recorded in two ways. First, under non-swelling conditions, i.e. with an appropriate bath-pipette osmotic pressure gradient, single-channel events corresponding to the kinetic behavior shown in Fig. 1 can be recorded (Worrell et al., 1989); that is, single channels inactivate in a step-wise manner during depolarizing voltage pulses. When a large number of such records are summed, the ensemble currents show the same inactivation time constants as the whole-cell currents recorded during cell swelling. Similar single channel events can be observed during cell-attached recording from swollen cells (Shoemaker et al.,1986). Canine tracheal cells swollen in KCl show a channel that inactivates during depolarizing voltage pulses.

The G^{swell}_{Cl} is most likely involved in the response of secretory cells to hypo-osmotic challenge. Since these cells maintain a basal K conductance, the increase in Cl conductance evoked during swelling would lead to cellular KCl loss and a regulatory volume decrease to return swollen cells toward their original volumes. In a number of cell types, parallel Cl and K conductance pathways are responsible for the RVD (Hoffmann et al., 1986). Interestingly, although an increase in cell Ca is thought to trigger RVD in some cell types, simultaneous recordings of whole-cell currents and fura-2 fluorescence showed no detectable change in cell Ca during the swelling response in colonic tumor cells.

The Calcium-Stimulated Chloride Conductance: G^{Ca}_{Cl}

The whole-cell currents in unstimulated secretory cells are low and remain stable for long periods under conditions that prevent cell swelling. The Ca ionophores, A23187 and ionomycin, markedly increase whole-cell currents in T84 and other Cl secreting epithelial cells. Typical Ca-activated Cl currents are shown in Fig. 2 (see also Cliff and Frizzell, 1990). Calcium ionophores cause increases in both Cl and K currents, but the Cl current response can be isolated using N-methyl glucamine (NMDG)-Cl pipette and bath solutions. Chloride currents are stablely elevated in the presence of ionophore and 1 mM bath Ca and can be reversed when the bath is exchanged for a low Ca or ionophore-free solution. The cells can be stimulated several times with no loss of sensitivity to Ca. The instantaneous I-V relation of G^{Ca}_{Cl} is outwardly-rectified over the voltage range ±100 mV (Fig. 2B). Reduction in bath Cl by replacement with glutamate indicates that the Ca-activated currents are Cl-selective. Ionophore-stimulated Cl currents also show characteristic voltage- and time-dependent kinetic behavior. There is time-dependent activation of currents at depolarizing clamp potentials greater

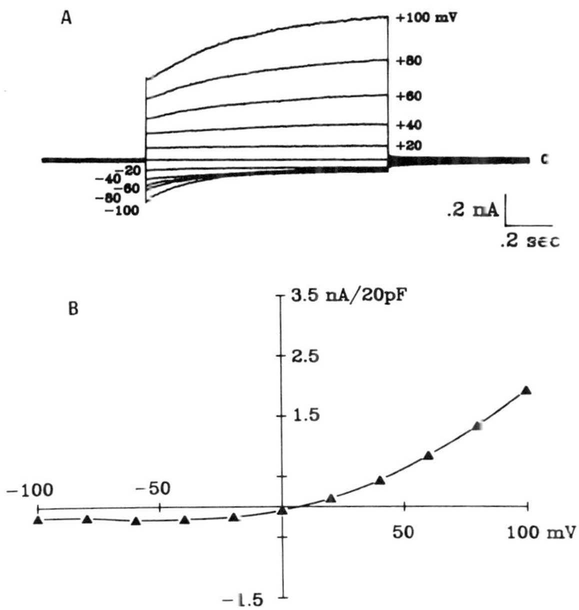

Figure 2. Calcium-stimulated whole-cell Cl currents. Pipette and bath solutions contained NMDG-Cl. (A) Overlay of cell currents recorded during pulses to ±100 mV after steady-state stimulation by 5 μM ionomycin in the presence of 1 mM bath Ca. (B) Instantaneous I-V relation of the currents shown in A.

than +40 mV and time-dependent inactivation of Cl currents at hyperpolarizing voltages greater than -40 mV. These time-dependent components of the Ca-activated Cl currents yield an even a more pronounced outward rectification of the I-V relation when the currents are sampled at later times during the voltage pulses. Thus, both G^{swell}_{Cl} and G^{Ca}_{Cl} are outwardly-rectified, but they show kinetically opposite voltage- and time-dependent properties.

Similar currents were evoked when pipette Ca was increased from 100 nM to 500 nM using a Ca-EGTA buffer. Under these conditions, currents developed progressively over several minutes after obtaining whole-cell recording conditions with the higher pipette Ca concentration. These currents show the same outward rectification and kinetic behavior as those stimulated by ionophores, indicating that the ionophore effects are due to increases in intracellular Ca. The receptor-mediated agonists, neurotensin and carbachol, elicited parallel increases in whole-cell Cl currents and cell Ca when these parameters were measured simultaneously in colonic tumor cells.

The cAMP-Stimulated Conductance: G^{cAMP}_{Cl}

Exposure of T84 and other secretory cells to forskolin, cAMP analogues or cAMP-mediated secretory agonists elicits sustained increases in whole-cell current that are reversible upon removing agonist from the bath (Cliff and Frizzell, 1990). Forskolin activates Cl conductance with little or no effect on K or Na conductances. The stimulated currents (Fig. 3) can be maintained for at least 30 min, indicating little run down of forskolin's effect during whole-cell recording. In addition, repetitive responses can be evoked by washing agonists into and from the bath. The properties of the Cl currents evoked by cAMP analogues, forskolin or receptor-mediated agonists are indistinguishable, indicating that increases in intracellular cAMP mediate the actions of forskolin on whole-cell Cl currents. The forskolin-stimulated I-V relation (Fig. 3B) is essentially linear over the voltage range ±100 mV. The current reversal potential is near 0 as expected for a Cl current and shifts in accordance with the Cl equilibrium potential as bath Cl concentration is changed. In addition, replacing Na in the bath by NMDG has no significant effect on the I-V relation. These results indicate that forskolin stimulates a Cl-selective current with properties different from the currents elicited by cell swelling or increased cell Ca.

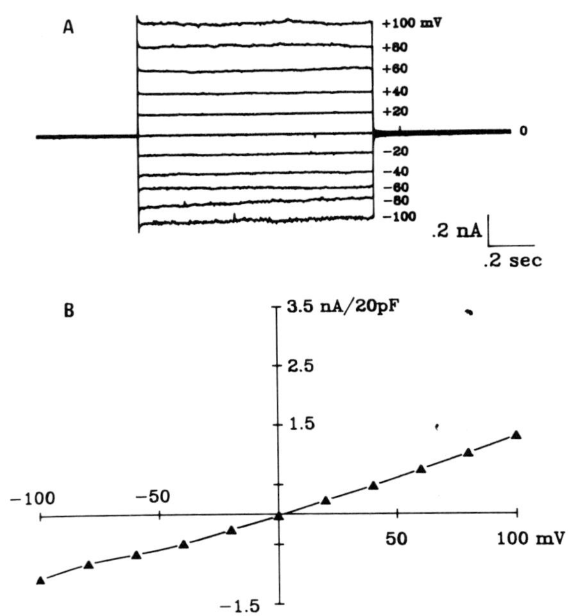

Figure 3. cAMP-activated whole-cell Cl currents. (A) Overlay of cell currents recorded during voltage pulses to ±100 mV after steady-state stimulation by 5 μM forskolin. (B) Instantaneous I-V relation for the currents shown in A.

Additivity of G^{Ca}_{Cl} and G^{cAMP}_{Cl}

Several experiments were performed to determine whether the increases in Cl conductance elicited by Ca and cAMP were additive. In response to forskolin, whole-cell Cl currents increased, with no changes in cation currents, as discussed previously. Subsequent addition of ionomycin evoked further increases Cl current and also elicited K currents, as when ionophore was added in the absence of a cAMP agonist. On average, the Cl conductance increase stimulated by ionomycin after forskolin was about 2-5 times the forskolin-stimulated current level. During the forskolin response, the Cl current showed no time-dependent properties, but following the addition of ionomycin, the currents took on the time-dependent inactivation at hyperpolarizing voltages characteristic of the G^{Ca}_{Cl}. Thus, the current components stimulated by cAMP and Ca appear to be additive.

Conclusions

According to the convention wisdom, Cl secretion across the apical membranes of secretory cells is mediated by a single Cl conductance pathway, on which the regulatory effects of cAMP and Ca converge. However, the results of these studies indicate that secretory cells possess three Cl conductances that can be distinguished by their different biophysical properties. The Cl conductance activated during cell swelling and that activated by increased cell Ca are outwardly rectified, but their time-dependent kinetics during voltage pulses differ. That is, G^{swell}_{Cl} inactivates during a depolarizing voltage pulse whereas G^{Ca}_{Cl} activates during a depolarizing voltage pulse. The conductance pathway elicited by cAMP, G^{cAMP}_{Cl}, shows even more distinct differences from the other Cl conductances. Its I-V relation is linear over the range ±100 mV and its currents are not time-dependent.

Thus, three distinct conductances are present in secretory epithelial cells, each with different conductive and kinetic properties and each of which responds to a different stimulus. Calcium- and cAMP-stimulated Cl currents with these properties have been observed in both epithelial and non-epithelial cells (Penner et al., 1988; Bahinski et al., 1989; Harvey and Hume, 1989). In particular, the Ca-activated Cl conductance of exocrine cells is similar to that observed here, although the single-channel events associated with this conductance have not been identified (Evans and Marty, 1986). Whole-cell Cl currents have also been recorded previously from secretory epithelial cells, but previous investigators failed to separate the volume- and agonist-induced Cl conductance components (Schoppa et al., 1989; McCann et al., 1989).

The relation of these conductances to single-channel events in secretory cells is somewhat uncertain. The single-channel basis of G^{swell}_{Cl} has been identified as an outwardly-rectifying Cl channel that inactivates with membrane depolarization in both whole-cell and single-channel records. In general, the conductance properties of the single-channel events that underlie the agonist-evoked whole-cell currents should be reflected most accurately by the properties of the instantaneous I-V relation of the whole-cell current, since time-dependent changes in channel kinetics at different voltages are minimal at the initiation of the voltage pulse. In this respect, the properties of the G^{Ca}_{Cl} are most consistent with those of the outward rectifier, identified in single-channel measurements. Both the single-channel and whole-cell currents show outward rectification. In addition, the open probability of the outward rectifier increases at depolarizing membrane voltages (Halm et al., 1987), and this Po-voltage relation is consistent with the time-dependence of G^{Ca}_{Cl}. While the activity of the outward rectifier in inside-out membrane patches is not reversibly altered by changes in bath Ca concentrations, the relevant Ca-dependent regulatory proteins may be lost after patch excision.

The whole-cell Cl currents activated by cAMP are difficult to reconcile with the properties of the outward rectifier, inasmuch as the instantaneous and steady-state I-V relations of G^{cAMP}_{Cl} are linear and the currents are time-independent. Two explanations for the discrepancies between the single-channel and whole-cell current measurements have occurred to us. First, the stimulation of apical membrane Cl conductance by cAMP could result from activation of a channel other than an outward rectifier. For example, the activity of a 5 pS Cl channel with a linear I-V relation and no voltage-dependence was stimulated by secretin in pancreatic duct cells (Gray et al., 1989). In addition, single Cl channels compatible with the properties of G^{cAMP}_{Cl} have been observed in human tracheal cells (Frizzell et al., 1986) and in _Necturus_ enterocytes (Giraldez et al., 1989). Second, if the outward rectifier is activated by cAMP, its behavior may be different in the whole-cell and single-channel measurements. There are certainly indications that the activity of this channel varies with measurement conditions. For example, depolarizing membrane voltages activate the outward rectifier (from a previously inactive state) in excised inside-out membrane patches; however, this does not occur under whole-cell or in cell-attached recording conditions. In addition, raising the bath temperature from 20 to 37°C can activate the outward rectifier in excised membrane patches, but this does not occur under whole-cell conditions. Thus, the outward rectifier may exist in different conductive and kinetic states, and perhaps the state routinely observed in the excised membrane patch corresponds to the properties of the whole-cell current that is elicited by increased cell Ca. Activation of the channel by cAMP in the intact cell may induce another state which is not routinely observed in single-channel studies. Regardless of the explanation of these discrepancies, it is now important to determine whether the activation of the outward rectifier in excised membrane patches assesses the physiological regulatory properties of secretory cell apical membranes or whether the results obtained from _in_ *vitro* phosphorylation experiments (e.g. with purified kinase) are unique to the excised membrane patch. Clearly, the answer to this question will be informative with respect to both the physiology of Cl secretion and the mechanistic basis of the CF defect.

Acknowledgements

[Supported by grants from the NIH, DK31091 and DK38518 and from the Cystic Fibrosis Foundation].

References

Bahinski, A., A.C. Nairn, P. Greengard, and D.C. Gadsby, 1989, Chloride conductance regulated by cyclic AMP-dependent protein kinase in cardiac myocytes, **Nature** 340:718.

Cliff, W.H., and R.A. Frizzell, 1990, Separate Cl conduct ances activated by cAMP and Ca in Cl-secreting epithelial cells, **Proc. Natl. Acad. Sci. USA**, in press.

Evans, M.G., and A. Marty, 1986, Calcium-dependent chloride currents in isolated cells from rat lacrimal glands, **J. Physiol.** 378:437.

Frizzell, R.A., 1987, Cystic fibrosis: A disease of ion channels? **Trends in Neurosci.** 10:190.

Frizzell, R.A., M. Field, and S.G. Schultz, 1979, Sodium-coupled chloride transport by epithelial tissues, **Am. J. Physiol.** 236:F1.

Frizzell, R.A., G.R. Rechkemmer, and R.L. Shoemaker, 1986, Altered regulation of airway epithelial cell chloride channels in cystic fibrosis, **Science** 233:558.

Giraldez, F., K.J. Murray, F.V. Sepulveda, and D.N. Sheppard, 1989, Characterization of a phosphorylation-activated Cl^--selective channel in isolated Necturus enterocytes, **J. Physiol.** 416:517.

Gray, M.A., A. Harris, L. Coleman, J.R. Greenwell, and B.E. Argent, 1989, Two types of chloride channel on duct cells cultured from human fetal pancreas, **Am. J. Physiol.** 257:C240.

Halm, D.R., G.R. Rechkemmer, R.A. Schoumacher, and R.A. Frizzell, 1988, Apical membrane chloride channels in a colonic cell line activated by secretory agonists, **Am. J. Physiol.** 254:C505.

Harvey, R.D., and J.R. Hume, 1989, Autonomic regulation of a chloride current in heart, **Science** 244:983.

Hoffmann, E.K., I.H. Lambert, and L.O. Simonsen, 1986, Separate Ca^{2+} activated K and Cl transport pathways in Ehrlich ascites tumor cells, **J. Membr. Biol.** 91:227.

Hwang, T.-C, L. Lu, P.L. Zeitlin, D.C. Gruenert, R. Huganir, and W.B. Guggino, 1989, Cl channels in CF: Lack of activation by protein kinase C and cAMP-dependent protein kinase, **Science** 244:1351.

Li, M., J.D. McCann, M.P. Anderson, J.P. Clancy, C.M. Liedtke, A.C. Nairn, P. Greengard, and M.J. Welsh, 1989, Regulation of chloride channels by protein kinase C in normal and cystic fibrosis airway epithelia, **Science** 244:1353.

Li, M., J.D. McCann, C.M. Liedtke, A.C. Nairn, P. Greengard, and M.J. Welsh, 1988, Cyclic AMP-dependent protein kinase opens chloride channels in normal but not cystic fibrosis airway epithelium, **Nature** 331:358.

McCann, J.D., M. Li, and M.J. Welsh, 1989. Identification and regulation of whole-cell chloride currents in airway epithelium, **J. Gen. Physiol.** 94:1015.

Penner, R., G. Matthews, and E. Neher, 1988, Regulation of calcium influx by second messengers in rat mast cells, **Nature** 334:499.

Schoppa, N., S.R. Shorofsky, F. Jow, and D.J. Nelson, 1989, Voltage-gated chloride currents in cultured canine tracheal epithelial cells, **J. Membr. Biol.** 108:73.

Schoumacher, R.A., R.L. Shoemaker, D.R. Halm, E.A. Tallant, R.W. Wallace, and R.A. Frizzel, 1987, Phosphorylation fails to activate chloride channels from cystic fibrosis airway cells, **Nature** 330:752.

Shoemaker, R.L., R.A. Frizzell, T.M. Dwyer, and J.M. Farley, 1986. Single-channel currents from canine tracheal epithelial cells, **Biochem. Biophys. Acta** 858:235.

Worrell, R.T., A.G. Butt, W.J. Cliff, and R.A. Frizzell, 1989, A volume-sensitive chloride conductance in the human colonic cell line, T84, **Am. J. Physiol.** 256:C1111.

DISCUSSION

GREGER: May I ask you one question regarding your new concept of different Cl^- conductances? If we take your classification we have the provocative situation that all of the PK-A experiments in excised patches might be misinterpreted. In these experiments, PK-A activation led to the appearance of outwardly rectifying Cl^- channels of intermediate conductance, yet you are suggesting from your whole cell measurements that the cAMP-dependent current is through a non rectifying channel. Do you agree with this consideration?

FRIZZELL: It's not really possible for me to do much more than repeat what I said at the end of the talk, in this respect. I don't really have additional insights into this possibility. In my mind, the biggest discrepancy that occurs results when one performs the kinase experiment. You excise a membrane patch, you routinely remain at inside negative voltages to avoid the activation by depolarizing voltages, add a kinase and ATP to the inside of the membrane and see a channel activate. Routinely, our diagnostic principle for determining whether the activation of the channel has occurred, is that we change the membrane voltage to a positive value to assess outward rectification. I think the question that we need to answer at this point in time is whether the outward rectifier can exist in different conductive and kinetic states, and whether in the excised membrane patch, the particular state that the outward rectifier happens to be in is different from that which occurs in the intact cell, for whatever reason.

GREGER: May I comment on this a little further. Seven years ago, when working with the rectal gland Cl^- channel, we noted that this conductance, unlike that of other Cl^- conductances of HT_{29}-cells or respiratory epithelial cell, was not rectifying. Furthermore we were able to show for this channel in a few experiments that it was PK-A activated. Both findings fit well to the conclusions drawn from your present study, namely that the PK-A activated channel is not rectifying.

AUSIELLO: Maybe I can add a comment. Ray, you're perhaps familiar with this data, but maybe it would help the interpretation. I think Doug Eaton is about to publish some data on an outwardly rectifying chloride channel using single channel recordings in the A6 cell, which can be converted from an outwardly rectifying channel to a linear channel in the presence of either insulin or dephosphorylation. So I think that supports very nicely what you've just said, that the possibility that we are perhaps statically looking at a single component of a channel which might be much be much more dynamic, either in the whole cell or, perhaps, under different patch conditions. And that's an example of some change in such conductance.

FRIZZELL: I think some really beautiful data in this respect is from Pancho Sepulveda's lab, where they have recently shown that the cell-attached appearance of single-channel events results in the activation of a single channel with a linear conductance, but in the excised mode, one sees predominantly an outwardly rectified conductance.

S.GUGGINO: I have just one comment, that I would like here to remember that you're measuring whole cell currents. This measures the channels on the apical membrane, the basolateral membrane and the lateral membranes. Lindeman has shown that there is a linear chloride channel on the basolateral membrane of colonic cells. So you cannot say, at this point, exactly what is going on in these cells without, again, looking by single channel recordings on the basolateral membrane.

FRIZZELL: I would only say that what we have tried to do is look at the specific regulator-mediated conductance pathways that, we believe, in cells of this type, are responsible for chloride secretion. This particular pattern which I've shown you, that shows up in the whole cell conductances is not peculiar to T84 cells, but is seen also in HT29 cells, and in human nasal, tracheal and sweat gland in primary culture. I agree with what you're saying, Sandy, that we obviously, I think, in some sense, have now to go even further back, to the Ussing chamber perhaps, or with microelectrode techniques, to look for the presence of conductances with specific properties of the type that have emerged from these studies, and try to localize them in apical or basolateral membranes.

WINE: Ray, is it possible to try to dissect apart the two possibilities that you raised at the end by looking for additivity effects in your whole-cell currents? I thought that you showed that they were additive, suggesting that they would be two separate conductances present in the cell. Is that over-interpreting the data?

FRIZZELL: I think it is, because all you can really say is that there is a further 2 to 7 fold increase in whole-cell current that you can measure after you add ionophore after cAMP. We generally have not tried the reverse sequence, because at least in T84 cells, many, many more of the cells, I think because of their differentiated state, are more sensitive to calcium than they are to cAMP. So we generally, in the addition experiments, have tried to make sure that the cell is first responsive to forskolin, then we look at the response to calcium. Now, since the forskolin effect is reversible, we could certainly go back and add the agonists in the other direction, and we need to do that.

GARDNER: It seems to me there are a lot of pitfalls, first in looking at whole-cell IV curves and trying to make any statements about single channel IV curves, because your current-voltage relationship is as much a part of both voltage- and time-dependent kinetics and steady-state kinetics (your probability of opening at various voltages), as it is your single channel conductance.

FRIZZELL: I agree.

GARDNER: And so I think to do that is dangerous.

FRIZZELL: That is why, first of all, we have focused entirely on the instantaneous currents recorded 6 milliseconds after a voltage pulse, in order to try as much as possible to minimize the time-dependent effects.

GARDNER: That minimizes time dependence but certainly not any kind of probability of opening. So I think it's very dangerous to go from one to the other. And the second thing, with additivity experiments, I also think that has a lot of pitfalls, because we have discussed doing these things to a great extent. First of all, what if your probability of opening under one pathway of activation is much greater than another? You're going to get additivity and I think it is overinterpretation to say that if you have additivity that it is a different channel, which I think you agree with.

FRIZZELL: Absolutely.

REGULATION OF EPITHELIAL CHLORIDE CHANNELS:

ROLES OF PROTEIN KINASES AND ARACHIDONIC ACID

Tzyh-Chang Hwang and William B. Guggino

Department of Physiology, School of Medicine
The Johns Hopkins University

INTRODUCTION

Epithelial Cl^- secretion plays a crucial role in controlling the content of airway fluid. The quantity and composition of airway fluid determine the effectiveness of mucociliary clearance, one of the natural lung defense mechanisms that protect the respiratory system from infection. Transcellular Cl^- secretion accompanied by Na^+ provides the osmotic force for water flow which hydrates the mucous coating on the surface of epithelia (see ref 1 for review).

A variety of neurohumoral and pharmacological agents regulate the rate of Cl^- secretion in tracheal epithelium. Those that increase intracellular cAMP concentration have been the most extensively investigated (2-8). Studies using membrane-permeable cAMP analogues provide further evidence supporting cAMP as an important second messenger in the regulation of epithelial Cl^- secretion (9). However, not all secretagogues increase intracellular cAMP concentration. For example, prostaglandin F2α (7,10,11) and phorbol ester (12,13) stimulate Cl^- secretion without inducing measurable changes in cAMP levels. In addition, Ca^{2+} ionophore A23187 also increases Cl^- secretion (13,14), suggesting that cytosolic Ca^{2+} may play a role in mediating Cl^- secretion. Therefore, second messengers other than cAMP are also important in regulating Cl^- secretion.

Cystic fibrosis (CF), the most common hereditary disease in the Caucasian population, is a multisystem disorder affecting children and young adults. An abnormally low Cl^- permeability was first demonstrated in sweat ducts isolated from CF patients (15). The same conclusion that the Cl^- permeability is low in CF is derived from studies in respiratory epithelial cells (16-18) and small intestinal mucosa (19). Since primary cultures (20-22) and established cell lines (23) of CF airway epithelia retain this abnormality, the defective electrolyte transport is localized in epithelia per se. Intracellular microelectrode measurements combined with ion substitutions further localize the CF Cl^- impermeability to the apical membrane (20, 24).

We used the patch clamp technique to understand the molecular mechanisms of normal Cl^- channel regulation and the decreased apical Cl^- permeability in CF.

APICAL Cl^- CHANNELS IN AIRWAY EPITHELIA

Several different types of Cl^- channels have been reported in airway epithelia (25-29). The single channel conductances in 150 mM NaCl solutions are 45, 20, 250 and 10 pS respectively. Among them, the outwardly rectifying Cl^- channel with single channel conductance of 45 pS at positive membrane potential is well characterized. This channel is thought to be responsible for the defect in CF, because β-adrenergic agonists can activate the channel in normal cells and the same channel is also found in other tissues affected by CF, such as colonic epithelia (30,31) and human sweat gland (32). In excised inside-out membrane patches, the channel can be activated by applying a sustained depolarizing voltage. This voltage activation is different from the classical voltage-gated channels found in excitable cells in that it requires a long period of time to occur and still remains even if the membrane potential is shifted to negative potentials. The outwardly rectifying Cl^- channel is the most abundant anion channel observed in the cells we used in our studies, including human fetal, adult, and CF airway epithelia. This channel is 7-10 times more selective for Cl^- over Na^+. Current is not carried by gluconate or sulfate, and substitution of Cs^+, Na^+ or K^+ has little effect on the conductive properties. In addition, experiments using whole-cell recording also demonstrate a cAMP-dependent Cl^- current with similar outward rectification in tracheal epithelia (33). In the remainder of this review, we will focus specifically on the properties of this outwardly rectifying Cl^- channel.

DIRECT ACTIVATION OF Cl^- CHANNELS BY PROTEIN KINASES

In recordings from cell-attached patches, β-adrenergic agonists or membrane-permeable cAMP analogues evoke Cl^- channel activity in membrane patches of normal airway cells (25,34). Experiments with excised inside-out membrane patches provide more insight into the actual regulatory mechanism of Cl^- channels. When membrane patches are excised into the physiological salt solution, there is no channel activity if the membrane is maintained at a physiological membrane potential of -50 mV. Application of the catalytic subunits of protein kinase A in the presence of ATP can open Cl^- channels (Fig. 1) in normal airway epithelial cells (35-37). Further experiments using a specific

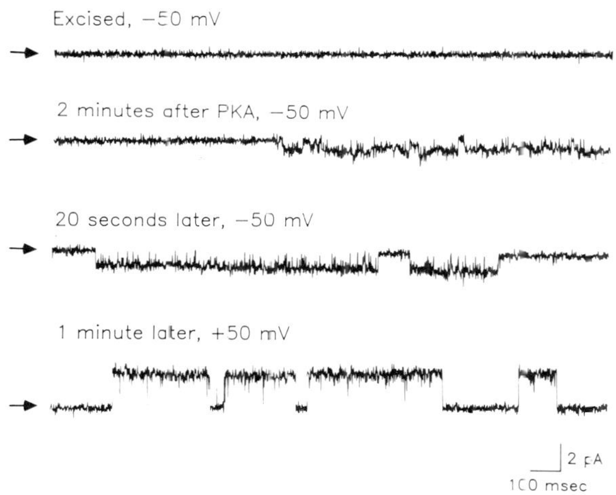

Fig. 1. Activation of epithelial Cl^- channels by PKA. There is no channel activity when the membrane patch is excised and maintained at negative membrane potential. Two minutes after addition of the catalytic subunit of PKA and ATP, Cl^- channel open. Shifting the voltage to the positive side confirms the outward rectification. Arrows indicate the closed state. Final concentrations of enzyme and chemical are 1 mM ATP and 50 nM PKA.

PKA inhibitor (37), the Walsh inhibitory peptide, confirm that phosphorylation by PKA is involved in activating outwardly rectifying Cl^- channels. The Walsh peptide has been shown to inhibit specifically substrate phosphorylation by PKA (38). A similar activation (Fig. 2) is also observed when protein kinase C (PKC) with the diacylglycerol analogue, dioctanoylglycerol (diC_8) and ATP are used (37,39). This activation can be prevented by a PKC inhibitor, staurosporine (40). Since the requirement of diacylglycerol for PKC activity can be overcome by high Ca^{2+} (41,42), PKC alone in the presence of 50μM Ca^{2+} was tested. Cl^- channels can be activated in this condition (43). In contrast, PKC in the presence of diolein (diacylglycerol analogue with long-chain fatty acid moieties), and high Ca^{2+} can inhibit Cl^- channel activity in both normal and CF cells (39). This inhibition appears to be reversible because channel activity returns when PKC is washed out. Taken together, the data suggests that PKC may be acting on two different phosphorylation sites, one activating, another inhibitory. This dual effect could depend upon the concentration on intracellular Ca^{2+}, or on the nature of the diacylglycerol analogue, or on different isozymes of protein kinase C (42) acting on separate phosphorylation sites. Since diacylglycerol has been shown to decrease calcium currents independently of protein kinase C activation (44), it is possible that type of diacylglycerol analogue may determine whether PKC has an activating or inhibitory effect.

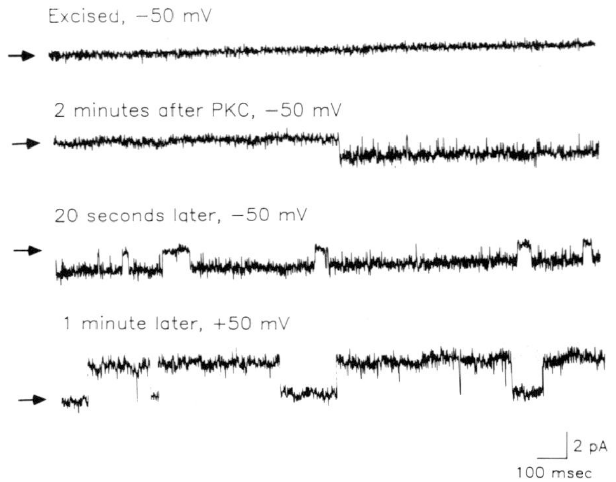

Fig. 2. Activation of epithelial Cl^- channels by PKC. The experimental protocol is the same as in Fig. 1. Final concentrations of enzyme and chemicals are 1 mM ATP, 2μM Walsh inhibitor, 1 μM DiC8 and 10 nM PKC.

DEFECTIVE REGULATION IN CF

As discussed above, β-adrenergic agonists fail to stimulate Cl^- secretion in CF epithelia. In cell-attached membrane patches from CF cells, Cl^- channels are not responsive to β-adrenergic agonists (25,34). Abnormal regulation of Cl^- channel does not appear to result from defective interactions between hormone, receptor, and second messenger pathway, because β-adrenergic agonists increase intracellular cAMP concentration and open basolateral K^+ channels in CF cells (34). CF Cl^- channels fail to open even when 8-bromo-cAMP or forskolin is used to bypass the receptor and adenylate cyclase. Therefore, the defective site (or sites) is beyond the second messenger cAMP. The failure to respond to β-adrenergic agents in CF cells does not result from the absence of Cl^- channels either, because in excised patches from CF cells sustained depolarization opens Cl^- channels and the conductive properties of CF channels are identical to those of normal cells. Experiments showing that purified PKC or the catalytic subunit of protein kinase A fails to activate Cl^- channels in excised inside-out membrane patches (Fig. 3) suggest that the channel itself or an associated regulatory protein is defective in CF. Cl^- channels in CF cells can be activated by manipulating intracellular Ca^{2+}. The Ca^{2+} ionophore A23187, when added to the bath solution of CF cells, induces Cl^- secretion (45,46) and channel opening (25). How Ca^{2+} controls Cl^- channel activity is unknown. Cl^- channel opening probability does not change with the Ca^{2+} concentration (47), suggesting the effect of Ca^{2+} is indirect. It will be important to investigate its mechanism in the hope of discovering an effective treatment for CF.

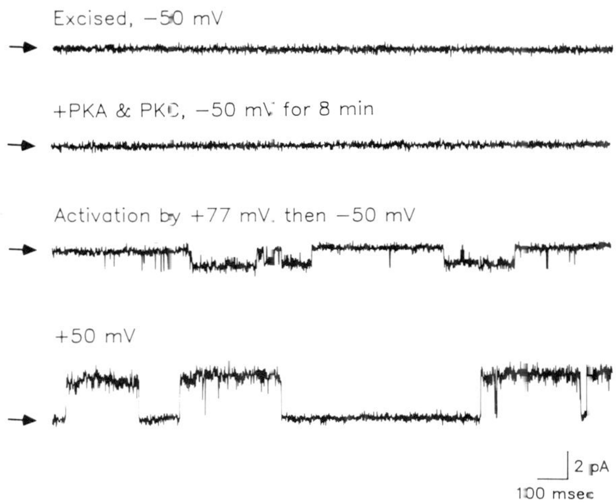

Fig. 3. Lack of activation by PKA and PKC in CF cells. The experimental conditions are the same as in Fig. 1 and 2 except Walsh inhibitor is not included.

INHIBITION OF Cl^- CHANNEL ACTIVITY BY ARACHIDONIC ACID

Arachidonic acid has been implicated as a secondary messenger in signal transduction by directly activating PKC (48), and as a retrograde messenger to carry information from the postsynaptic to presynaptic terminals (49,50). Arachidonic acid, as well as other fatty acids, can activate a variety of K+ channels in cardiac and smooth muscle cells (51,52). In contrast, in lymphocytes, linoleic acid inhibits Cl^- efflux induced by cell swelling (53). The structure of these fatty acids is similar to that of synthetic organic acids which have been well characterized as Cl^- channel blockers (54). They all have a carboxyl group and a hydrophobic domain. Therefore, fatty acids can act as an inhibitory factors in the regulation of epithelial Cl^- secretion by directly blocking the Cl^- channel. We tested this hypothesis by studying the effect of various fatty acids on single channel current.

Figure 4 shows the dose-dependent block of Cl^- channel by arachidonic acid (55). As the concentration of arachidonic acid is increased, the number of flickerings indicating the block increases and the channel open time decreases. However, the current interruptions induced by arachidonic acid are similar in duration to normal closing events. Kinetic analysis shows the reciprocal of the mean open time increases linearly with the concentration of blocker, whereas the reciprocal of the closed time does not change with the drug concentration (Fig. 5). These results are consistent with a bimolecular process of a channel alternating between closed, open and blocked states. Other cis-unsaturated fatty acids like linoleic, oleic, and ricinoleic

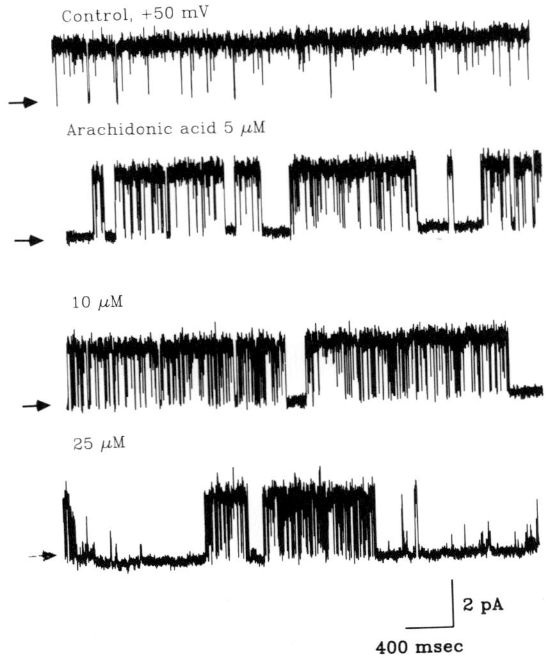

Fig. 4. Dose-dependent block of epithelial Cl^- channels by arachidonic acid. Arachidonic acid induces a flickering block. The number of flickers increases as the concentration of arachidonic acid is increased.

acid have the same potency of block (Fig. 5A). This inhibition of single channel current by arachidonic acid is also observed in the presence of inhibitors of both lipoxygenase and cyclooxygenase, indicating arachidonic acid and not its metabolites, is the effective agent. The possibility that cis-unsaturated fatty acids exert a nonspecific effect on the membrane is ruled out by experiments showing that the nonselective cation channel in the same epithelia is not affected by any of these fatty acids (43,55). Figure 5A also shows that trans-unsaturated or saturated fatty acids have no effect on channel kinetics. Arachidonyl alcohol, with the carboxyl group replaced by a hydroxyl group, does not affect channel kinetics. Thus, the structural requirement for fatty acids as Cl^- channel blockers are a carboxyl group and a kinked long carbon chain.

Further biophysical studies reveal the potency of arachidonic acid blockade depends on the membrane potential. The association rate between the channel protein and the blocker does not change with the voltage, whereas the dissociation rate is faster as the membrane is depolarized (43). Therefore, the inhibitory effect of arachidonic acid is expected to be stronger at negative membrane potentials. The inhibitory effect of cis-

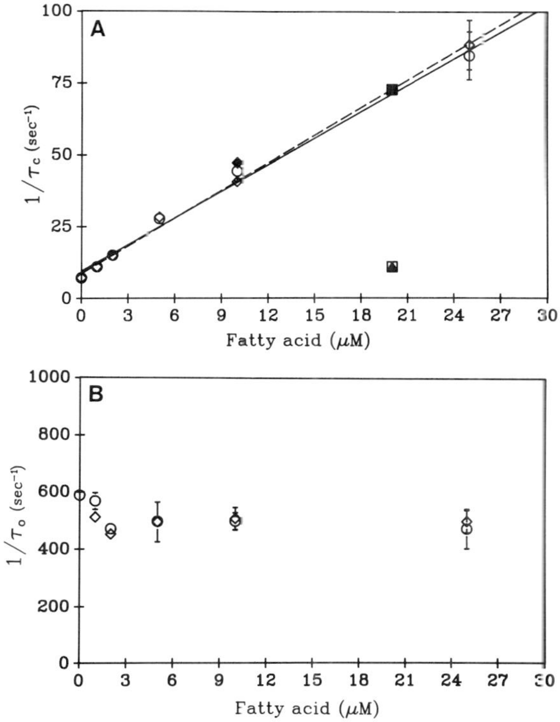

Fig. 5. A. Concentration dependence of the reciprocal of mean open time, and B. concentration independence of the reciprocal of mean closed time. Mean open time and mean closed time are determined by fitting open time and closed time histograms with single exponential. Each point represents mean ± SEM (n = 3-7). Fatty acids are: arachidonic acid (open diamond), linoleic acid (open circle), oleic acid (filled diamond), ricinoleic acid (filled square), stearic acid (open square) and elaidic acid (filled triangle).

unsaturated fatty acids on Cl^- secretion is demonstrated in short circuit current experiments (Fig. 6). After the Cl^- secretion is stimulated in T84 cells, a colonic carcinoma cell line that secretes Cl^- in response to secretagogues (56), the Cl^- channel blocker diphenylamine-2-carboxylate (DPC) and ricinoleic acid inhibit I_{sc} to a similar extent.

Although the physiological significance of arachidonic acid blockade is unknown, several speculations can be made. Since Cl^- channel activation by protein kinases appears to act in an "all or none" fashion, arachidonic acid, generated from the membrane phospholipids, may serve as a fine-tuning modulator for Cl^- secretion by inhibiting open channel current. Arachidonic acid and its metabolites may also be involved in the pathophysiology of

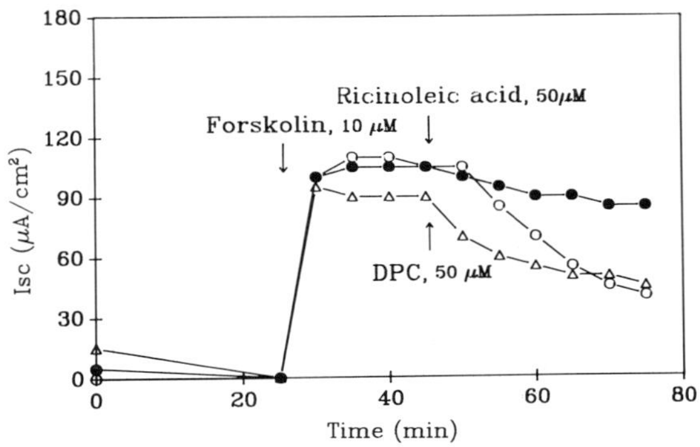

Fig. 6. Representative experiments of short circuit current inhibition by ricinoleic acid and DPC in T84 cells. I_{sc} is stimulated by applying forskolin in the mucosal bath at a final concentration of 10μM. DPC (open triangle) or ricinoleic acid (open circle) is added 25 min after forskolin. 30 min after addition of ricinoleic acid, I_{sc} decreases to 39 ± 2% (n = 5) of the initial value, whereas, in control experiments (filled circle), I_{sc} is maintained at 77 ± 6% (n = 6).

airway diseases. For example, leukotrienes C_4, D_4, and E_4 are thought to be the major mediators of allergic bronchoconstriction (57-59), and increased release of arachidonic acid is observed in lymphocytes of CF patients (60). It has been shown that these chemical mediators can be generated by a variety of cells in the respiratory tracts (61-63). Although this study does not demonstrate that arachidonic acid is "the" responsible chemical mediator, the data suggest that arachidonic acid or other fatty acid derivatives may serve as Cl^- channel blockers and this blockage may contribute to the abnormal mucociliary transport seen in several disease states. Cl^- channels exist not only in epithelia, but in excitable cells (64). Cl^- channel blockers used to study epithelial Cl^- channels can block Cl^- channels in other tissues (65). One recent report demonstrates similar regulation of cardiac Cl^- channels by PKA and block by 4,4'-dinitrostilbene-2,2'-disulphonic acid (66). If Cl^- channels in neuronal cells can be blocked by arachidonic acid, the resulting increase of neuronal excitability would account for the effects of arachidonic acid on transmitters release in presynaptic terminals (49).

OVERALL SCHEME OF Cl^- CHANNEL REGULATION

Phosphorylation sites are present on several different types of ion channels and are important in channel regulation (67). cAMP-dependent phosphorylation has been shown to regulate cardiac Ca^{2+} channel (68,69) and several different types of K^+

channels in neuronal and smooth muscle cells (70-72). Biochemical studies also display phosphorylation of purified dihydropyridine receptor (73) and acetylcholine nicotinic receptor (74, 75) by PKA. Protein kinase C, as well, modulates a number of K^+ channels (70,76) and Ca^{2+} channels (77-79). In cardiac cells both delayed rectified K^+ channels (76) and L-type Ca^{2+} channels (68,69,78) are activated by PKC and PKA. Both kinases accelerate the rate of desensitization of the nicotinic receptor (80). The epithelial, outwardly rectifying Cl^- channel is another example that protein phosphorylation plays a regulatory role in channel activity. Phosphorylation produces its effects by several different molecular mechanisms. The membrane current I, carried by a given population of ion channels can be described by the equation

$$I = N * p * i$$

where N is the number of functional channels, p is the channel open probability, and i is the single channel current amplitude. Theoretically, it is possible to regulate I by modulating either N, p, or i. For example, it has been shown that β-adrenergic agonists modulate the activity of L-type Ca^{2+} channels in cardiac cells by increasing their probability of opening (69). Whereas, intracellular pressure injection of PKC increases Ca^{2+} currents by recruiting a novel type of Ca^{2+} channel in Aplysia bag cell neurons (79). In the present study, before kinases are added, the membrane is quiet without Cl^- channel activity. Phosphorylation by added kinases causes irreversible channel

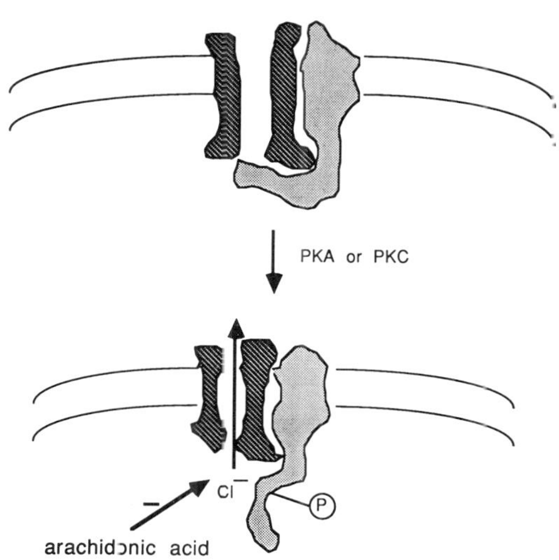

Fig. 7. Schematic picture of epithelial Cl^- channel regulation.

activation, suggesting that the mechanism of activation is an increase of the number of functional channels. Figure 7 is the schematic representation of Cl^- channel regulation. Phosphorylation by either PKA or PKC activates the channel from a inactive state. Once the channel is active, arachidonic as well as other cis-unsaturated fatty acids decreases the time channels stay in open state by directly blocking the channel, thereby reducing the total current passing through the channel.

Acknowledgement

We wish to thank Ms. E. Rohland for technique assistance with cell cultures, Mr. C. Layton for some of the short circuit current experiments, Dr. M. Egan for growing T84 cells, Dr. R. Huganir for generously providing kinases. We also express our gratitude to Drs. G. Yellen and D. Raben for their helpful discussion. This work is supported by National Heart Lung Institute Grant No. HL40178 and Cystic Fibrosis Foundation Grant No. CFR020.

References

1. Welsh, M.J., 1987, Electrolyte transport by airway epithelia, **Physiol. Rev.**, 67:1143.
2. Al-Bazzaz, F., Cheng, E., 1979, Effect of catecholamines on ion transport in dog tracheal epithelium, **J. Appl. Physiol.**, 47:397.
3. Boucher, R.C., Gatzy, J.T., 1982, Regional effects of autonomic agents on ion transport across excised canine airways, **J. Appl. Physiol.**, 52:893.
4. Davis, B., Marin, M.G., Yee, J.W., Nadel, J.A.,1979, Effect of terbutaline on movement of Cl^- and Na^+ across the trachea of the dog in vitro, **Am. Rev. Respir. Dis.**, 120:547.
5. Lazarus, S.C., Basbaum, C.B., Gold, W.M., 1984, Localization of cAMP in dog and cat trachea: effects of β-adrenergic agonists, **Am. J. Physiol.**, 247:C327.
6. Liedtke, C.M., Tandler, B., 1984, Physiological responsiveness of isolated rabbit tracheal epithelial cells, **Am. J. Physiol.**, 247:C441.
7. Smith, P.L. Welsh, M.J., Stoff, J.W., Frizzell, R.A., 1982, Chloride secretion by canine tracheal epithelium. I. Role of intracellular cAMP levels, **J. Membr. Biol.**, 70:217.
8. Welsh, M.J., 1986, Adrenergic regulation of ion transport by primary cultures of canine tracheal epithelium: cellular electrophysiology, **J. Membr. Biol.**, 91:121.
9. Al-Bazzaz, F.J., 1981, Role of cyclic AMP in regulation of chloride secretion by canine tracheal mucosa, **Am. Rev. Respir. Dis.**, 123:295.
10. Al-Bazzaz, F.J., Yadava, V.P., Westen-Felder, C., 1981, Modification of Na^+ and Cl^- transport in canine tracheal mucosa by prostaglandins, **Am. J. Physiol.**, 240:F101.
11. Liedtke, C.M., Boat, T.F., Rudolph, S.A., 1982, Neurohormo-

nal receptors and cyclic AMP-binding proteins in rabbit tracheal mucosa-submucosa, **Biochim. Biophys. Acta** 719:169.
12. Barthelson, R.A., Jacoby, D.B., Widdicombe, J.H., 1987, Regulation of chloride secretion in dog tracheal epithelium by protein kinase C, **Am. J. Physiol.**, 253:C302. 13. Welsh, M.J., 1987, Effect of phorbol ester and calcium ionophore on chloride secretion in canine tracheal epithelium, **Am. J. Physiol.**, 253:C828.
14. Eling, T.E., Danilowicz, R.M., Henke, D.C., 1986, Arachidonic acid metabolism by canine tracheal epithelial cells. Product formation and relationship to chloride secretion, **J. Biol. Chem.**, 261:12841.
15. Quinton, P.M., 1983, Chloride impermeability in cystic fibrosis. Nature 301:421.
16. Knowles, M., Gatzy, J., Boucher, R., 1981, Increased bioelectric potential difference across respiratory epithelia in cystic fibrosis, **N. Engl. J. Med.**, 305:1489.
17. Knowles, M., Gatzy, J., Boucher, R., 1983, Relative ion permeability of normal and cystic fibrosis nasal epithelium, **J. Clin. Invest.**, 71:1410.
18. Knowles, M.R., Stutts, M.J., Spock, A., 1983, Abnormal ion permeation through cystic fibrosis respiratory epithelium, **Science**, 221:1067.
19. Berschneider, H.M., Knowles, M.R., Azizkhan, R.G., 1988, Altered intestinal chloride transport in cystic fibrosis, **FASEB J.**, 2:2625.
20. Widdicombe, J.H., Welsh, M.J., Finkbeiner, W.E., 1985, Cystic fibrosis decreases the apical membrane chloride permeability of monolayer cultured from cells of tracheal epithelium, **Proc. Natl. Acad. Sci.**, 82:6167.
21. Yankaskas, J.R., Cotton, C.U., Knowles, M.R., 1985, Culture of human nasal epithelial cells on collagen matrix support: a comparison of bioelectric properties of normal and cystic fibrosis epithelia, **Am. Rev. Respir. Dis.** 132:1281.
22. Yankaskas, J.R., Gatzy, J.T., Knowles, M.R., Boucher, R.C., 1985, Persistence of abnormal chloride ion permeability in cystic fibrosis nasal epithelial cells in heterologous culture, **Lancet**, 8435:954.
23. Jetten, A.M., Yankaskas, J.R., Stutts, M.J., 1989, Persistence of abnormal chloride conductance regulation in transformed cystic fibrosis epithelia, **Science** 244.1472.
24. Cotton, C.U., Stutts, M.J., Knowles, M.R., 1987, Abnormal apical cell membrane in cystic fibrosis respiratory epithelium. An in vitro electrophysiological analysis, **J. Clin. Invest.**, 79:30.
25. Frizzell, R.A., Rechkemmer, G., Shoemaker, R.L., 1986, Altered regulation of airway epithelial cell chloride channels in cystic fibrosis, **Science**, 233:558.
26. Shoemaker, R.L., Frizzell, R.A., Dwyer, T.M., Farley, J.M., 1986, Single chloride channel currents from canine tracheal epithelial cells, **Biochim. Biophys. Acta**, 858:235.
27. Welsh, M.J., 1986, An apical-membrane chloride channel in human tracheal epithelium, **Science**, 232:1648.

28. Welsh, M.J., 1986, Single apical membrane anion channels in primary cultures of canine tracheal epithelium, **Pfluegers Arch.**, 407:S116.
29. Duszyk, M., French, A.S., Paul Man, S.F., 1989, The 20-pS chloride channel of the human airway epithelium, **Biophys. J.**, 57:223.
30. Halm, D.R., Rechkemmer, G.R., Schoumacher, R.A., Frizzell, R.A., 1988, Apical membrane chloride channels in a colonic cell line activated by secretory agonists, **Am. J. Physiol.**, 254:CSOS.
31. Hayslett, J.P., Goegelein, H., Kunzelmarm, K., Greger, R., 1987: Characteristics of apical chloride channels in human colon cells (H19), **Pfluegers Arch.**, 410:487.
32. Krouse, M.E., Hagiwara, G., Chen, J., 1989, Ion channels in normal human and cystic fibrosis sweat gland cells, **Am. J. Physiol.**, 257: C129.
33. McCann, J.D., Li, M., Welsh, M.J., 1989, Identification and regulation of whole-cell chloride currents in airway epithelium, **J. Gen. Physiol.**, 94:1015.
34. Welsh, M.J., Liedtke, C.M., 1986, Chloride and potassium channels in cystic fibrosis airway epithelia, **Nature**, 322:467.
35. Li, M., McCann, J.D., Liedtke, C.M., 1988, Cyclic AMP-dependent protein kinase opens chloride channels in normal but not cystic fibrosis airway epithelium, **Nature**, 331:358.
36. Schoumacher, R.A., Shoemaker, R.L. Halm, D.R., 1987, Phosphorylation fails to activate chloride channels from cystic fibrosis airway cells, **Nature**, 330:752.
37. Hwang, T.C., Lu, L., Zeitlin, P., 1989, Cl^- channels in CF: lack of activation by protein kinase C and cAMP-dependent protein kinase, **Science**, 244:1352.
38. Walsh, P.A., Ashby, C.D., Gonzalez, C., 1985, Purification and characterization of a protein kinase inhibitor of adenosine 3',5'-monophosphate-dependent protein kinase, **J. Biol. Chem.**, 246:1977.
39. Li, M., McCann, J.D., Anderson, M.P., 1989, Regulation of chloride channels by protein kinase C in normal and cystic fibrosis airway epithelia, **Science**, 244:1353.
40. Tamaoki, T., Nomoto, H., Takahashi, I., 1986, Staurosporine, a potent inhibitor of phospholipid/Ca^{++} dependent protein kinase, **Biochem. Biophys. Res. Commun.**, 135:397.
41. Nishizuka, Y., 1984, : The role of protein kinase C in cell surface signal transduction and tumor promotion, **Nature**, 308:693.
42. Nishizuka, Y., 1988, The molecular heterogeneity of protein kinase C and its implications for cellular regulation, **Nature**, 334:661.
43. Hwang, T.C., 1990, Ph.D. thesis: Regulation of epithelial Cl^- channels: roles of protein kinases and arachidonic acid. The Johns Hopkins University, School of Medicine.
44. Hockberger, P., Toselli, M., Swandulla, D., Lux, H.D., 1989, A diacylglycerol analogue reduces neuronal calcium currents independently of protein kinase C activation, **Nature**, 338:340.

45. Willumsen, N.J., Boucher, R.C., 1989, Activation of an apical Cl^- conductance by Ca^{2-} ionophores in cystic fibrosis airway epithelia, **Am. J. Physiol.**, 256:C226.
46. Widdicombe, J.H., 1986, Cystic fibrosis and β-adrenergic response of airway epithelial cell culture, **Am. J. Physiol.**, 251:R818.
47. Clancy, J.P., McCann, J.D., Li, M., Welsh, M.J., 1990, Calcium-dependent regulation of airway epithelial chloride channels, **Am. J. Physiol.**, 258:L25.
48. McPhail, L.C., Clayton, C.C., Snyderman, R., 1984, A potential second messenger role for unsaturated fatty acids: activation of Ca^{2+}-dependent protein kinase, **Science**, 224:622.
49. Williams, J.H., Errington, M.L., Lynch, M.A., 1989, Arachidonic acid induces a longterm activity-dependent enhancement of synaptic transmission in the hippocampus, **Nature**, 341:739.
50. Barbour, B., Szatkowski, M., Ingledew, N., Attwell, D., 1989, Arachidonic acid induces a prolonged inhibition of glutamate uptake in glial cells, **Nature**, 342:918.
51. Kim, D., Clapham, D.E., 1989, Potassium channels in cardiac cells activated by arachidonic acid and phospholipids, **Science**, 244:1174.
52. Ordway, R.W., Walsh, J.V., Singer, J.J. Jr., 1989, Arachidonic acid and other fatty acids directly activate potassium channels in smooth muscle cells, **Science**, 244:1176.
53. Sarkadi, B., Cheung, R., Mack, E., 1985, Cation and anion transport pathways in volume regulatory response of human lymphocytes to hyposmotic media, **Am. J. Physiol.**, 248:C480.
54. Landry, D.W., Reitman, M., Cragoe, E.J. Jr., Al-Awqati, Q., 1987, Epithelial chloride channel: Development of inhibitory ligands, **J. Gen. Physiol.**, 90:779.
55. Hwang, T.C., Guggino, S.E., Guggino, W.B., Direct modulation of secretory Cl^- channels by arachidonic acid, Accepted for publication by **Proc. Natl. Acad. Sci. USA.**
56. Dharmsathaphorn, K., McRoberts, J.A., Mandei, K.G., 1984, A human colonic tumor cell line that maintains vectorial electrolyte transport, **Am. J. Physiol.**, 246:G204.
57. Dahlen, S.-E., Hansson, G., Hedqvist, P., 1983, Allergen challenge of lung tissue from asthmatics elicits bronchial contraction that correlates with the release of leukotriene C4, D4 and E4, **Proc. Natl. Acad. Sci. USA**, 80:1712.
58. Weiss, J.W., Drazen, J.M., Coles, N., 1982, Bronchoconstrictor effects of leukotriene C in humans, **Science**, 216:196.
59. Hardy, C.C., Robinson, C., Tattersfield, A.E., Holgate, S.T., 1984, The bronchosonstrictor effect of inhales prostaglandin D2 in normal and asthmatic men, **N. Engl. J. Med.**, 311:209.
60. Carlstedt-Duke, J., Broennegard, M., Strandvik, B., 1986, Pathological regulation of arachidonic acid release in cystic fibrosis: the putative basic defect, **Proc. Natl. Acad. Sci. USA**, 83:9202.
61. Lewis, R.A., Soter, N.A., Diamond, P.T., 1982, Prostaglandin D_2 generation after activation of rat and human mast cells with anti-IgE, **J. Immun.**, 129:1627.

62. Churchill, L., Chilton, F.H., Resau, J.H., 1989, Cyclooxygenase metabolism of endogenous arachidonic acid by cultured human tracheal epithelial cells, **Am. Rev. Respir. dis.**, 140:449.
63. MacDermot, J., Barnes, P.J., 1980, Activation of guinea pig pulmonary adenylate cyclase by prostacyclin, **Eur. J., Pharmacol.**, 67:419.
64. Hille, B., 1984, "Ionic Channels in Excitable Membranes," Sinauer Association Inc., Massachusetts.
65. White, M.M., Miller, C., 1981, Probes of the conduction process of a voltage-gated Cl^- channel from Torpedo electroplax, **J. Gen. Physiol.**, 78:1.
66. Bahinski, A., Nairn, A.C., Greengard, P., Gadsby, D.C., 1989, Chloride conductance regulated by cyclic AMP-dependent protein kinase in cardiac myocytes, **Nature**, 340:718.
67. Levitan I.B., 1988, Modulation of ion channels in neurons and other cells, **Ann. Rev. Neurosci.**, 11:119
68. Bean, B.P., Nowycky, M.C., Tsien, R.W., 1984, β-adrenergic modulation of calcium channels in frog ventricular heart cells, **Nature** 307:371.
69. Cachelin, A.B., de Peyer, J.E., Kokubun, S., Reuter, H., 1983, Ca^{2+} channel modulation by 8-bromocyclic AMP in cultured heart cells, **Nature**, 304:462.
70. Soliven, B., Szuchet, S., Arnason, B.G.W., Nelson, D.J., 1988, Forskolin and phorbol esters decrease the same K^+ conductance in cultured oligodendrocytes, **J. Membr. Biol.**, 105:177.
71. Sims, S.M., Singer, J.J., Walsh, J.V. Jr., 1988, Antagonistic adrenergic-muscarinic regulation of M current in smooth muscle cells, **Science**, 239:190.
72. Kume, H., Takai, A., Tokuno, H., Tomita, T., 1989, Regulation of Ca^{2+}-dependent K^+ channel activity in tracheal myocytes by phosphorylation, **Nature**, 341:152.
73. Nastainczyk, W., Roehrkasten, A., Sieber, M., 1987, Phosphorylation of the purified receptor for calcium channel blockers by cAMP kinase and protein kinase C, **Euro. J. Biochem.**, 169:137.
74. Huganir, R.L., Greengard, P., cAMP-dependent protein kinase phosphorylates the nicotinic acetylcholine receptor, **Proc. Natl. Acad. Sci. USA**, 80:1130.
75. Miles, K., Anthony, D.T., Ruben, L.L., 1987, Regulation of nicotinic acetylcholine receptor phosphorylation in rat myotubes by forskolin and cAMP, **Proc. Natl. Acad. Sci. USA**, 84:6591.
76. Walsh, K.B., Kass, R.S., 1988, Regulation of a heart potassium channel by protein kinase A and C, **Science**, 242:67.
77. Hammond, C., Paupardin-Tritsch, D., Nairn, A.C., 1987, Cholecystokinin induces a decrease in Ca^{2+} current in snail neurons that appears to be mediated by protein kinase C, **Nature**, 325:809.
78. Lacerda, A.E., Rampe, D., Brown, A.M., 1988, Effects of pro-

tein kinase C activators on cardiac Ca^{2+} channels, **Nature**, 335:249.
79. Strong, J.A., Fox, A.P., Tsien, R.W., Kaczmarek, L.K., 1987 Stimulation of protein kinase C recruits covert calcium channels in Aplysia bag cell neurons, **Nature**, 325:714.
80. Miles, K., Huganir, R.L., 1988, Regulation of nicotinic acetylcholine receptors by protein phosphorylation, **Mol. Neurobiol.** 2:91.

DISCUSSION

FRIZZELL: Maybe I missed it, but where was the ricinoleic acid added: to the apical side?

W. GUGGINO: We added it to both sides in that experiment.

FRIZZELL: And the time constant - the rate constant - for its effect is on the order of several minutes?

W. GUGGINO: Several minutes, yes.

(Unidentified): What kind of media did you use in the T84 experiments, because some of the media and especially the serum contains arachidonic acid, or similar substitutes, so what was the basal state before you added arachidonic acid?

W. GUGGINO: We grow T84 cells in 10% serum. But there is of course no serum in the short-circuit current experiments. This is just the bicarbonate-free Ringer solution. But the cells have been pre-exposed to serum, 10% serum.

(Unidentified: Bill, were the effects reversible for ricinoleic acid, when you washed it out?

W. GUGGINO: Yes, the effects were reversible.

AL-AWQATI: Bill, were there experiments (my memory is not really good on this) weren't there experiments on saturated fatty acids including arachidonic acid that actually inhibit potassium channels in a variety of cells?

GUGGINO W.: Arachidonic acid activates K^+ channels in the heart, and I could mention that we also see non-selective cation channels in the same patches where we see outwardly-rectifying chloride channels, and those non-selective cation channels are not affected by arachidonic acid.

(Unidentified): Could I just clarify one thing? Did you say that it didn't matter which side you added the arachidonic acid to, or the different analogues?

GUGGINO W.: In the excised patch experiments, of course we were adding the arachidonic acid to the intracellular side of the membrane. In the short circuit current experiments we put ricinoleic acid, which is the laxative, on both sides of the T84 cell cultures. We have not done outside-out patches on the airway cells to determine if arachidonic acid is effective from the outside. We don't know if the arachidonic acid is getting in the cell and then blocking from the inside.

BEAR: I was wondering if you had any explanation for the observation that a laxative-blocked chloride channel should be involved in fluid secretion?

GUGGINO W.: Yes, it's a good question. We have also used some leukotrienes and we also show that leukotrienes can also cause this flickery block of the chloride channel, even though they stimulate Cl^- secretion in the whole tissue. Perhaps what is happening, at least with the leukotrienes, is that the leukotrienes are binding to a receptor and then activating the channel in another way, for example via cAMP. If the fatty acid concentration builds up to a high enough concentration, then you would get channel block. Thus there must be an alternative pathway for laxative stimulation of chloride secretion in the intestine and colon.

CYTOSOLIC INHIBITION AND EXCISION ACTIVATION OF EPITHELIAL CHLORIDE CHANNELS

Karl Kunzelmann, Monika Tilmann, and Rainer Greger

Physiologisches Institut der Albert-Ludwigs-Universität Freiburg, Hermann-Herder-Straße 7, D-7800 Freiburg, FRG

Epithelial chloride channels with intermediate conductance (20-80 pS) have been identified in respiratory epithelia in primary culture (4,10,12, 13,19,21),in colonic carcinoma cells (9,11), sweat ducts (1,22), in the rectal gland of Squalus acanthias (8) and in the thick ascending limb of Henle's loop (6). The properties of these channels have recently been reviewed (5). In short, it appears likely that most of these channels share in common the following properties: They are activated by the cAMP pathway (7,16,17) and by proteinkinase C (12,15), and they are indirectly activated by increases in cytosolic Ca^{2+} activity (20,21). Furthermore, these channels are insensitive to cytosolic pH, ATP, and they are inhibited by chloride channel blockers such as NPPB, DIDS, amidine and indanylphenoxyacetic acids (2,14,18). In addition, it has been shown by several laboratories (3,13,19) that this type of chloride channel is absent in respiratory cells from cystic fibrosis (CF) patients, but that it appears when membrane patches from these cells are ripped off. This channel activation was caused by long lasting depolarization of excised membrane patches from CF cells (12,16, 17). On the other hand, we have shown more recently (13) that no depolarization was necessary for channel activation if, unlike the previous studies, excision was carried out not at room temperature but at 37°C. The present short review will focus on our recent observations on the excision activation of chloride channels and in a second part on our attempts to characterize the cytosolic inhibitor responsible for chloride channel inactivation in intact cells.

Excision activation of chloride channels occurs immediately and is found in normal as well as in CF respiratory cells and in colonic carcinoma cells. Fig. 1 summarizes two patch clamp experiments in CF respiratory cells. In these, like in all other

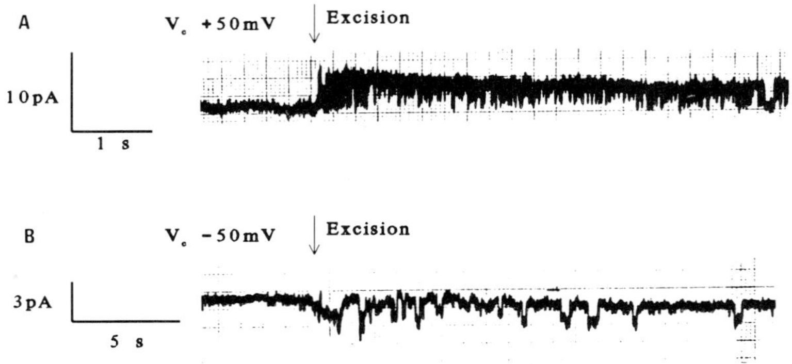

Fig.1. Excision activation of chloride channels in cystic fibrosis (CF) respiratory cells. Respiratory epithelial cells from adenoids in primary culture were examined by the patch clamp technique at 37°C in a bath containing NaCl-Ringers. For details of methods see ref. 13. The patch pipette was filled with KCl-Ringers. Original current traces are displayed. A: Membrane patch was clamped at a voltage (V_c) of +50 mV in cell attached mode. No channel openings were noted over more than 1 min. The arrow indicates the time of excision of the patch. Note that channel openings appear instantaneously after the excision. The channel was identified as a chloride channel of intermediate conductance. Note also that the open probability of the channel shows a slight run-down with time. B: Membrane patch was clamped at a voltage of -50 mV prior to excision and no channel openings were noted. Channel openings in the downward direction were observed shortly after the excision of the patch. The current events were caused by a chloride channel of intermediate conductance.

experiments in CF cells, no chloride channels were found in the cell attached configuration. When the pipette was pulled off the cell, chloride channel openings occurred immediately and irrespective of the clamp voltage (13). Hence, our data do not support the view that depolarization is required for chloride channel opening in excised patches. Long lasting depolarization may, in fact, have the opposite effect: We usually observe a slow run-down of the open probability, if patches are kept for some time at depolarizing voltage (also c.f. Figs. 1 and 2). A brief hyperpolarizing spell usually "reactivates" channels which have been subject to a run-down phenomena.

Excision activation of chloride channels is not unique to CF respiratory cells. We have made comparable observations in normal respiratory as well as in colonic carcinoma cells. Fig. 2 shows a patch clamp experiment in a HT_{29} colonic carcinoma cell. In the cell attached configuration no chloride channel was found. After excision several chloride channels were immediately

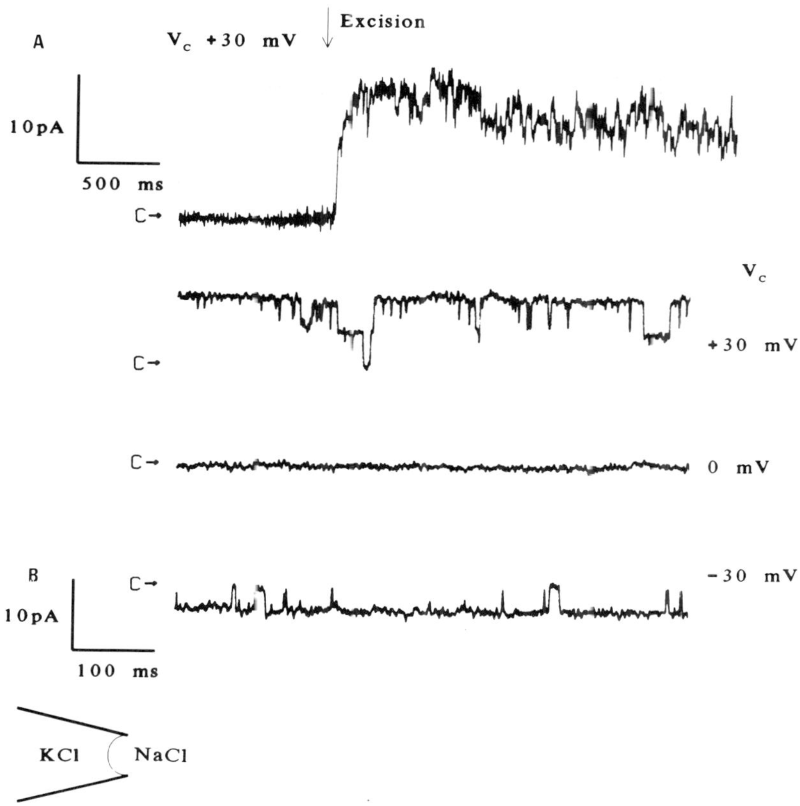

Fig.2. Excision activation of chloride channels in a colonic carcinoma cell line (HT_{29}). The cells were cultured in low glucose media and examined at 37°C in NaCl-Ringers. No hormonal agonist was added to the bath solution. For details of methods see ref. 11. The patch pipette was filled with KCl-Ringers. A: Current trace at low time resolution. No current events were apparent in cell attached configuration at a clamp voltage (V_c) of +30 mV. C- indicates the zero current level. Excision of the patch induced immediate current events caused by the superposition of many channels. Note that the mean current shows a run-down phenomena with time. B: Typical current traces of the same patch at higher time resolution, approximately 2 min after the excision. C- indicates the zero current level. The current amplitude for positive clamp voltages (V_c) exceeds that for the same negative clamp voltages (slight outward rectification of this kind of chloride channel). Note also that the open state probability is high at depolarized and low at hyperpolarized voltages.

activated. The lower panel of this Fig. documents that the general properties of these channels (voltage dependence of current amplitude and open state probability) are identical to those of hormone induced chloride channels in these cells. These studies in four different preparations (normal and CF respiratory cells, HT_{29} and T_{84} cells) indicate that cytosolic inhibition may be one regulatory component of chloride channels. This down-regulation by cytosolic inhibitor antagonizes the activating pathways through proteinkinase A, proteinkinase C, and Ca^{2+}-calmodulin. The pathophysiology of CF, could therefore not only be caused by a defective activation pathway. It might, alternatively or in addition, be caused by an increased cytosolic inhibition of chloride channels. Our further attempts have been directed towards the characterization of this cytosolic inhibitory pathway.

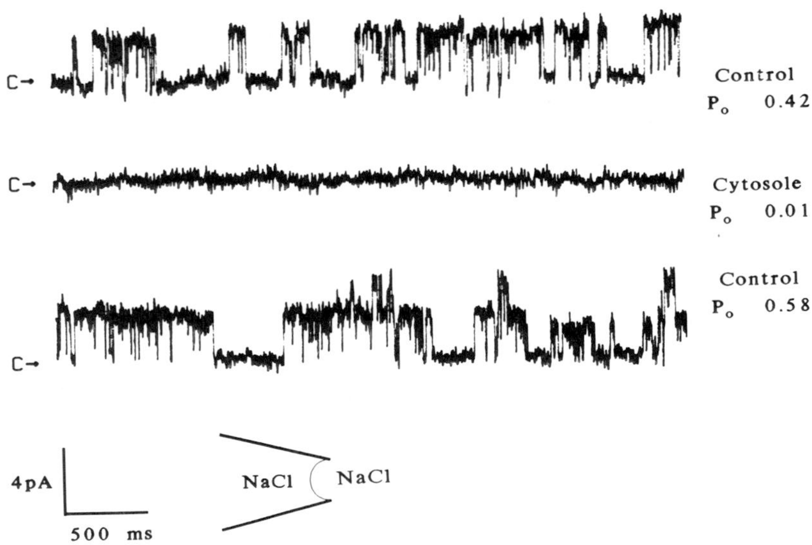

Fig.3. Effect of cytosol extracted from colonic carcinoma cells (HT_{29}) on a chloride channel in an excised membrane patch from a cystic fibrosis (CF) respiratory cell. For methods see Fig.1 and ref. 13. The cytosol was prepared from HT_{29} cells by homogenization and ultracentrifugation. Its pH was titrated to 7.2 and its ionic composition was adjusted to achieve a chloride concentration of approximately 150 mmol/l. The dilution was approximately twofold. The bath contained NaCl-Ringers (control), and the pipette was filled with the same NaCl-Ringers. The chloride channel was activated by excision. Its orientation was inside/ out. C- indicates the zero current level. The clamp voltage was +60 mV. Replacing the bath by diluted cytosol led to a rapid and reversible reduction of the open state probability of this channel (p_o).

Cytololic inhibitor of chloride channels can be extracted from HT_{29} and placenta cells. Cytosol from HT_{29} cells or placenta was prepared by ultracentrifugation. The cytosol was recovered as the supernatant, its pH and ionic composition were adjusted. This cytosolic fraction was examined on the cytosolic surface of excised chloride channels of HT_{29} and respiratory cells. Fig. 3 shows an example of a chloride channel in a CF respiratory cell membrane which was activated by excision. Cytosol, in a dilution of approximately 1:20, inhibited this channel reversibly. The inhibition was dependent on the dilution and was characterized by a flicker block. This is shown more clearly in Fig. 4 for a

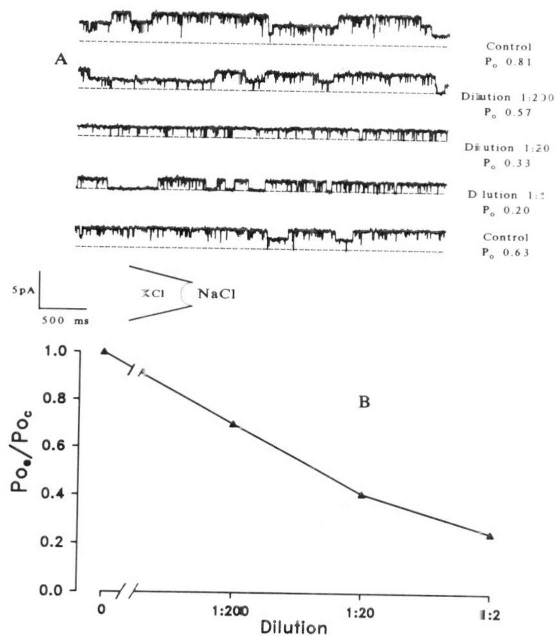

Fig.4. Effect of cytosol extracted from human placenta on chloride channels of colonic carcinoma cells (HT_{29}). For details of methods see refs. 11 and 13. The cytosol was prepared from human placenta and adjusted as stated in the legend to Fig. 3. Different dilutions with invariant chloride concentrations of 150 mmol/l were tested. The bath contained NaCl-Ringers (control). The pipette was filled with KCl-Ringers. The channels were activated by excision and kept at a clamp voltage of +30 mV. A: Typical current traces are shown. The open state probability (p_c) was determined from 30 s time samples obtained under the different conditions. The dashed line indicates the zero current level. Cytosol in a dilution of 1:200 or less reduced p_o reversibly. B: Dose response curve for the inhibitory effect of cytosol.

HT_{29} chloride channel. In this experiment cytosol from human placenta was used. This Fig. also contains a dose response curve for this inhibitory effect. The inhibitor acted down to a dilution of 1:200. In a larger series the mean half maximal inhibition was achieved at a dilution of around 1:20.

In a next series we have used the same method to prepare cytosol extracts from skeletal muscle and fibroblasts. Even if undiluted these extracts did not inhibit HT_{29} chloride channels.

In further series we have used excision activated HT_{29} chloride channels as a bioassay system for cytosol fractions. When placenta cytosol was filtered through a filter with a pore size corresponding to < 10 kD, we found inhibitory activity in the supernatant as well as in the filtrate. This may indicate that the cytosolic inhibitor has a small molecular mass, but that some of the inhibitor may have escaped filtering because of binding to larger molecules or because of incomplete filtration. In a separate series we have used gel filtration and found no inhibitory activity in the fractions > 5 kD. Furthermore, we were able to separate the inhibitor from the cytosol through a dialysis membrane with pores corresponding to 10 kD. These three series indicate that the inhibitor has a molecular mass of > 5 kD.

Next we have boiled the cytosol for 5 min., and found no loss in inhibitory activity. This renders it rather unlikely that the inhibitor is a protein. In a next series we have used chloroform to extract lipophilic components from the cytosol and found that the inhibitory activity was mostly in the hydrophobic fraction. These preliminary data suggest that the cytosolic inhibitor has a molecular mass of <5kD and that it is lipophilic. The mode of action of this inhibitor suggests that the inhibitor itself is responsible for a direct effect on the channel, inasmuch as its effect is immediate, easily reversible, and can be reproduced in simple in vitro systems such as the excised patch.

Acknowledgements

This study has been supported by BMFT 01 GA 8816, DFG Gr 489/10 and by GIF I-86-100.1/88.

References

1. Bijman, J., Scholte, B., De Jonge, H.R., Chloride transport in cystic fibrosis: chloride channel regulation in cultured sweat duct and cultured nasal epithelium, In: "Cellular and Molecular Basis of Cystic Fibrosis," Mastella, G. and Quinton, P.M., eds., San Francisco Press, San Francisco, 1988.
2. Dreinhöfer, J., Gögelein, H., Greger, R., 1988, Blocking kinetics of Cl^- channels in colonic carcinoma cells (HT_{29}) as revealed by 5-nitro-2-(3-phenylpropylamino)-benzoic acid (NPPB), **Biochim. Biophys. Acta**, 956:135.

3. Frizzell, R.A., 1987, Cystic fibrosis: a disease of ion channels? **Trends Neurosci.**, 10:190.
4. Frizzell, R.A., Halm, D.R., Rechkemmer, G.R., Shoemaker, R.L., 1986, Chloride channel regulation in secretory epithelia, **Fed. Proc.**, 45:2727.
5. Gögelein, H., 1988, Chloride channels in epithelia, **Biochim. Biophys. Acta**, 947:521.
6. Greger, R., Bleich, M., Schlatter, E., 1990, Ion channels in the thick ascending limb of Henle's loop, **Renal. Physiol. Biochem.**, 13:37.
7. Greger, R., Schlatter, E., Gögelein, H., 1985, Cl^--channels in the apical cell membrane of the rectal gland "induced" by cAMP, **Pflügers Arch.**, 403:446.
8. Greger, R., Schlatter, E., Gögelein, H., 1987, Chloride channels in the luminal membrane of the rectal gland of the dogfish (Squalus acanthias). Properties of the "larger" conductance channel, **Pflügers Arch.**, 409:114.
9. Halm, D.R., Rechkemmer, G., Schoumacher, R.A., Frizzell, R.A., 1989, Biophysical properties of a chloride channel in the apical membrane of a secretory epithelial cell, **Comp. Biochem. Physiol.**, 90A:597.
10. Halm, D.R., Rechkemmer, G.R., Schoumacher, R.A., Frizzell, R.A., 1988, Apical membrane chloride channels in a colonic cell line activated by secretory agonists, **An. J. Physiol.**, 254:C505.
11. Hayslett, J.P., Gögelein, H., Kunzelmann, K., Greger, R., 1987, Characteristics of apical chloride channels in human colon cells (HT_{29}), **Pflügers Arch.**, 410:487.
12. Hwang, T.-C., Lu, L., Zeitlin, L., Gruenert, D.C., Huganir, R., Guggino, W.B., 1989, Cl- channels in CF: lack of activation by protein kinase C and cAMP-dependent protein kinase, **Science**, 244:1351.
13. Kunzelmann, K., Pavenstädt, H., Greger, R., 1989, Properties and regulation of chloride channels in cystic fibrosis and normal airway cells, **Pflügers Arch.**, 415:172.
14. Landry, D.W., Reitman, M., Cragoe, E.J., Al-Awqati, Q., 1987, Epithelial chloride channel, **J. Gen. Physiol.**, 90:779.
15. Li, M., McCann, J.D., Anderson, M.P., Clancy, J.P., Liedtke, C.M., Nairn, A.C., Greengard, P., Welsh, M.J., 1989, Regulation of chloride channels by protein kinase C in normal and cystic fibrosis airway epithelia, **Science**, 244:1353.
16. Li, M., McCann, J.D., Liedtke, C.M., Nairn, A.C., Greengard, P., Welsh, M.J., 1988, Cyclic AMP-dependent protein kinase opens chloride channels in normal but not cystic fibrosis airway epithelium, **Nature** 331:358.
17. Schoumacher, R.A., Shoemaker, R.L., Halm, D.R., Tallant, E.A., Wallace, R.W., Frizzell, R.A., 1987, Phosphorylation fails to activate chloride channels from cystic fibrosis airway cells, **Nature**, 330:752.
18. Tilmann, M., Kunzelmann, K., Greger, R., 1990, Different types of blockers for epithelial chloride channels, **Pflügers Arch.**, 415:R19 (Abstract).
19. Welsh, M.J., 1986, An apical-membrane chloride channel in human tracheal epithelium, **Science**, 232:1648.

20. Welsh, M.J., 1987, Effect of phorbol ester and calcium ionophore on chloride secretion in canine tracheal epithelium, **Am. J. Physiol.**, 253:C828.
21. Welsh, M.J., 1987, Electrolyte transport by airway epithelia, **Physiol. Rev.**, 67:1143.
22. Welsh, M.J., Mc Cann, J.D., Dearborn, D.G., 1987, Chloride and sodium channel currents in normal and cystic fibrosis sweat duct epithelium, **Fed. Am. Soc. Expt. Biol.**, 46:5567.

DISCUSSION

(Unidentified): What role would the cytosolic inhibitor play in the calcium- and swelling-activated chloride currents in these cells? How do your findings fit with the model of defective Cl^--channel regulation in CF?

KUNZELMANN: I do not know. At the present we can only speculate that maybe this lipophylic cytosolic inhibitor is yet another regulatory pathway of Cl^--channels.

FRIZZELL: Did you look into the voltage dependence of the inhibition by cytosol?

KUNZELMANN: Yes, we did not find any impressive voltage dependence of the block by cytosolic inhibitor over the narrow voltage range (usually ± 30 mV of our experiments.

RECHKEMMER: My question regards the effect of temperature which you examined. You did not observe any change in open channel probability when you compared 4° to 37°. Now the question arises whether the blocking effect of linoleic acid or arachidonic acid were the same at both temperatures.

KUNZELMANN: We have not done the respective experiments.

BIJMAN: Is the effect of cytosolic inhibitor rapidly reversible when you remove this inhibitor?

KUNZELMANN: Yes, the inhibition is completely and rapidly, i.e. within less than 5 sec, reversible.

QUINTON: In view of the fact that the Cl^--channel appears to be constitutively on in the sweat duct, which is a purely absorptive epithelium, I wonder if the cytosolic factors would be the same in purely absorptive epithelia and if perhaps you might try the cytosol, say, from something like the urinary bladder.

GREGER: May I respond to this. I have been looking into the Cl^--channel of the basolateral membrane of thick ascending limb of Henle lately, and I found an excision activation of Cl^--channels in this preparation too. Hence, I conclude that cytosolic inhibitor is also present in the kidney. In fact, Dr. Frömter has shown yesterday that his and Dr. Burckhardt's groups are presently extracting cytosolic inhibitor from the kidney.

KUNZELMANN: I just wanted to add that our cytosolic fraction was obtained from kidney.

PURIFICATION OF THE EPITHELIAL Cl CHANNEL

Qais Al-Awqati, Donald Landry, Myles Akabas,
Christopher Redhead, Aleksander Edelman,
and John Edwards

Departments of Medicine and Physiology
College of Physicians and Surgeons of
Columbia University, New York, NY, USA

Chloride channels are present in plasma membranes and in intracellular organelles. In a variety of epithelial cells these channels mediate NaCl absorption and secretion. The aim of our work was to purify the Cl channel and to develop molecular reagents that could be used to study the biochemical basis of cystic fibrosis.

Our strategy for accomplishing this task started with development of a simple but specific assay for voltage-driven ^{36}Cl uptake. Using this assay we identified a tissue source enriched in this activity and then screened a large number of compounds that might be useful as high affinity ligands. A potent inhibitor was found, titrated and used to study binding in the membranes and to follow the channel through chromatographic purification. The structure of such a compound also suggested a possible affinity matrix. Finally, after the "drug-binding" protein was purified, we reconstituted into artificial lipid bilayers to demonstrate that it was indeed a Cl channel. The purified protein was used to generate antibodies and sequences for the purposes of cloning.

Development of Inhibitory Ligands (Ref. 1)

We devised a transport assay for the chloride conductance in membrane vesicles derived from bovine renal cortex. The vesicles were loaded with KCl, stripped of external Cl via a gluconate-loaded anion exchange resin and the resulting electrochemical gradient was used to drive ^{36}Cl uptake with aliquots collected and counted on a filter. Using this assay a large number of varied compounds were screened for inhibition. The indanyloxyacetic acids (IAA) proved to be a potent class of agents and one of these, IAA94, inhibited 33Cl uptake with an IC50 of approxi-

The Identification of the CF (Cystic Fibrosis) Gene
Edited by L.-C. Tsui *et al.*, Plenum Press, New York, 1991

mately 1 μM. We titrated IAA-94 by acid catalyzed $^1H^+$-$^3H^+$ exchange and obtained a reagent with specific activity 0.6 Ci/mmol. Scatchard analysis of 3H-IAA-94 binding showed that this drug bound to the kidney microsomes with a Kd of 0.6 μM, similar to the inhibitory potency of IAA-94 against ^{36}Cl transport. For a large number of compounds, we compared the ability to inhibit ^{36}Cl transport and the ability to displace 3H-IAA-94 from its binding site and we found a good concordance between these two parameters suggesting that the drugs bound to the chloride channel. These studies were extended to trachea and skeletal muscle.

Purification of the Cl Channel (Ref. 2)

To purify the Cl channel from kidney we initially prepared microsomal membrane vesicles from bovine kidney cortex which demonstrate a voltage sensitive ^{36}Cl uptake. We synthesized an affinity column based on the IAA structure using a derivative of the most potent ligand, IAA-23. We found that this matrix depleted the IAA-94 binding activity. Extensive washing of the column followed by elution with excess IAA-94 resulted in the release of 4 polypeptides with molecular masses of 96, 64, 40 and 27 kDa.

Reconstitution of the Cl Channel

The purified proteins were concentrated and excess soybean phospholipid was added to the mixture. The proteoliposomes were then dialyzed for one or two days to remove the detergents. The resultant vesicles were incorporated into planar lipid bilayers and electrophysiological measurements showed the presence of several anion selective channels that had single channel conductances of 30 pS and were ideally anion selective. We also saw channels that had a single channel conductance of 400 pS and which showed different substrates. None of these channels showed any marked voltage dependence.

We also co-reconstituted these channels with bacteriorhodopsin, a light sensitive, electrogenic proton pump. Exposure to light results in a membrane potential, positive inside. When the purified proteins were incorporated into these vesicles, exposure to light resulted in a voltage-driven uptake of ^{36}Cl. These results demonstrate that the purified proteins are indeed the chloride channel.

Identification of the 64 kDa Protein as the Cl Channel

Which of the four proteins purified is the Cl channel? Because of the relatively low affinity of the drug (only 1 μM), it was possible that some of the proteins purified were actually drug binding proteins rather than channel proteins. We prepared antibodies to these proteins by injecting slices from SDS-PAGE of large scale purifications. The antiserum to the 64 kDa protein showed a high titer and stained only the 64 kDa protein on western blot of purified and unpurified membranes. It also stained the same protein in western blots of kidney, colon, fibroblasts and other cells from a variety of mammals.

We also were able to immunoprecipitate the Cl channel activity using this antibody. Crude microsomal kidney cortex vesicles were solubilized by using n-octyl glucoside and the total proteins were incubated with anti-64 kDa serum or non-immune serum. The antigen-antibody complex was precipitated using protein A-Sepharose beads. The total proteins were then incorporated into soybean phospholipid and the voltage dependent uptake of ^{36}Cl was measured. When the solubilized proteins were exposed to sepharose beads to which protein A was coupled, we did not find any reduction in the tracer uptake. However, if the beads were loaded with the immune serum, the supernatant was completely depleted of ^{36}Cl uptake activity. Similar experiments, in which the beads were loaded with non-immune globulin did not result in any depletion of the transport activity. These results demonstrate that the 64 kDa protein is the Cl channel.

Further studies were performed to confirm this observation. Solubilized membranes were separated by gel filtration before purification on affinity columns. We found a fraction of apparent Mr of above 200,000 Da which contained the transport activity. When this fraction was incubated with the IAA affinity column, we found that it contained the 64 kDa protein.

Cloning the 64 kDa Protein

We prepared a cDNA library from bovine kidney cortex poly A mRNA in an expression vector. Two clones were identified using the anti-64 kDa antibody. The fusion proteins expressed in E. coli had molecular masses on western blots of approximately 70 and 40 kDa. Polyclonal antibodies were affinity purified against the expressed fusion proteins from each clone. Both affinity-purified antibodies recognized both fusion proteins indicating that the two clones code for the same or related proteins. More importantly they recognized the 64 kDa protein present in crude kidney vesicles and in the chloride channel proteins purified by affinity chromatography. These results lend strong support to the hypothesis that these two clones actually code for the chloride channel. These clones are being sequenced and further characterized.

In an independent approach to isolate clones for the 64 kDa protein, we obtained an N-terminal sequence from purified 64 kDa protein. The sequence was back-translated to obtain oligonucleotide sequences which code for the protein to it. The oligonucleotide is being used with the polymerase chain reaction to generate larger sequences from the corresponding mRNA.

The Role of Intracellular Cl Channels in Regulation of Intravesicular pH. (Ref. 3)

We (and others) had previously shown that several intracellular compartments contain an electrogenic proton-translocating ATPase. Recent studies have shown, however that the pH of these compartments is variable; the golgi having a pH of 6.57.0, the endosome is 5.5-6.0 and the lysosome around 4.5-5.0. Since these ATPases are similar, we had proposed that the reason for the pH difference is the presence of a regulatable ion conductance which we later discovered to be a Cl channel. In one cell type, the parafollicular cell of the thyroid, we found that the secretory granule was not acid even though it contained the proton-translocating ATPase. The pH of the isolated vesicle could be induced to go down to its minimal level by shunting the membrane potential generated by the ATPase using the ionophore valinomycin. When these cells were stimulated by TSH to secrete, we found that the granules became acid. The reason the granules became acid was due to the opening of a Cl channel in the granule.

References

1. D.W. Landry, M. Reitman, E.J. Cragoe and Q. Al-Awqati, 1987, Epithelial Chloride Channels; Development of inhibitory ligands, J. Gen. Physiol., 90:779.
2. D.W. Landry, M.H. Akabas, C.Redhead, A. Edelman, E.J. Cragoe, Jr., and Q. Al-Awqati, 1989, Purification and reconstitution of chloride channels from kidney and trachea, Science, 244:1469.69.
3. J. Barasch, M.D. Gershon, E.A. Nunez, H. Tamir and Q. AlAwqati, 1988, Thyrotropin induces the acidification of the secretory granules of parafollicular cells by increasing the chloride conductance of the granular membrane, J. Cell Biol., 107:2137.

DISCUSSION

CORCELLI: I would like to know if the binding of the ligand in the solubilized extract requires some specific lipids.

AL-AWQATI: No. We have tried many kinds of added lipids but, because we do the solubilization in very small volumes, there is a large amount of the native lipids - endogenous lipids - and we have prepared, actually, endogenous lipids and they seem to be necessary for reconstitutability, but we really haven't done extensive work on this.

IANNUZZI: Do you know the size of the messenger RNA?

AL-AWQATI: Yes. We have probed messenger RNA from the kidney. First, several messages are recognized: three messages are recognized and they are all in the 4 kb range (3.5, 4.0 and 7.0 kb). There is no homology to anything that is in the data base, in particular no homology to other chloride transporters, such as band 3, GABA receptors and CFTR.

GREGER: We have been looking at a couple of preparations which possess Cl^- channels, and you seem to pick up such membranes with your antibody. On the other hand, we are unable to see Cl^--channels, comparable to those of Cl^--transporting epithelia, in smooth muscle cells, yet smooth muscle cells are positive with your antibody. What is your interpretation? Are you in the process of digging out something which may turn out to be a superfamily of choride channels?

AL-AWQATI: Well that's what we hope we'll end up doing - that all of the chloride channels will at least be related to each other. That would be the optimistic view, I think, similar to the potassium channel story, where minor modifications in the protein give various kinds of electro-physiological behavior. I might add that it is really impressive that this antibody recognizes so many channels (well, so many proteins) in so many cells. We have not had (and this is rather bad for us) a cell that is negative. It would be nice to find negative cells, but as all of you know, there is no cell that doesn't have chloride channels.

GREGER: Have you looked specifically at the acini of the pancreas?

AL-AWQATI: Not with immunocytochemistry.

(Unidentified): I wasn't completely clear about the gel filtration profile where you had the Westerns. I mean what your interpretation was. Did you say there were no other proteins there?

AL-AWQATI: That was a Western blot, not an immunoprecipitate. As I mentioned, we will change the detergents and the salt concentration to see if now proteins will appear and our sort of fuzzy first result shows that sometimes other proteins are immunoprecipitated in addition to the 64 kDa protein.

(Same): So was the suggestion that there would be different oligomers at different points along the gel filtration profile?

AL-AWQATI: Yes, that's what I think actually happens, that if one has a trimer, or a dimer, or a monomer, it cannot reconstitutively produce a chloride channel.

GREGER: Can I ask one more question? As you have already some results on the sequence, does it match with that of the band three protein?

AL-AWQATI: Yes. There is no homology at all here.

GRUENERT: The doublet that you see, how often do you see that and have you done anything to try and sequence it to see if there is any difference between the bands?

AL-AWQATI: No we haven't done sequencing. I think the thing that has prevented us from actually pursuing this effectively is the fact that it is unpredictable. From the same preparation (we essentially do most of our work on the kidney) sometimes we see just one band and sometimes we see two. We eventually have to purify the antibody against one band and study its behavior.

GRUENERT: When you see one band, is it always the same band of the doublet pair that you see, or does it vary in terms of the top one vs. the bottom one?

AL-AWQATI: Well, I think you will get different answers from the 4 people who do these experiments - that's all I can tell you.

DEARBORN: In your silver-stain gels there was always a fairly dark, thin band at high molecular weight. Does that have any significance?

AL AWQATI: No, they were not really reproducible. We sometimes found something at around 150,000, but it was never really reproducible.

REGULATION OF EXPRESSION OF CFTR IN HUMAN INTESTINAL EPITHELIAL CELLS

M. Buchwald[1], R. Sood and W. Auerbach

Research Institute - Hospital for Sick Children
Departments of Medical Genetics and Medical Biophysics
University of Toronto[1], Toronto, Ontario, Canada

ABSTRACT

As a first step in our efforts to delineate the role of CFTR in cellular phenotypes we have studied its expression in cultured human intestinal epithelial cells. In particular we have examined the effect of cellular differentiation on CFTR gene expression. CFTR mRNA was measured by quantitative densitometry of Northern blots and normalized to the amounts of pyruvate dehydrogenase message. We have found that in T84 cells the levels of CFTR mRNA do not change as the cells grow to confluence. In contrast, levels of CFTR mRNA increase by a factor of 10-20 as Caco2 cells grow after subculture. This change in the levels of CFTR mRNA is correlated with the morphological differentiation that occurs in Caco2 cells during culture. The potential significance of this observation is discussed.

INTRODUCTION

The initial experiments characterizing the expression of the CF gene (CFTR) showed that it is preferentially expressed in exocrine tissues, such as lung, pancreas and sweat gland, that are also the site of the disease. Furthermore, expression of CFTR appears to be much lower in non-exocrine tissues such as kidney and brain (Riordan et al., 1989). These results are consistent with the view that the regulation of expression of CFTR plays an important role in its cellular and organismic function and that defining the types of regulatory mechanisms will be an integral aspect in understanding the function of CFTR.

The initial description of the expression of CFTR also indicated that CFTR was expressed at a high level in the colonic adenocarcinoma cell line T84 (Riordan et al., 1989). This cell line consists of highly differentiated epithelial cells that produce monolayers with resistances on the order of 1500 ohm•cm^2 (Dharmasathaphorn et al., 1984). The cell line appears to express the cellular function (chloride-mediated transport that can be stimulated by elevations of intracellular cyclic AMP) thought to be the site of the CF defect (Quinton, 1989). Thus, one could hypothesize that expression of CFTR is a prerequisite for the existence of a cyclic AMP stimulated chloride transport pathway. One method to examine this question would be to determine if variation in the expression of CFTR is correlated with the transport function. We have therefore examined the regulation of expression of CFTR in T84 cells as a function of growth in culture. In addition, since other human intestinal epithelial cell lines are also available, these were examined as well. In particular we have focused on Caco2 cells because these undergo differentiation in culture (Rousset, 1986) and we reasoned that such differentiation might involve the expression of CFTR. This manuscript presents our initial results of these studies. We have found that while the expression of CFTR does not change in T84 cells, it increases by a factor of 10-20 fold in Caco2 cells. The possible implications of this change in expression are also discussed.

MATERIALS AND METHODS

<u>Cells</u>: T84 (ATCC # 248-CCL) and Caco2 (ATCC # HTB 37) cells were obtained from the American Type Tissue Culture Collection. Cells are certified free of mycoplasma. T84 cells were cultured in a 1:1 mixture of Ham's F12 medium and alpha MEM (both from Gibco) + 5% Nu serum (Collaborative Research Inc.). Caco2 cells

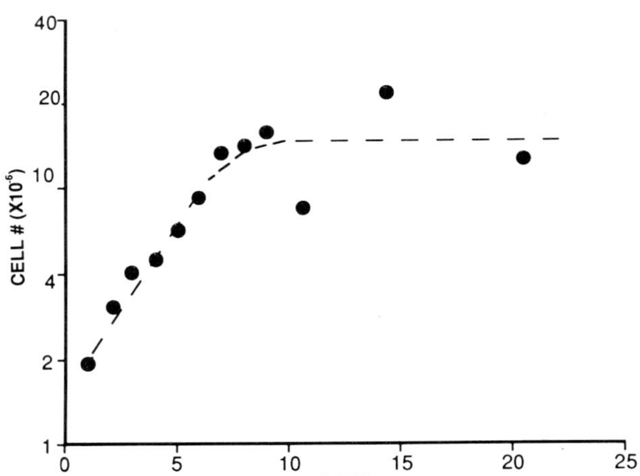

Fig. 1. Growth of T84 cells following subculture. Cells were seeded at approximately 2.5 x 10^6 per 100 mm dish and counted on the indicated days.

were cultured in alpha MEM with 10% fetal bovine serum (Flow Laboratories). Cells were passaged by trypsinization every 7 days, T84 as a 1:2 and Caco2 as a 1:4 or 1:6 subculture.

Growth curves: To determine the effect of growth on the expression of CFTR, cells were seeded at a density of 2.5 x 10^6 per 100 cm plate (approximately a 1:5 subculture) and counted daily until the end of the experiment using a Coulter counter. Cells were fed every other day until they reached confluence (T84- 7 days; Caco2- 4 days) and daily after that.

RNA isolation: An appropriate number of plates was chosen at each time point to yield approximately 0.1 mg of RNA. Total RNA was isolated by the method of Chirgwin et al. (1979). Briefly, the procedure involved lysing the cells with denaturing solution (4 M guanidinium thiocyanate, 25 mM sodium citrate, pH 7, 0.5% N-lauryl sarcosine and 0.1 M 2-mercapto-ethanol) and pelleting the RNA by centrifugation at 30,000 rpm, 20°C for 18-24 hr over a cushion of 5.7 M CsCl, 0.1 M EDTA, pH 8, in SW41 ultra-clear tubes. The pellet was then rinsed twice with ethanol, resuspended in diethyl pyrocarbonate-treated TE (10 mM Tris HCl, 1 mM EDTA, pH 7.5) and precipitated with 2 volumes of ice-cold ethanol.

Northern blotting and hybridization: Total RNA (20 µg/lane) was fractionated on 0.66 M formaldehyde/1.2% agarose gels (Fourney et al., 1988) in 1 X MOPS/EDTA buffer [0.2 MOPS (3-(N-morpholine) propanesulfonic acid), 50 mM sodium acetate and 10 mM EDTA, pH 7]. Gels were equilibrated in 10 X SSC (1 X SSC = 0.15 M NaCl, 0.015 M Na citrate, pH 7.0) for 20 min and then transferred onto Gene Screen plus membranes by capillary action. Blots were baked at 80° under vacuum for 2 hr and prehybridized in 1% SDS, 1 M NaCl and 10% dextran sulfate at 65° for 2-6 hr. Probes were labelled with ^{32}P-dCTP by random priming (Feinberg and Vogelstein, 1983) to specific activities of 5-15 x 10^8 cpm/mg and blots hybridized at 65°C overnight with denatured probes. Blots were then washed twice at room temperature in 2 X SSC, 0.1% SDS, followed by a wash at 65°C in 0.1 X SSC, 0.1% SDS for 30 min and then autoradiographed on X-ray film at -70°C.

Quantitation of CFTR and PDH message levels: Appropriately exposed autoradiograms were scanned using a Laser densitometer (Molecular Dynamics). Comparative data are expressed in relationship to the signal produced by a probe for pyruvate dehydrogenase (PDH), a marker chosen because it is not expected to change during the culture and differentiation of cells.

Electron microscopy: Cells were grown on nucleopore PC membranes attached to 35 mm petri dishes with rat tail collagen. Cells were fixed at various points along the growth curve in universal fixative and stored at 4°C until all samples were ready. For scanning EM the fixed cells were washed with phosphate buffer, dehydrated in graded series of ethanol, critical point

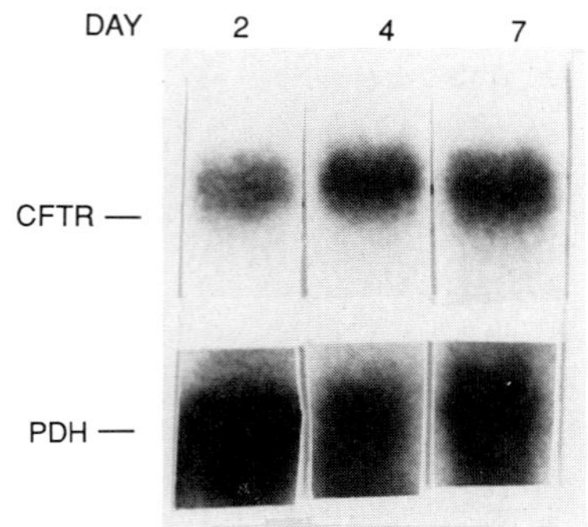

Fig. 2. Expression of CFTR and PDH mRNA in T84 cells. At the indicated days, total RNA samples from the cells were separated by formaldehyde-agarose gel electrophoresis, blotted onto Gene Screen plus membranes and hybridized sequentially to ^{32}P-labelled CFTR and PDH cDNA.

dried, mounted on metal stubs, palladium coated and examined with a Jeol 820 scanning electron microscope. For transmission EM the fixed cells were washed with phosphate buffer and post-fixed with osmium tetroxide. After dehydration through a graded acetone series the cells were infiltrated with Epon resin and polymerized overnight at 60°C in a flat embedded mold. Appropriate areas were selected by examining 0.5-1 micron thick sections under a light microscope. Thin sections were then cut on an ultramicrotome, collected on copper grids, counterstained with Sato's lead and uranyl acetate and examined using a Phillips 201 transmission electron microscope.

RESULTS

To determine whether the expression of CFTR was regulated in T84 cells we measured the steady-state levels of CFTR mRNA during the growth of T84 cells following subculture. Total RNA was isolated as described in Materials and Methods, separated by size on formaldehyde-agarose gels and hybridized to radioactively labelled CFTR cDNA. As is shown in Figure 1, T84 cells increase in number by approximately a factor of 10 following subculture. However, there does not appear to be any change in the amounts of CFTR mRNA that can be detected on Northern blots (Figure 2). The levels at days 2, 4, and 7 after subculture are similar to each other. To ensure that the amounts of RNA loaded onto each lane of the gel were the same, the blot was separately hybridized to radioactively labelled pyruvate dehydrogenase cDNA. This metabolic enzyme is not expected to vary during the growth of cells. It can be seen in Figure 2 that approximately equal signals with pyruvate dehydrogenase mRNA are seen in the three different lanes. We have similarly detected unaltered amounts of CFTR mRNA at other days throughout the growth cycle (data not shown).

We then examined the levels of CFTR mRNA in Caco2 cells. These cells differ from T84 in that they undergo morphological and functional differentiation in culture following subculture and growth to confluency. Representative data are shown in Figure 3. As was the case with T84, Caco2 cells grow by a factor of about 10 following subculture. However, unlike T84 cells there is a dramatic increase in the amounts of CFTR mRNA that can be detected. Levels of the message increase by 10-20 fold during the growth of Caco2 cells (Figure 4). The amounts of RNA loaded on the gel were approximately the same in all cases as determined by hybridization to pyruvate dehydrogenase cDNA.

Caco2 differentiation has been observed morphologically by a marked increase in the numbers of microvilli on the apical cell surface of the cells and by a change in the shape of cells to more polarized form. We therefore confirmed that Caco2 cells had differentiated during this growth in culture by examining the cells by scanning and transmission electron microscopy. Figure 5 illustrates a representative scanning electron micrograph of Caco2 cells at days 1 and 14 after subculture. A marked increase in microvilli is observed. As well, the size of the surface outline of the cells decreases significantly. This change is a reflection of the fact that the cells have become more cuboidal and elongated. This is seen in Figure 6 which illustrates representative transmission electron micrographs. Cells on day 1 are flat and spread over a large distance on the plastic and have few, short microvilli. On day 14, in contrast, the cells are taller and have a profusion of long microvilli. Thus, the Caco2 cells that we have analyzed are undergoing a process of morphological differentiation.

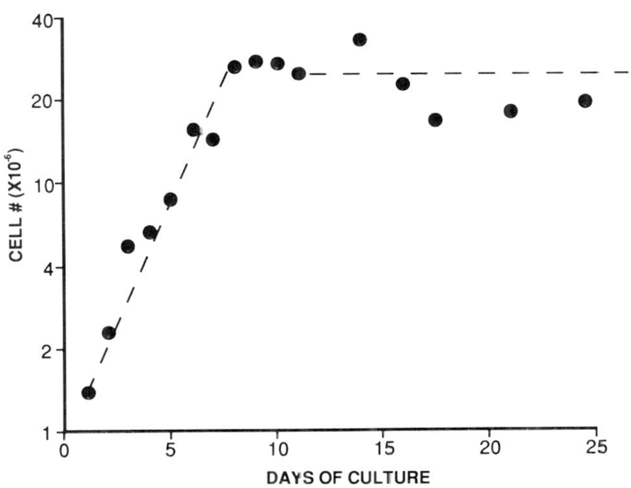

Fig. 3. Growth of Caco2 cells following subculture. Cells were seeded at approximately 2.5 x 10^6 per 100 mm dish and counted on the indicated days.

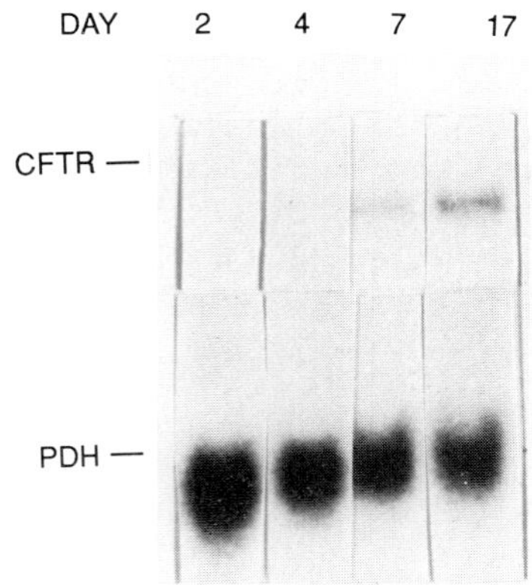

Fig. 4. Expression of CFTR and PDH mRNA in Caco2 cells. Total RNA samples were harvested from cells growing as described in Figure 3 on the indicated days, separated on formaldehyde-agarose gels, blotted on Gene Screen plus membranes and hybridized sequentially to ^{32}P-labelled CFTR and PDH cDNA.

DISCUSSION

We have determined that CFTR mRNA abundance is correlated with the state of differentiation of Caco2 cells. The levels of CFTR mRNA increase by an order of magnitude as the cells grow to confluence and undergo morphological and functional differentiation. This increase in CFTR mRNA abundance is not observed in T84 cells that do not undergo differentiation in culture. It appears therefore that in T84 cells CFTR expression is always maximal whereas in Caco2 cells it is modulated. This modulation could be mediated by increased transcription or by alterations in message stability. We are currently exploring these two possibilities.

As detected in these experiments, the amounts of CFTR mRNA seem to be proportional to the number of cells in the dish, in that the increase in CFTR mRNA abundance starts shortly after subculture. However, Caco2 cells tend to grow as small islands in which cell division occurs at the edges while the centre of the island contains non-growing cells (Pinto et al., 1983). It is possible that the cells in the centre of the islands have already differentiated and that the low levels of CFTR mRNA detected two and three days after subculture represent CFTR mRNA produced by the differentiated cells in the centre of the islands. We are currently examining this aspect of CFTR expression by *in situ* hybridization on growing Caco2 cells.

These results raise the question of the functional significance of the regulated CFTR expression. However, since the precise function of CFTR is not yet known, it is difficult to assess this point. Current assays for CFTR are based on differences in cellular transport functions detected when comparing

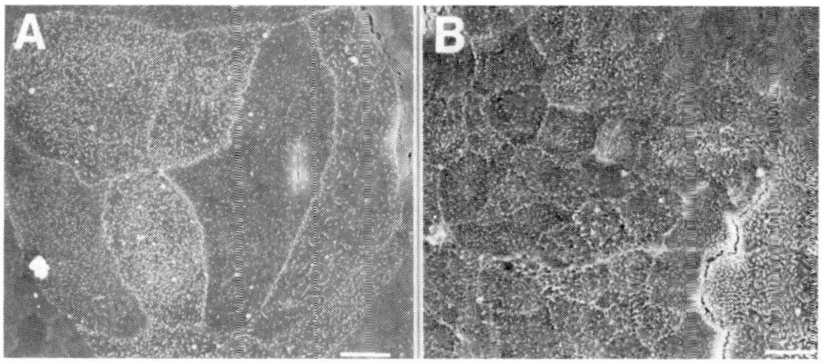

Fig. 5. Scanning electron micrographs of Caco2 cells. Cells on day 1 (panel A) or day 14 (panel B) were analyzed on a scanning electron microscope as described in Materials and Methods. Magnification X 1200. The bar represents 10 micro M.

normal and CF cells. These include the measurement of altered transcellular electric potentials (Knowles et al., 1981; Quinton, 1983), changes in transcellular resistance following stimulation with agents that increase intracellular cyclic AMP levels (Sato and Sato, 1984; Widdicombe et al., 1985), or the measurement of chloride channels in cells (Li et al., 1986; Frizzell et al., 1986) or isolated membranes (Schoumacher et al., 1987; Li et al., 1988). It is believed that the changes in transcellular resistance observed in normal cells upon stimulation with agents that raise intracellular cyclic AMP levels represent increased chloride transport (Quinton, 1989). T84 cells form highly resistant monolayers (1500 ohm·cm^2) and following treatment with agents such as vasoactive intestinal peptide there is an increase in net chloride secretion (Dharmasathaphorn et al., 1985). Given the expression of CFTR mRNA in T84 cells, one would have to assume that the cellular response to VIP occurs as a result of CFTR action inside these cells.

If this line of reasoning were applied to Caco2 cells, one would expect that those expressing the highest levels of CFTR mRNA would also show the largest changes in transcellular chloride movement. However, two difficulties arise in assessing this possibility. First, monolayers of Caco2 cells have much lower resistances than those of T84 cells (150 ohm·cm^2) and chloride-specific changes in electrical properties yield smaller differences (Grasset et al., 1984; 1985). Secondly, it is technically more complex to compare subconfluent (low expression) and confluent (differentiated, highly expressing) Caco2 cells using the same technique. Since measurements in Ussing chambers, that require intact monolayers cannot be used on the subconfluent cells, one would have to use single cell assays to compare the two sets of cells. However, single cell assays on confluent monolayers are not generally available and will have to be developed further. Thus, it is currently difficult to assess the cellular consequences of the altered CFTR mRNA expression. Nonetheless, provided this technical problem can be overcome it

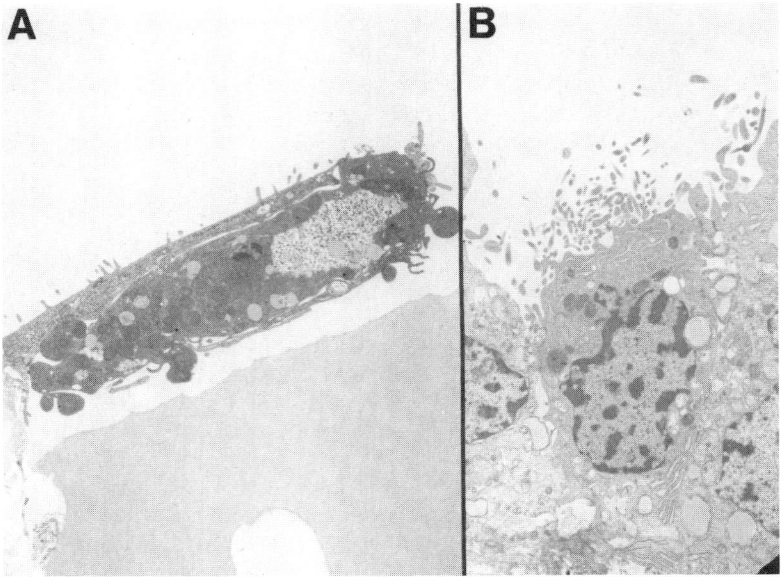

Fig. 6. Transmission electron micrographs of Caco2 cells. Cells on day 1 (panel A) and on day 14 (panel B) were analyzed on transmission electron microscope as described in Materials and Methods. Magnification X 7,600.

should be possible to determine if the regulated expression of CFTR mRNA has functional consequences.

Even though Caco2 cells were derived from an adult adenocarcinoma of the colon, on the basis of the patterns of enzyme expression and on their morphological differentiation, they are considered to represent a developmental precursor of enterocytes (Rousset, 1986). Thus, it is possible that the altered expression of CFTR mRNA that is observed in these cells has an *in vivo* counterpart in the differentiation that occurs in the crypt-villus progression of enterocytes. We are currently examining this possibility.

Acknowledgements

This research was supported by grants from the Canadian and U.S. Cystic Fibrosis Foundations and the National Institutes of Health (U.S.A) and represents part of the Cystic Fibrosis Research Development Programme at the Hospital for Sick Children. R.S. was supported by a Fellowship from the Canadian Cystic Fibrosis Foundation. We thank our colleagues Jack Riordan and Lap-Chee Tsui for their continued assistance. We also thank Ron Buick for introducing us to Caco2 cells and for assistance in starting this research.

References

Chirgwin, J.M., Przybyla, A.E., MacDonald, R.J., Rutter, W.J., 1979, Isolation of biologically active ribonucleic acid from sources enriched in ribonucleases, **Biochemistry**, 18:5294.

Dharmsathaphorn, K., Mandel, K.G., McRoberts, J.A., Tisdale, L.D. and Masui, H., 1984, A human colonic tumor cell line that maintains vectorial electrolyte transport, **Am. J. Physiol.**, 246:G204.

Dharmasathaphorn, K., Mandel, K.G., Masui, H., and McRoberts, J.A., 1985, **J. Clin. Invest.**, 75:462.

Feinberg, A.P. and Vogelstein, B., 1983, A technique for radiolabelling DNA restriction endonuclease fragments to high specific activity, **Anal. Biochem.**, 132:6.

Fourney, R.M., Miyakoshi, J., Day, R.S., and Paterson, M.C., 1988, Northern blotting: efficient RNA staining and transfer, **Focus**, 10:5.

Frizzell, R.A., Rechkemmer, G., and Shoemaker, R.L., 1986, Altered regulation of airway epithelial cell chloride channels in cystic fibrosis, **Science**, 233:558.

Grasset, E., Pinto, M., Dussaulx, E., Zweibaum, A., and Desjeux, J.-F., 1984, Epithelial properties of human colonic carcinoma cell line Caco-2: electrical parameters, **Am. J.Physiol.**, 247:C260.

Grasset, E., Bemabeu, J., and Pinto, M., 1985, Epithelial properties of human colonic carcinoma cell line Caco-2: effect of secretagogues, **Am. J.Physiol.**, 248:C410.

Knowles, M., Gatzy, J., and Boucher, R., 1981, Increased bioelectric potential difference across respiratory epithelia in cystic fibrosis, **New England J. Med.** 305:1489.

Li, M., McCann, J.D., Liedke, C.M., Naim, A.C., Greengard, P., and Welsh, M.J., 1988, Cyclic AMP-dependent protein kinase opens chloride channels in normal but not in cystic fibrosis airway epithelium, **Nature**, 331:358.

Pinto M., Robine-Leon, S., Appay, M.-D., Kedinger, M., Triadou, N., Dussaulx, E., Lacroix, B., Simon-Assmann, P., Haffen, K., Fogh, J., Zweibaum, A., 1983, Enterocyte differentiation and polarization of the human colon carcinoma cell line Caco-2 in culture, **Biol. Cell,** 47:323.

Quinton, P.M., 1983, Chloride impermeability in cystic fibrosis, **Nature**, 301:421.

Quinton, P., 1989, Defective epithelial ion transport in cystic fibrosis, **Clin. Chem.**, 35:726.

Riordan, J.R., Rommens, J.M., Kerem, B.-S., Alon, N., Rozmahel, R., Grzelczak, Z., Zielenski, J., Lok, S., Plavsic, N., Chou. J.-L., Drumm, M.L., Iannuzzi, M.C., Collins, F.S., Tsui, L.-C., 1989, Identification of the cystic fibrosis gene: cloning and characterization of complementary DNA, **Science**, 245:1066.

Rousset, M., 1986, The human colon carcinoma cell lines HT-29 and Caco-2: two in vitro models for the study of intestinal differentiation, **Biochimie**, 68:1035.

Sato, K. and Sato, F., 1984, Defective beta-adrenergic response

of cystic fibrosis sweat glands <u>in vivo</u> and <u>in vitro</u>, **J. Clin. Invest.**, 73:1763.
Schoumacher, R.A., Shoemaker, R.L., Halm, D.R., Tallant, E.A., Wallace, R.W., and Frizzell, R.A., 1987, Phosphorylation fails to activate chloride channels from cystic fibrosis airway cells, **Nature**, 330:752.
Widdicombe, J.H., Welsh, M.J., and Finkbeiner, W.E., 1985, Cystic fibrosis decreases the apical membrane chloride permeability of monolayers cultured from cells of tracheal epithelium. **Proc. Natl. Aca. Sci. USA** 82: 6167.

DISCUSSION

WINE: Dr. Buchwald, do you think the increasing levels of CFTR message you see during time in culture could mean that the density of CFTR mRNA is tracking the total membrane surface area.

BUCHWALD: I guess it's possible. The question is how would one test that. We are in the midst of trying to repeat these experiments with the P-glycoprotein probe that we have, which is also a membrane enzyme, to see whether this is peculiar to CFTR or whether it is characteristic of P-glycoprotein as well.

GREGER: Could one not argue in the following way: The cells increase in number after they are seeded; they are by no means confluent then, yet the sum of their surface areas is already large. If you now measure a low expression of CFTR and only an increase of this expression some time after seeding, when they start to become confluent, would one not have to conclude that the area of plasma membrane has little to do with the expression of CFTR?

BUCHWALD: I guess the question is, you know, some of the cells are changing their shape quite significantly and I don't know that anyone has ever looked at what the membrane density is in highly differentiated cells vs. relatively undifferentiated cells.

GREUNERT: Have you looked at any other factors, such as extracellular matrix components or soluble factors that might enhance differentiation in some of the other transformed lines that you have?

BUCHWALD: No. That's an experiment that we plan to do now that we know that expression may be related to the stage of differentiation. We're going to go back to our transformed sweat gland and nasal polyp cells and see whether we can influence the degree of differentiation. We haven't done it yet.

AUSIELLO: Yes. I had a question, but perhaps I can comment on Dr. Wine's question. We have conducted similar studies for apical membrane G-proteins and the sodium channel. What Dr. Greger says is quite true. There is stabilization of total membrane area, even though there is some microvilli development. But the lability of apical membrane proteins is greater than that for basolateral membrane proteins. W see an identical pattern of development of message on day 4, 5, 6 and 7 of cultured epithelia for certain G-proteins that regulate sodium channels as Dr. Buchwald presented for CFTR mRNA. The reverse is also true. When you take the fully developed polar cells, and either trypsinize them or even simply lift them off the plate and replate them, they rapidly lose their mRNA for G proteins and then redevelop the mRNA. Thus,the cell develops a lability with regard to apical membranes mRNA and protein. The data that Dr. Buchwald presented then are consistent with the possibility that CFTR is an apical membrane protein in these cells.

BUCHWALD: Right, which is why I suggested maybe message stability might also be part of this mechanism.

RECHKEMMER: I have a technical question on something I might have missed: are these cells grown on permeable supports?

BUCHWALD: No, they're grown on plastic.

RECHKEMMER: And you have no experience with these cells grown on permeable supports as far as differentiation?

BUCHWALD: No. We haven't done these experiments. You know, CaCo2 differentiation was reported 7 or 8 years ago by French workers. So we just basically followed their protocols and that was growing on plastic and feeding them every day once they got to a certain density.

FRIZZELL: It's certainly worth saying that Panc-1, if it doesn't express CFTR, certainly has a lot of the rectified chloride channels, as we've seen and as John Hanrahan has reported as well. Also, we have performed some preliminary experiments in T84 cells which sort of show the opposite side of the coin. We find, as you do, that transport is a function of time after plating. The amount of apparent CFTR message isn't changing in time but one thing that is changing is that the cells after plating are becoming more and more responsive to forskolin. The cAMP-induced chloride permeability change is essentially absent about 6 hours after plating T84 cells and increases in time up to a maximum at 2 or 3 days. During this time, CFTR expression by RNA blot shows only a small increase in CFTR message.

BUCHWALD: Are these monolayer experiments, Ray?

FRIZZELL: That's right.

BUCHWALD: Because there was a report that I read somewhere that said that the development of tight junctions takes place over a period of days, following subculture of T84 cells.

SLICHTER: We also find, with T84 cells, after you plate them down that the responsiveness to cAMP agonists does increase with time up to about 4 days and then is stable for the next few, and that this correlates very well with resting levels of cAMP which are very high early, and then decrease, and then can be stimulated, so that the channels may be somehow prestimulated or downregulated by the high cAMP levels, perhaps.

GREGER: Can the conclusion now be that CFTR and cAMP stimulation have little to do with each other?

BUCHWALD: Oh, no.

GREGER: But weren't you saying that the cAMP response comes prior to CFTR expression?

BUCHWALD: No, no - the other way around.

AUSIELLO: Maybe I can comment on that, which supports what Ray said. We see the sodium channel appear before the regulatory protein, the G-protein, so that for 2 or 3 days, in the A6 cells you have a sodium channel that is not pertussis-toxin-sensitive, which then gains pertussis-toxin sensitivity, corresponding to the message development for the G-protein. So I think one of the things we have to pay attention to when we're looking at the developmental message is that if these are multi-protein regulatory components there might very well be some asymmetry to that. I think that is what Ray is alluding to, and the asymmetry may go in different directions.

W. GUGGINO: We have done some preliminary experiments with HT29 cells in collaboration with Chip Montrose and Chahryad Montrose. We have shown that if you grow the cells in glucose or galactose, that cells grown in the more differentiating galactose growing medium, that CFTR mRNA will be expressed in greater quantity. This confirms what you find.

BUCHWALD: One of the things that I should make clear is that when you grow any epithelial cell, they don't grow as single cells. They always grow as patches. So a monolayer that is half confluent may in fact contain undifferentiated and differentiated cells depending on whether they are growing in patches of different sizes. So these measurements of message levels are really average, of some cells that have a lot and some cells that have a little, and you're just counting the mixture, in a sense.

CYSTIC FIBROSIS, THE CFTR, AND RECTIFYING Cl^- CHANNELS

J.J. Wine, D.J. Brayden, G. Hagiwara, M.E. Krouse, T.C. Law, U.J. Müller, C.K. Solc, C.L. Ward, J.H. Widdicombe*, and Y. Xia

Cystic Fibrosis Research Laboratory
Stanford University, Stanford, California

*Cardiovascular Research Institute
University of San Francisco

ABSTRACT

The human genetic disease cystic fibrosis is caused by a single defective gene on chromosome 7 that codes for a 1480 amino acid protein called the cystic fibrosis transmembrane conductance regulator (CFTR). The defect causes a profound reduction of Cl^- permeability in several tissues, which in turn impairs salt absorption and fluid secretion. A 25-80 pS, rectifying Cl^- channel has been targeted as the exclusive or primary channel affected in CF. However, we have found no evidence for significant activation or spontaneous activity of this channel in cell-attached patches of normal lymphoblasts or dog tracheal cells. However, in dog tracheal cells, we find lower conductance, linear Cl^- channels that are spontaneously active in unstimulated cells and may show increased activity in stimulated cells. Attempts to correlate the expression of mRNA for the CFTR protein in various types of cells with the presence of the rectifying Cl^- channel show a lack of correlation: i.e., depolarization-activated rectifying Cl^- channels have been found in excised, inside-out patches from all cell types that we have examined to date, but the CFTR mRNA has so far only been detected in a subset of epithelial cells.

INTRODUCTION

The human genetic disease cystic fibrosis is caused by a single defective gene on chronosome 7 that codes for a 1480 amino acid protein called the cystic fibrosis transmembrane conductance regulator (CFTR, ref. 1). A single amino acid deletion or

The Identification of the CF (Cystic Fibrosis) Gene
Edited by L.-C. Tsui *et al.*, Plenum Press, New York, 1991

any of several point mutations causes a profound reduction in Cl^- permeability of the sweat ducts (2-5) and airway cells (6-8) from subjects with cystic fibrosis. Sweat secretion following intradermal isoproterenol injections is abolished in CF subjects and reduced to 1/2 normal in heterozygotes, although secretion to cholinergic agonists is normal (9,10). In intestinal tissues, Cl^- secretory responses to agents that act via cAMP, cGMP and Ca^{2+} are abolished (11-13). Na^+ permeability in airway cells is enhanced (e.g. 14). The goal now is to understand exactly how the CFTR protein regulates normal membrane conductance and how the mutations interfere with that function.

A leading hypothesis is that the CFTR mutation renders a specific, distinctive, rectifying Cl^- channel incapable of being switched from a totally non-conductive state to an activated state by a phosphorylation event mediated by cAMP-dependent protein kinase (15). The hypothesis is based on Ussing chamber and microelectrode experiments in several tissues that provide convincing evidence for a Cl^- secretory mechanism that is controlled, in part, by regulation of apical Cl^- permeability (e.g. 16,17), and on patch-clamp studies that show that dog and human tracheal cells possess a distinctive, rectifying Cl^- channel that is thought to be either the sole channel (18-19) or one of two channels (15,20) mediating Cl^- secretion in response to adrenergic stimulation. The channel may (20) or may not (19) be affected by Ca^{2+}, but in both studies was found to be non-responsive to adrenergic stimulation in cells from CF subjects. Subsequent studies of this channel in excised, inside-out patches indicated that it could be activated with catalytic subunit of cAMP-dependent protein kinase, and perhaps by protein kinase C, in patches from normal but not CF subjects (e.g. 21-24).

The hypothesis can take 3 forms (Fig. 1). The most restrictive form equates the rectifying Cl^- channel with the CFTR protein. This hypothesis is tenable because the CFTR has predicted structural properties consistent with its being a channel, and has abundant sites for regulation by both PKA and PKC (1). A less restrictive form of the hypothesis postulates that the rectifying Cl^- channel is a different protein, but is uniquely associated with the CFTR protein, i.e., only the rectifying Cl^- channel is affected in CF, and the CFTR and the channel will always be found together. The least restrictive form of the hypothesis is that the rectifying channel is a separate protein, and is one of several channels that are regulated, in some as yet unspecified way, by the CFTR. Three other possibilities are that the CFTR may be a different Cl^- channel, or may be associated with a different Cl^- channel either uniquely or not.

In our laboratory we are attempting to assess these hypotheses. In the absence of antibodies to the CFTR protein, and without the ability at present to express the CFTR protein in cells that normally lack it, we have resorted to a comparative approach, in which we are attempting to study the regulation of

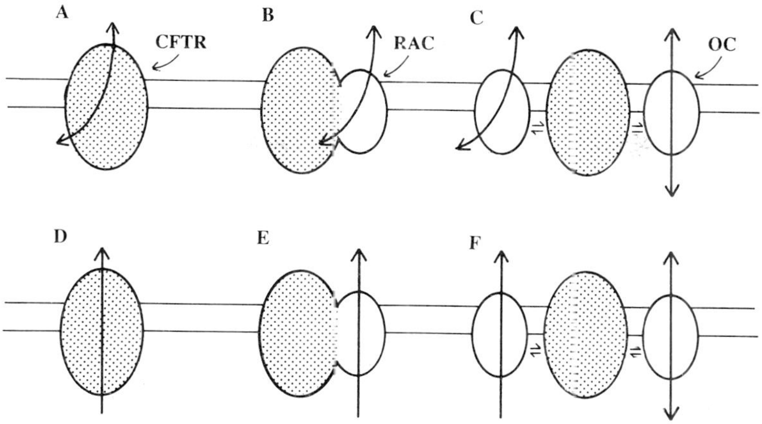

Fig. 1. How is the CFTR protein related to the rectifying Cl^- channel? Six possibilities are shown. A. The CFTR and the rectifying Cl^- channel may be the same protein. B. The CFTR and the rectifying Cl^- channel may be obligatory components of a multi-protein transmembrane transporter. C. The CFTR may more loosely regulate several channels, including the rectifying Cl^- channel. D - F. The CFTR may be associated with a different channel and not with the rectifying Cl^- channel, in any of the three ways just presented.

Cl^- conductance in a variety of tissues that show differential expression of the CFTR protein, as assessed by measurements of the CFTR mRNA. Our goal is to provide a context that will aid the interpretation of expression studies of the CFTR protein.

METHODS AND RESULTS

Defective Apical Membrane Chloride Permeability in CF Tissues

We have confirmed the Cl^- permeability defect in sweat ducts and the secretory defect in sweat glands, and find that each is so profound that these measures can detect a significant reduction of function even in heterozygotes (10, 25, 26). Since then, we have focused our efforts on the single-channel basis for Cl^- permeability.

Patch-Clamp Studies of Sweat Gland Cells

We initially surveyed channel populations in primary cultures of sweat secretory coil and sweat ducts, where we hoped we would see dozens of Cl^- channels in every patch, as predicted by the conductance values of intact, microperfused ducts. Our results were not what we expected (27), and led us to suspect that the properties of sweat gland cells were being compromised in culture. Subsequent work makes that seem more likely (28). When we turned to assessing the rectifying Cl^- channel hypothesis, we concentrated our efforts on cell-attached patches, since the channel has an inherent tendency to activate in excised patches from either normal or CF subjects, with such activation apparently being irreversible. Since activation by phosphorylation can take many minutes, attribution of channel activity to the phosphorylation event is subjective, and rigorous controls are required to avoid false-positives. Assessing activation of rectifying channels in cell-attached patches is also non-trivial, and we had no success in activating the rectifying Cl^- channel in a small sample of sweat gland cells.

Studies with Lymphocyte and Lymphoblast Cell Lines

As these experiments were proceeding, it was discovered that lymphocytes and lymphoblasts contain high densities of a rectifying Cl^- channel that is nearly indistinguishable from the channel in epithelial cells. Crucially, this channel could be activated with almost 100% efficacy in normal cells (70% of cell-attached patches could be activated with 8-Br-cAMP, vs. 78% of excised patches with strong depolarization) but was never observed to be activated by cAMP in CF cells (29). Given that report, it isn't surprising that we chose to study lymphoblasts, which are seductively convenient to obtain and culture compared with primary cultures of epithelial cells.

Unfortunately, although we did find rectifying Cl^- channels in lymphoblasts (30), we observed no significant agonist-induced activation of the rectifying Cl^- channels using a variety of stimuli and conditions, including our best efforts to reproduce the conditions of the original study (31-33). A blind study of normal and CF B-lymphoblasts also did not reveal any significant differences (unpublished results). Subsequent attempts to detect message for the CFTR in lymphocyte cell lines were negative (1), and to date we have been unable to detect any evidence for CFTR message in these cells, using the PCR technique with primers directed to the C-terminus or middle regions of the CFTR coding sequence.

Studies of Dog Tracheal Cells

The work with lymphocytes left us in a quandary. We had now failed to observe activation of the rectifying Cl^- channel in two cell types, and although our activation experiments with sweat gland cells had been cursory, the experiments with lymphocytes were extensive, and many maneuvers that might have made a difference had been tried. So we decided to renew our studies of epithelial cells. We chose dog tracheal cells because they are readily available and have been extensively studied in Ussing chambers both as native tissues and as cultures. More importantly, it was in dog tracheal cells that activation of rectifying Cl^- channels was first reported (17).

It is important to review what was reported in that study. Active rectifying Cl^- channels were detected in 43 patches; the proportion of patches having such channels varied from 2-10% of patches in different dogs. Using the highest percentage, the total of 43 patches with rectifying Cl^- channels came from a minimum of 430 patches. Rectifying Cl^- channel activity following isoproterenol addition was seen in 8 patches, with the channel spontaneously active in 4 patches and refractory until excision in 31 patches. In sum, activation of rectifying Cl^- channels was seen in 2% of patches tested, a proportion 35 times less than that reported for lymphocytes.

The rarity of the regulatory event that is postulated to be the basis for Cl^- secretion in airway cells has not generally been appreciated, and raises the issue of whether activation of the rectifying Cl^- channel is necessary and sufficient to account for macroscopic currents in these tissues. Simply reexamining the efficacy of activation of rectifying Cl^- channels did not seem likely to be informative, but we thought it important to determine the full range of channel activity in which the postulated regulatory event occurs, to see if activation of the rectifying Cl^- channel is the only significant channel event following stimulation.

Single-channel recordings were made from the apical membranes of confluent, primary cultures of canine tracheal cells grown on human placental collagen in both cell-attached and excised, inside-out modes. From 551 attempts, we obtained 221 usable patch recordings (40%). Of these, 87 were recorded at 22°C, and 134 at 32-35°C. All patches were checked for high conductance channels, and a subset of 104 patches was analyzed in more detail in an attempt to characterize the lower conductance, active channels. Evidence for at least 6 types of channels was obtained. Linear cation channels of 10-15 pS, a 40 pS cation non-selective channel, and a K^+ channel of 10 pS were recorded in 20-40% of the patches (cell-attached and excised), provided conditions were appropriate for seeing the channel. Two linear anion channels of 7-12 pS and c. 300 pS and a rectifying anion channel of c. 24 pS at 0 mV were recorded in 7-12% of patches. (The lower conductance anion channel was only seen in certain cell-attached patches, and the two higher conductance channels were seen almost exclusively in excised patches.)

Cell-attached patches from unstimulated cells at 22°C were typically silent or near silent, but at 32-37°C active channels were recorded in almost 80% of the patches, the great majority of active channels having conductances smaller than 10 pS. Following excision at either temperature, activity in higher conductance channels greatly increased, with the most common higher conductance channel being a cation non-selective channel of 22 pS at 22°C and 40 pS at 32-37°C. When excised patches were held for long periods, openings of other higher conductance channels gradually emerged, including the rectifying Cl^- channel and the very high conductance, linear, multi-level Cl^- channel.

Cells that were preincubated or acutely stimulated with isoproterenol did not differ qualitatively from unstimulated cells, but we have preliminary evidence for increased activity in smaller conductance anion channels. We saw only 4 instances of possible activation of higher conductance channels: 3 of these were linear channels and had properties suggesting they were cation non-selective, one was a rectifying Cl^- channel. In 2 of the 4 examples, the data suggest that the patch was inadvertently excised soon after addition of isoproterenol, but in one patch where activation of the linear channel occurred, its large offset allowed us to establish unequivocally that the patch remained attached throughout the response (Fig. 2).

It is important to emphasize that our data with regard to the rectifying Cl^- channel agree about as closely as could be expected with the original study by Welsh (18), as Table 1 makes clear. Both studies of dog tracheal cells find the rectifying Cl^- channels in about the same proportions overall; both find they are rarely spontaneously active in cell-attached patches, and in both they rarely respond to isoproterenol.

Although the data on the rectifying Cl^- channel agree, the significance of the rectifying Cl^- channel appears quite different when it is placed in the context of other channel activity. When observations are restricted to patches that are definitely

Table 1

COMPARISON OF DOG CELL STUDIES

	Welsh, 1986	Müller et al.
Rectifying Cl^- channel:		
Total:	43/430 (10%)	16/221 (7%)
spontaneously active c/a:	4/430 (1%)	1/52 (2%)
iso activated:	8/430 (2%)	1/52 (2%)
excision induced:	31/430 (7%)	15/221 (7%)
other anion channels active cell-attached:	not discussed	10/52 (19%)

Two studies of primary cultures of dog tracheal cells. The denominator in Welsh study was set to n = 430 patches because at most 10% of patches had active rectifying Cl^- channels, with a total of 43 active patches. We now know that the proportion of active patches is strongly influenced by the degree of depolarization. Müller et al. held 221 excised patches long enough to check for rectifying Cl^- channels, but these were usually not depolarized beyond 75-100 mV, which may account for the low proportion of patches with active rectifying Cl^- channels. Only 51 patches were maintained in the cell-attached configuration throughout the testing period.

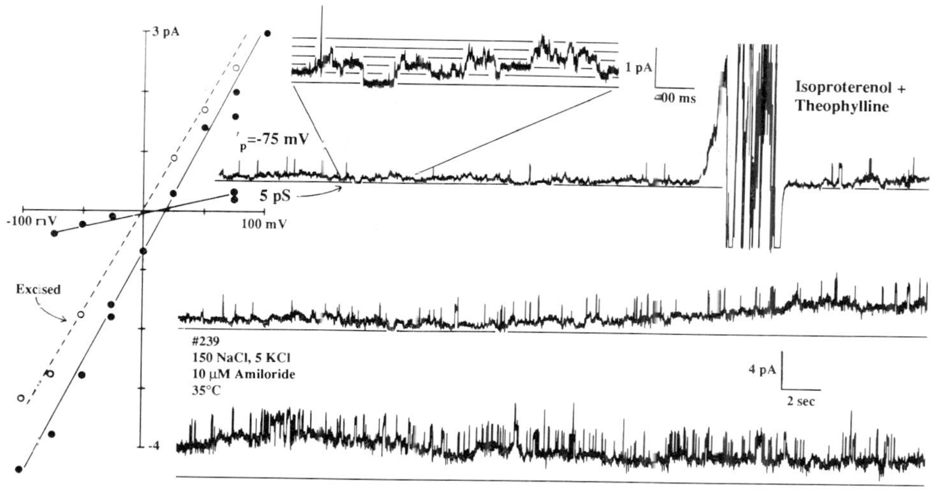

Fig. 2. Increased activity in 5 pS channels and 34 pS cation nonselective channels following bath addition of isoproterenol (50 μM) and theophylline (0.5 mM). The trace is a continuous current recording from a cell-attached patch held at V_p = -75 mV. Prior to addition of isoproterenol, single brief openings of the 34 pS channel and multiple copies of the 5 pS channel are seen (amplification and time base are shown expanded in the inset). After isoproterenol addition, a gradual increase in activity was observed for both channels. I(V) relations for both channels are shown at left. Following excision later in the experiment, the I(V) curve for the 34 pS channel shifted to the left and reversed at V_p = 0. The lower conductance channels became inactive or were obscured by the increased activity in the higher conductance channel following excision. Pipet and bath contained 150 mM NaCl, 2 mM Mg^{2+} and 2 mM Ca^{2+} . The nonselective cation channel typically has a large offset of V_p = -25 to -45 mV when cell-attached. While the 5 pS channels reverses at about V_p= -10 mV, which is consistent with either a Cl^- channel, or a cation channel with greater permeability to K^+ than to Na^+.

cell-attached, and when cells are observed at close to their normal temperatures, the rectifying Cl^- channel was seen to make a trivial contribution to the apical membrane permeability in our studies. Instead, the permeability was dominated by smaller conductance channels, and one larger conductance cation channel.

Increased activity in these channels following isoproterenol was much more common than increased activity in the rectifying Cl^- channel, but the effects were nevertheless small and variable. The small anion channels are thus candidates for mediating the increased Cl^- conductance of the apical membrane, but we do not yet consider the evidence to be compelling. Nevertheless, since the apical Cl^- permeability is essentially abolished in CF airway cells (8), these other anion channels, which are active in cell-attached patches, must be considered when assessing the consequences of the mutation in the CFTR.

The activity we see in cation channels does not fit with expectations for a Cl^--secreting epithelium, but tracheal airway cells, even in dogs, have reabsorptive as well as secretory functions, and display both amiloride-sensitive and amiloride insensitive Na^+ currents.

What, then, is the role of the rectifying Cl^- channel? We aren't sure, but during the course of our experiments, we have been struck by many points of similarity between the rectifying Cl^- channel and another unusual anion channel of unknown function: the very high conductance, linear anion channel. Both channels occur with similar frequency (in excised patches from dog cells, we recorded the high conductance anion channel in 14% of patches and the rectifying anion channel in 7% of patches); both channels were seen almost exclusively in excised patches; for both kinds of channel it was often the case that they appeared only after the patch had been excised for several minutes and subjected to non-physiological levels of depolarization. Finally, both channels appear to occur ubiquitously in many different types of cultured cells (for the very high conductance linear anion channel, see ref. 34). It seems clear that rectifying Cl^- channels participate in cell volume regulation (35), but it is not yet clear that the voltage-induced and swelling-induced Cl^- channels are identical.

Studies of T84 and HT29 Colonic Tumor Cells

The above work was done without benefit of the new knowledge that came with the discovery of the CF gene. As soon as it became possible to look for CFTR mRNA in cells, it became clear that, all other things being equal, it was probably wise for people interested in CF to study cells for which there was some evidence of CFTR expression. We are presently studying the single channel basis of apical membrane permeability in T84 cells (35, 36) and HT_{29} cells (e.g. 37) using an approach similar to the one used with dog tracheal cells. Preliminary results indicate that the rectifying Cl^- channel is no more prevalent in these cells than in any other cell types we have tested, even though I_{sc} responses in Ussing chamber experiments are, as previously reported, very large (Fig. 3C), and even though levels of CFTR expression in these cells are higher than any other cell types we have tested. We have not yet observed activity of the rectifying Cl^- channel in cell-attached patches of T84 cells, even in cells stimulated at elevated temperatures.

We are also using channel blockers to investigate the channel basis of the I_{sc} response in T84 cells. We find that DNDS at concentrations up to 10 mM does not have an observable effect on the I_{sc} produced by either VIP or carbachol (Fig. 3C). However,

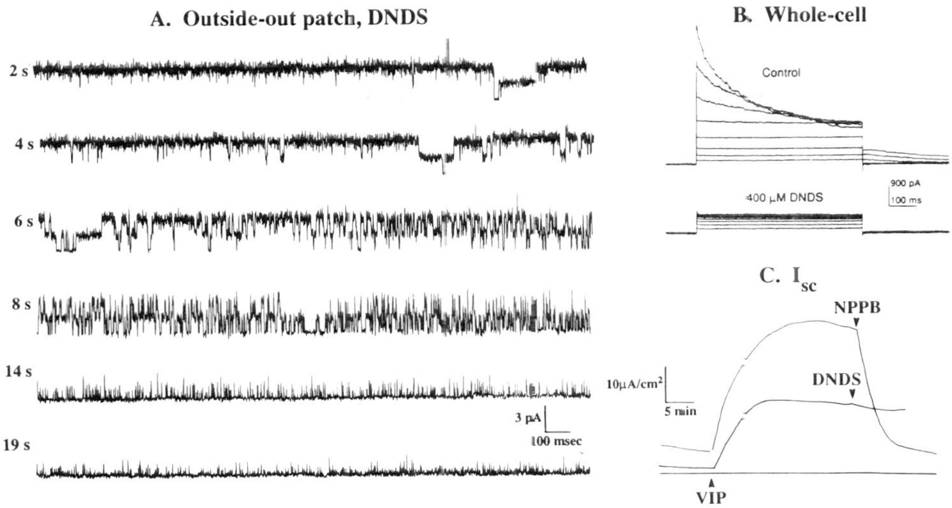

Fig. 3. The stilbene DNDS blocks outwardly rectifying Cl^- channels induced by depolarization or by swelling, but does not block Cl^- secretion of T84 cells in Ussing chamber experiments. A. Current recording from an outside-out patch excised from a T84 cell and depolarized until outwardly rectifying Cl^- channels appeared. Records shown at +50 mV while washing in bath solution containing 0.4 mM DNDS; trace is continuous for first 8 sec. The block was fully reversible and was obtained with concentrations as low as 0.1 mM (lower concentrations not tried); the effect was also strongly voltage dependent, as is shown in Fig. 4B. Whole-cell Cl^- currents recorded in response to voltage steps between -40 and +100 mV from an osmotically swollen Tn T84 cell, before and immediately after the introduction of DNDS to a final concentration of ~0.4 mM. Block of whole cell Cl^- currents is also rapidly reversible and voltage dependent. C. Isc responses of two T84 monolayers to VIP, followed by application of either NPPB or DNDS. DNDS at concentrations as high as 10 mM did not affect Isc (or PD in open circuit conditions).

as previously reported (37), DNDS does block swelling-induced whole cell currents in T84 cells (Fig. 3B), and 0.1 mM also blocks single-channel currents from depolarization-induced rectifying Cl^- channels recorded in outside-out patches from non-swollen T84 cells (Fig. 3A). The inconsistency between DNDS block of rectifying Cl^- channels reconstituted into lipid bilayers and its failure to affect I_{sc} was noted previously (37). One problem with interpreting the results is that DNDS block is voltage-dependent, and so would be expected to be less in I_{sc} experiments. We are presently attempting to determine if that explanation is sufficient.

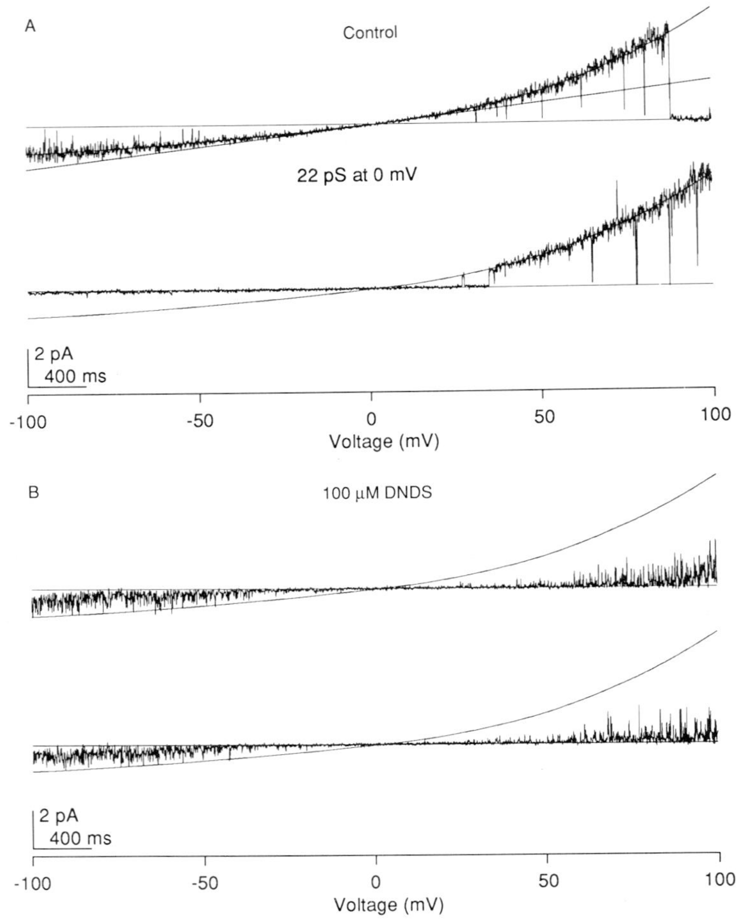

Fig. 4. Currents from a voltage-induced rectifying Cl^- channel in response to voltage ramps between −100 and +100 mV in control (A) and in the presence of 0.1 mM DNDS (B). The curved line is a best-fit to the open-channel in control solutions. Note that even though the block is strongly voltage-dependent, the open channel current at negative voltages is still reduced by ~50%. If the apical Cl^- channels are mostly open during Isc measurements, and if the current through these channels is the rate-limiting step in the Isc response, these results indicate that DNDS should be partly effective at blocking Isc associated with Cl^- secretion and very effective at blocking Isc associated with Cl^- absorption.

Survey of Rectifying Cl^- Channels and CFTR mRNA in Different Cell Types

We are attempting to correlate the presence of rectifying Cl^- channels with mRNA for the CFTR protein, both by surveying the literature and with our own data. Preliminary studies have not shown a correlation between detectable CFTR expression and the presence of rectifying Cl^- channels (Table 2). Rectifying Cl^- channels that typically require excision and depolarization to activate have been found at similar densities in lymphoblasts (29), fibroblasts (40), keratinocytes (13), glia (41), the Hep G2 liver cell line (unpublished observations, see ref.42 for review of these cells), and PANC-1 pancreatic tumor cells (43), in which we or others have been unable to find evidence for CFTR message (see Table 2), as well as in T84 colon carcinoma cells, HT_{29} colon carcinoma cells and primary cultures of tissues from sweat secretory coil, sweat duct, airway and pancreas, in which CFTR message has been found (Table 2). One problem with this attempted correlation is that we have not yet found a cell which definitively lacks the depolarization-activated rectifying Cl^-

Table 2

Cell Type	Rectifying Cl^- Channels	(Ref)	CFTR mRNA	(Ref)
Lymphoblasts	++	Chen et al., 1989	?	Riordan et al., 1989,*
Fibroblasts	+	Bear, 1988	?	Riordan et al., 1989
Keratinocytes	++	Bijman et al., 1988	?	*
Glia	+	Barres et al., 1988	?	Riordan et al., 1989
PANC-1	++	Tabcharani et al.,1989	?	*
Hep G2	+	*	?	*
—				
T_{84}	+	Halm et al., 1988	+	Riordan et al., 1989*
HT_{29}	+	Hayslett et al., 1987	+	*
1° Sweat gland	+	Krouse et al., 1989	+	Riordan et al., 1989
1° Sweat duct	+	Joris et al., 1989	+	Riordan et al., 1989
1° Airway	+	Welsh & Liedtke, 1986 Frizzell et al., 1986	+	Riordan et al., 1989
1° Pancreas	+	Gray et al., 1989	+	Riordan et al., 1989

* = Unpublished results from Stanford CF Research Lab

Presence of depolarization-activated rectifying Cl channels is not predictive for detection of CFTR mRNA. Rectifying Cl^- channels appear to be ubiquitously distributed in cells. Double + indicates that these channels were found to be present in most patches, often as multiple copies. Asterisks in reference column for mRNA studies indicate unpublished work based on PCR studies using primers to the C-terminus. Cells assessed for mRNA were not always identical lines used for electrophysiology, and in some cases PCR or Northern blots were done from native tissues whereas electrophysiology was always from cultured cells. A question mark in column for CFTR studies means no signal has yet been detected for these cell types using 30 or 2x30 round of PCR amplification, or Northern blots.

channel. Unfortunately, the loose linkage between mRNA and protein levels does not allow a definite conclusion; the CFTR protein may have a long half life and may be sustained at sufficient levels with either very low levels of expression or with sporadic expression. Even so, we think the pattern of the above data makes it less likely that the CFTR and the rectifying Cl^- channel are identical.

DISCUSSION

Until recently, the linkage between phosphorylation-dependent regulation of the rectifying Cl^- channel and the CF Cl^- permeability defect seemed sufficiently secure to support the hypothesis that the rectifying Cl^- channel might be the protein affected in CF. One part of the evidence for that linkage has been reexamined here: the ability of normal secretagogues to increase activity of rectifying Cl^- channels in cell-attached patches of normal epithelial and non-epithelial cells. Existing evidence for activation is equivocal. More importantly, the rectifying Cl^- channel is shown to be just one of several anion channels in the membranes of these cells, and in our experiments it was active almost exclusively in excised patches.

If the rectifying Cl^- channel is not the CFTR, is it uniquely associated with the CFTR protein? That is, does cystic fibrosis result exclusively from the defective operation of a CFTR-rectifying Cl^- channel complex? This possibility, once so attractive, is undermined both by the weakness of the direct evidence for activation of the rectifying Cl^- channel, and by evidence that cAMP-dependent Cl^- conductances do not have the properties predicted for the rectifying Cl^- channel. For example, in whole-cell measurements of T84 cells, forskolin caused a large increase in a Cl^- conductance with a linear I(V) relation (44), whereas rectifying whole-cell currents are seen after cell-swelling (37). Experiments with channel blockers are also difficult to reconcile with the idea that the rectifying Cl^- channel is uniquely responsible for the Cl^- conductances affected in CF. In addition to the evidence presented above, blockers that are effective against the rectifying Cl^- channel in excised patches do not affect the conductance of the eccrine sweat duct (45), suggesting that the duct conductance is mediated by a different Cl^- channel. These problems are outlined in Table 3.

If the rectifying Cl^- channel is not uniquely associated with the CFTR, can other channels be identified whose operation might be regulated by the CFTR? The possibility that other channels might be involved was first suggested by Frizzell et al. (15, 20), who identified a 20 pS anion channel in airway cells that was defectively regulated in CF cells. Gray and Argent et al. (46) have found copious quantities of a linear, 5-7 pS, secreting and cAMP-regulated Cl^- channel in primary cultures of fetal pancreatic duct cells. Since they rarely saw rectifying

Cl^- channels, they propose the small conductance channel as a candidate for the channel that is defective in CF. As indicated above, we see a similar channel in both tracheal and T84 cells, but we have not established its function. More recently, in experiments with normal nasal turbinate cells that were cultured in the presence of cholera toxin, the most common channel encountered in excised inside-out patches at 37°C was a 20 pS anion channel seen in 117/495 (24%) of patches; the rectifying Cl^- channel was seen in 4 patches (1%) and the very high conductance anion channel in 8 (2%) patches (47). The authors report that the 20 pS anion channel is defectively regulated in cystic fibrosis tissues, and that the number of Na^+ channels is abnormally high, but the experiments are not reported in sufficient detail to be evaluated (48).

Table 3

DO RECTIFYING Cl^- CHANNELS PLAY A CENTRAL ROLE IN THE PATHOPHYSIOLOGY OF CYSTIC FIBROSIS?

We found that rectifying Cl^- channels:

1. are found widely in epithelial and non-epithelial cells.
2. are not correlated with expression of CFTR mRNA.
3. were rarely activated by isoproterenol.
4. were rarely active cell-attached in our experiments.
5. require excision and sustained depolarization to activate.
6. may mediate swelling-induced Cl^- conductance increases.
7. are less active cell-attached than other apical Cl^- channels.

CONCLUSION

In conclusion, the evidence pointing to a unique or central role for the rectifying Cl^- channel in the pathophysiology of cystic fibrosis is inconclusive. The behavior of the channel is complicated and multidetermined - properties that require considerable subjectivity in the interpretation of results. Some investigators have used the complexity of the channel to dismiss negative evidence, but the argument cuts both ways: the channel's complexity may also lead to false positive results. The accepted method for dealing with complex and multidetermined outcomes is to conduct the experiments in blinded fashion. The use of such controls, plus powerful new approaches made possible by the cloning of the CF gene, should allow these questions to be resolved in the near future. If the rectifying Cl^- channel is uniquely responsible for Cl^- secretion and absorption in the tissues affected by CF, it will be important to design paradigms that allow its normal and abnormal function to be demonstrated reproducibly. If it is not uniquely responsible for these functions, it is critical that the other molecular species involved be identified soon.

Acknowledgements

Supported by NIH grants R01 DK 39659 and HL42368; by Cystic Fibrosis Research, Inc.; and by the Cystic Fibrosis Foundation. We thank Jan Ruby for help in preparing the manuscript, and Ron Kopito for collaborative efforts with PCR studies.

References

1. Riordan, J.R., J.M. Rommens, B.-S. Kerem, N. Alon, R. Rozmahel, Z. Grzelczak, J. Zielenski, S. Lok, N. Plavsic, J.-L. Chou, M.L. Drumm, M.C. Iannuzzi, F.S. Collins, and L.-C. Tsui, 1989, Identification of the cystic fibrosis gene: Cloning and characterization of complementary DNA, **Sci.**, 245:1066.
2. Quinton, P.M., 1983, Chloride impermeability in cystic fibrosis, **Nature**, 301:421.
3. Bijman, J., and P.M. Quinton, 1984, Influence of abnormal Cl^- impermeability on sweating in cystic fibrosis, **Am. J. Physiol.**, 247:C3.
4. Quinton, P.M., 1986, Missing Cl conductance in cystic fibrosis, **Am. J. Physiol.** 251:C649.
5. Bijman, J., and E. Frömter, 1986, Direct demonstration of high transepithelial chloride conductance in normal human sweat duct which is absent in cystic fibrosis, **Pflug. Arch.**, 407:S123.
6. Knowles, M.R., M.J. Stutts, A. Spock, N. Fischer, J.T. Gatzy, and R.C. Boucher, 1983, Abnormal ion permeation through cystic fibrosis respiratory epithelium, **Sci.** 221:1067.
7. Widdicombe, J.H., M.J. Welsh, and W.E. Finkbeiner, 1985, Cystic fibrosis decreases the apical membrane chloride permeability of monolayers cultured from cells of tracheal epithelium, **Proc. Natl. Acad. Sci. USA**, 82:6167.
8. Willumsen, N.J., and R.C. Boucher, 1989, Shunt resistance and ion permeabilities in normal and cystic fibrosis airway epithelia, **Am. J. Physiol.**, 256:C1054.
9. Sato, K., and F. Sato, 1984, Defective beta adrenergic response of cystic fibrosis sweat glands in vivo and in vitro, **J. Clin. Invest.**, 73:1763.
10. Behm, J.K., G. Hagiwara, N.J. Lewiston, P.M. Quinton, and J.J. Wine, 1986, Hyposecretion of beta-adrenergically induced sweating in CF heterozygotes, **Pediatr. Res.**, 22:271.
11. Taylor, C.J., P.S. Baxter, J. Hardcastle, and P.T. Hardcastle, 1988, Failure to induce secretion in jejunal biopsies from children with cystic fibrosis, **Gut**, 29:957.
12. Berschneider, H.M., M.R. Knowles, R.G. Azizkhan, R. Boucher, N.A. Tobey, R.C. Orlando, and D. W. Powell, 1988, Altered intestinal chloride transport in cystic fibrosis, **FASEB J.**, 2:2625.
13. Bijman, J., M. Kansen, A.H. Hoogeveen, B. Scholte, A. van der Kamp, and H. de Jonge, 1988, Electrolyte transport in normal and CF epithelia, in: "Exocrine Secretion," P.Y.D.

Wong and J.A. Young, eds., Hong Kong University Press, Hong Kong.

14. Boucher, R.C., C.U. Cotton, J.T. Gatzy, M.R. Knowles, and J.R. Yankaskas, 1988, Evidence for reduced Cl^- and increased Na^+ permeability in cystic fibrosis human primary cell cultures, **J. Physiol.**, 405:77.
15. Frizzell, R.A., 1987, Cystic fibrosis: a disease of ion channels? **Trends in Neurosci.**, 10:190.
16. Frizzell, R.A., M. Field, and S.G. Schultz, 1979, Sodium-coupled chloride transport by epithelial cells, **Am. J. Physiol.**, 236:1.
17. Welsh, M.J., 1987, Electrolyte transport by the airway epithelia, **Physiol. Rev.**, 67:1143.
18. Welsh, M.J., 1986, Single apical membrane anion channels in primary cultures of canine tracheal epithelium, **Pflug. Arch.**, 407:S116.
19. Welsh, M.J., and C.M. Liedtke 1986, Chloride and potassium channels in cystic fibrosis airway epithelia, **Nature** 322:467.
20. Frizzell, R.A., G. Rechkemmer, and R.L. Shoemaker, 1986, Altered regulation of airway epithelial cell chloride channels in cystic fibrosis, **Sci.**, 233:558.
21. Schoumacher, R.A., R.L. Shoemaker, D.R. Halm, E.A. Tallant, R.W. Wallace, and R.A. Frizzell, 1987, Phosphorylation fails to activate chloride channels from cystic fibrosis airway cells, **Nature**, 330:752.
22. Li, M., J.D. McCann, C.M. Liedtke, A.C. Nairn, P. Greengard, and M.J. Welsh, 1988, Cyclic AMP-dependent protein kinase opens chloride channels in normal but not cystic fibrosis airway epithelium, **Nature**, 331:358.
23. Hwang, T.-C., L. Lu, P.L. Zeitlin, D.C. Gruenert, R. Huganir, and W.B. Guggino, 1989, Cl^- channels in CF: Lack of activation by protein kinase C and cAMP-dependent protein kinase, **Sci.**, 244:1351.
24. Li, M., J.D. McCann, M.P. Anderson, J.P. Clancy, C.M. Liedtke, A.C. Nairn, P. Greengard, and M.J. Welsh, 1989, Regulation of chloride channels by protein kinase C in normal and cystic fibrosis airway epithelia, **Sci.**, 244:1353.
25. Wine, J.J., N.J. Lewiston, and G. Hagiwara, 1988, Reduced chloride permeability in sweat ducts of cystic fibrosis heterozygotes, **Ped. Pulmon.**, suppl. 2:110.
26. Wine, J.J., 1988, The heterozygote phenotype in cystic fibrosis, **in:** "Cellular and Molecular Basis of Cystic Fibrosis," R. G. Mastella, and P.M. Quinton, eds., San Francisco Press, San Francisco.
27. Krouse, M.E., G. Hagiwara, J. Chen, N.J. Lewiston, and J.J. Wine, 1989, Ion channels in normal human and cystic fibrosis sweat gland cells, **Am. J. Physiol.**, 257:C129.
28. Brayden, D.J., A.W. Cuthbert, and C.M. Lee, 1988, Human eccrine sweat gland epithelial cultures express ductal characteristics, **J. Physiol.**, 405:657.
29. Chen, J.H., H. Schulman, and P. Gardner, 1989, A cAMP-regulated chloride channel in lymphocytes that is affected in cystic fibrosis, **Sci.**, 243:657.

30. Hagiwara, G., M.E. Krouse, U.J. Müller, and J.J. Wine, 1989, Is regulation of a chloride channel in lymphocytes affected in cystic fibrosis? **Sci.**, 246:1049.
31. Krouse, M.E., G. Hagiwara, U.J. Müller, C.L. Ward, T.C. Law, and J.J. Wine, 1989, Comparison of Cl channels from lymphocytes and primary sweat gland cultures, **Ped. Pulmon.**, suppl. 4:11.
32. Hagiwara, G., M.E. Krouse, U.J. Müller, C.L. Ward, T.C. Law, and J.J. Wine, 1989, Attempts to demonstrate cAMP-dependent regulation of Cl^- channels in T lymphocytes, **Ped. Pulmon.**, suppl. 4:124.
33. Müller, U.J., M.E. Krouse, G. Hagiwara, C.L. Ward, T.C. Law, and J.J. Wine, 1989, Attempts to regulate Cl^- channels in T lymphocytes with cholera toxin and PGE_1, **Ped. Pulmon.**, suppl. 4:125.
34. Woll, K.H., M.D. Leibowtiz, B. Neumcke, and B. Hille, 1987, A high-conductance anion channel in adult amphibian skeletal muscle, **Pflug. Arch.**, 410:632.
35. Worrell, R.T., A.G. Butt, W.H. Cliff, and R.A. Frizzell, 1989, A volume-sensitive chloride conductance in human colonic cell line T84, **Am. J. Physiol.**, 256:C1111.
36. Dharmsathaphorn, K., J.A. McRoberts, K.G. Mandel, L.D. Tisdale, and H. Masui, 1984, A human colonic tumor cell line that maintains vectorial electrolyte transport, **Am. J. Physiol.**, 246:G204.
37. Halm, D.R., G. Rechkemmer, R.A. Shoumacher, and R.A. Frizzell, 1988, Apical membrane chloride channels in a colonic cell line activated by secretory agonists, **Am. J. Physiol.**, 254:C505.
38. Hayslett, J.P., H. Gogelein, K. Kunzelmann, and R. Greger, 1987, Characteristics of apical chloride channels in human colon cells (HT29), **Pflug. Arch.**, 410:487.
39. Bridges, R.J., R.T. Worrell, R.A. Frizzell, and D.J. Benos, 1989,Stilbene disulfonate blockade of colonic secretory Cl^- channels in planar lipid bilayers, **Am. J. Physiol.**, 256:C902.
40. Bear, C.E., 1988, Phosphorylation-activated chloride channels in human skin fibroblasts, **FEBS Lett.**, 237:145.
41. Barres, B.A., L.L.Y. Chun, and D.P. Corey, 1988, Ion channel expression by white matter glia: I. Type 2 astrocytes and oligodendrocytes, **Glia**, 1:10.
42. Javitt, N.B., 1990, Hep G2 cells as a resource for metabolic studies: lipoprotein, cholesterol, and bile acids, **FASEB J.**, 4:161.
43. Tabcharani, J.A., T.J. Jensen, J.R. Riordan, and J.W. Hanrahan, 1989, Bicarbonate permeability of the outwardly rectifying anion channel, **J. Mem. Biol.**, 112:109.
44. Cliff, W.H., and R.A. Frizzell, 1989, cAMP- and Ca-mediated regulation of Cl conductances in the human colonic cell line T84, **Ped. Pulmon.**, suppl. 4:123.
45. Bijman, J., H.C. Englert, H.J. Lang, R. Greger, and E. Frömter, 1987, Characterization of human sweat duct chloride conductance by chloride channel blockers, **Pflug. Arch.**, 408:511.

46. Gray, M.A., A. Harris, L. Coleman, J.R. Greenwell, and B.E. Argent, 1989, Two types of chloride channels on duct cells cultured from human fetal pancreas, **Am. J. Physiol.**, 257:C240.
47. Duszyk, M., A.S. French, and S.F.P. Man, 1990, The 20-pS chloride channel of the human airway epithelium, **Biophys. J.**, 57:223.
48. Duszyk, M., A.S. French and S.F.P. Man, 1989, Cystic fibrosis affects chloride and sodium channels in human airway epithelia, **Can. J. Physiol. Pharmacol.**, 67:1362.

DISCUSSION

FRIZZELL: I wonder about the Ussing chamber experiments, only from the point of view about the possible voltage-dependence of the block by the stilbene. What are your thoughts about that?

WINE: It's a good point. What Ray is alluding to is that there is a voltage dependence of the DNDS block and in fact, when the cell is hyperpolarized, DNDS can actually increase the activity in the channels under some circumstances. What we've tried to do is get around that, as I said, just by concentration. So that's the only thing that we can say: we've gone to very high concentrations in the Ussing chamber and very low concentrations in the single channel and whole cell experiments. But every experiment, again, has an objection to it.

GUGGINO: If the outwardly rectifying chloride channel is not the one that's involved, how do you reconcile the clearly different pattern that is seen with the normal channels vs. CF channels in that normal channels are activated by kinases and CF channels aren't? How do you think that fits in with the whole picture?

WINE: That's right, that's the real problem, isn't it? It seems to be the one central thing we have so far to hang our hats on. I believe that this is a treacherous channel. I think that it's very easy to be fooled in those kinase experiments. I can tell you, and you probably know yourself: Many people who have done these experiments have told me, when we discussed them, that they often don't work. We didn't get it to work. There are a lot of reasons why it might not work. So what we need, I think, to settle the issue once and for all, is to do a blind, multicenter study, the way people do with drugs. You wouldn't start selling amiloride without a blind study - why do we want to sell this channel without a blind study.

GUGGINO: We are talking about already 4 or 5 groups of individuals who have published positive results. We're not talking about one individual who has seen this difference. We're talking about several groups that have seen it in airway cells, leucocytes, fibroblasts...in multi-country, international experiments that have been done.

WINE: I agree. Usually when something is replicated in a great many laboratories you feel very secure about it, and that's one of the reasons why I aged 10 years in the last few years, because I thought how could all these people possibly be barking up the wrong tree? However, when cold fusion hit and it was quickly replicated in 4 or 5 laboratories including by a colleague of mine at Stanford, I began to realize that these are complex phenomena that can trick all of us. All I can tell you, Bill, is that I have not done enough of those experiments so that I can say, in our hands we definitely couldn't make it work. So I have no data disagreement with people who have done kinase experiments, but if people who have tried it and haven't had any success would be willing to say so publicly, it would be a great help. I don't know if anybody is willing to do that, or not.

(Unidentified): One additional question, if you try to compare channels, either between the Ussing preparation or in isolated patches, or you compare them between different tissues, must you not also consider the time course, in which state an individual cell exists, what channels are expressed at a given time? It seemed to me as if you had taken this as constant.

WINE: Thanks very much for pointing that out. I didn't give you the time scale over which we did our patch experiments. All the patch experiments were done on cells that were confluent in their area. The mean age of the cells that were used for patch clamp experiments was 7 days of culture. The great difference that we worry about is that all of our patch experiments were done on cells that were grown on collagen-coated plastic, which isn't the same as growing them on permeable substrates. I don't think that is a difference, because of Dr. Frizzell's whole cell results and also because of his and other people's results with efflux, which suggest that cells grown in that way still have an activatable chloride conductance. So I think it's a good point you raise, but one of the reasons we used dog cells is so that we could track them with Ussing chambers, unlike human cells, which are too scarce. We've done these same studies on T84 cells. They're just not as far along, but the answer is not different, so far.

QUINTON: I think you've got a little bit of consolation from the sweat duct. That is, I think we all agree that the chloride channel, or at least the chloride conductance, is not present in the sweat duct of cystic fibrosis patients, and it is, clearly, present in normal sweat ducts. So even though it's unfortunate that the duct has not been patched, so that the actual channels in the apical membrane of the duct are not available for description, one might expect that the tissue should exhibit some of the properties of the basic chloride conductance unit. So I've had some consternation with the properties that have been described for the rectifying chloride channel in comparison to the duct, and I would just like to point out that there are three discrepancies, which I think occur, and which perhaps should be noted. One is that the duct is notoriously insensitive to stilbenes. Secondly, it has the wrong anion selectivity conductance sequence, at least as far as we can measure it. That is, in the patch it appears that iodide is more conductive than chloride, but the duct is clearly much more conductive to chloride than it is to iodide. Thirdly, the point I was making in my talk yesterday, it does not seem that the duct is easily regulated by the cAMP pathway. Thank you.

AUSIELLO: I'd like to follow up on a part that I find intriguing and I think bears on this issue. You said that most of the people who have done the protein kinase activation find it takes 10 to 20 minutes to activate?

WINE: No, no. Up to 10. The median or average time is more like 3 minutes.

AUSIELLO: All right. Still, I assume you are concerned about the relatively long time. I think for anyone who has worked with protein kinase activation in other systems, that is a long time. Is that the point you were trying to make?

WINE: Well, the real point I was trying to make, Dennis, is that because there is a long interval between the addition of the kinase and then something else that happens, the linkage between those two is looser, and it makes you start to worry. What happens if you have a CF patch and you add kinase and 10 minutes later a channel comes in? There is a great temptation to say "There's one of those spontaneously activating patches". But that's not really what my talk was about. Maybe there's something funny here, or maybe the rectifying chloride channel is one of the things that is affected in cystic fibrosis. My point is just that it is very unlikely to be a key player.

GUGGINO: I'd like to just comment, having worked with protein kinase in a number of other systems, that we usually look at seconds to a minute in terms of the ability for enzyme activation etc. So if indeed the time limit is prolonged, and I am not familiar with these studies, but I'm saying if the half-time is that long, then what you say is indeed true. There could be multiple events occurring in that time period.

GREGER: I think I should make a brief comment because I have had an opinion on that for some time. The difficulty with the kinase experiment, in my eyes, is that we have been unable to do it because the channel is spontaneously activated just by excision. I think that is the real dilemma. If you work at low temperature, this excision activation may or may not occur. I am aware, Bill, that you and others have worked with really much accuracy to find the right controls, like the voltage at which you clamp, and Ray Frizzell has done the same, Mike Welsh has done the same, but this is not really more than, still, circumstantial. I think the real evidence that I would ask for is someone who tells me that he can de-activate this phosphorylated channel and then re-activate again, or something of that kind. I think this is the real experiment. If anyone could tell me this, then I would be very happy.

S. GUGGINO: But how can you still change the effects in time? You know, even if you activate with kinase, de-activate and then re-activate, you still have the effect of time.

GREGER: But it's much better - it's double-paired. You go through your phases and you see exactly what you do. That's number one. Number two is, I want to have a satisfactory explanation, that if you work at the biologically appropriate temperature, that this phenomenon is not required at all, because the patches are active and you can throw on those patches whatever you want, you don't get additional activation. So the excision by itself, at the respective temperature, has done it. So my view of it is that at the low temperature you excise a patch which has very complex biophysical properties. It's a dumpling of some kind with a lot of things in it, and the biochemistry is very complex and it may well work, but if you do it at 37°, it's more, in a way, simplified. So I think the real dilemma for us is if there were not the excision activation at the normal temperature, it would be much easier for us.

REGULATION OF ION CONDUCTANCE IN HUMAN SKIN FIBROBLASTS

Christine E. Bear

Division of Cell Biology, Research Institute
Hospital for Sick Children, Toronto, Canada

INTRODUCTION

The function of the CFTR (Cystic Fibrosis Transport Regulatory) protein is unknown but based on ion flux and electrophysiological data in epithelial cells it appears probable that this protein functions either directly or indirectly in the regulation of ion transport. Current hypotheses suggest that this protein may be more specifically related to the activation of chloride channels via protein kinase A or C mediated phosphorylation since abnormal chloride channel regulation by these enzymes has been observed in airway epithelial cells obtained from cystic fibrosis patients (1,2,3). However, conductance pathways for other ions, ie. sodium ion, may also be affected in cystic fibrosis (4). In an attempt to further define the role for the CFTR protein we tested the hypothesis that cell types which express only low levels of CFTR, i.e. nonepithelial cells, will fail to exhibit one or more important features of ion transport regulation known to exist in epithelial cells. As cyclic AMP and calcium mediated ion conductance has been described quite well in epithelial cells we assessed the role of these regulatory pathways in ion conductance in skin fibroblasts.

In this set of experiments we assessed the regulation of ion conductance in human skin fibroblasts using the patch clamp technique. Changes in total membrane ion conductance evoked by intracellular calcium or cAMP accumulation were studied first using the whole-cell patch clamp configuration. Subsequently, the single ion channels underlying the changes in whole cell current were assessed using both the cell-attached and excised patch configurations. The results showed that intracellular calcium accumulation caused an increase in 200 pS potassium channel activity leading to transient outward whole cell currents at a cell potential of 0 mV. The effects of intracellular cyclic AMP accumulation were less consistent. In approximately half of the

The Identification of the CF (Cystic Fibrosis) Gene
Edited by L.-C. Tsui *et al.*, Plenum Press, New York, 1991

cells studied, cAMP accumulation caused small, transient inward currents at -60 mV membrane potential, possibly due to the activation of 30 pS outwardly- rectifying, chloride-selective channels.

METHODS

Cell Culture

Primary cultures of human skin fibroblasts from 4 different non-CF subjects were obtained from collaborators in the Cystic Fibrosis Research Development Program (HSC-Toronto). Cells were grown in alpha medium (Princess Margaret Hospital, Toronto) plus 10% bovine calf serum on 35 mm plastic cell culture dishes. Whole cell and single channel currents were recorded from cells in subconfluent and confluent monolayers (7-14 days after plating). Because of the difficulty in forming high resistance seals between patch pipettes and flattened fibroblasts in monolayers the cells were treated with a trypsin plus EDTA containing medium (Gibco) followed by 5-6 washes in Na^+-hepes buffer. After this treatment, most of the fibroblasts were still adherent to the culture plates but were more rounded.

Current recording

Membrane currents were recorded according to Hamill et al. (5) using a List EPC-7 patch-clamp amplifier (Medical Systems Corp., Great Neck, NY). Pipettes were fabricated from borosillicate glass type 7052 (Garner Glass Co.) using a two-stage Narishige pipette puller (PP-83). Pipette resistances ranged between 2-5 MΩ. The bath electrode was a Ag-AgCl wire connected to the bathing solution via an agar bridge. Current output was monitored on a Tektronix oscilloscope and recorded on video tape after A/D conversion by a video adaptor (PCM2, Medical Systems).

Whole Cell Recordings: The whole cell configuration was formed by application of abrupt pulses of suction to cell-attached membrane patches. The successful formation of the whole-cell patch was indicated by: i) appearance of capacitance currents which were then compensated with the analog circuit of the patch-clamp amplifier and ii) measurement of whole cell membrane potential in current-clamp mode (>-20 mV). A tri-level stimulator (S100, Medical Systems) was used to generate the voltage clamp protocols. In all whole cell experiments the pipette was filled with a salt solution containing (mM): KCl (140), HEPES (10), glucose (10), $MgCl_2$ (2), EGTA (5), ATP (4), GTP S (0.2). The bath contained a NaCl or choline Cl (140 mM) salt solution with (mM): HEPES (10), glucose (10), $MgCl_2$ (2) and $CaCl_2$ (2).

Single Channel Recordings: The salt solutions employed for single channel studies were basically the same as those described above except that KCl salt solutions did not contain ATP and GTP S. In some excised patch studies, cytosolic Ca^{2+} concentration was buffered using EGTA. For a final free Ca^{2+} concentration of approximately 10 nM; 1 mM EGTA was added to a nominally Ca^{2+} free KCl salt solution, 100 nM: 1 mM EGTA was added to a KCl salt solution with 0.5 mM $CaCl_2$ added, 1000 nM: 1 mM EGTA was added to a KCl salt solution containing 0.87 mM $CaCl_2$. Single channel analysis was performed using pCLAMP version 5 (Axon Inst. Burlingame, CA). All experiments were performed at room temperature.

Materials

A23187 was obtained from Calbiochem (La Jolla, CA) and isobutyl methylxanthine and dibutyryl cyclic AMP from Sigma (St. Louis, Missouri).

RESULTS

Calcium Mediated Whole Cell Ion Currents

Following formation of the whole-cell patch-clamp configuration, the resting or control current-voltage relationship was determined. Once the control I-V had been determined, a continual pulse protocol was established with the Vp alternatively switched from 0 to -60 mV for 600 and 400 ms respectively (in some cases hyperpolarizing potentials of -40, -50 or -80 mV were chosen). This protocol is similar to that employed by several investigators, including Findlay (6), in studies of exocrine gland cells, to differentiate between changes in total membrane current driven by K^+ and Cl^-. For the present studies, as in the above publications, the cell interior was perfused through the patch pipette with the pH and Ca^{2+}-buffered KCl (140 mM) solution described in METHODS and the cell exterior bathed with a buffered NaCl (140 mM) solution (METHODS). At Vp=0, there was no driving force for Cl^- transport but an outward electrochemical driving force for K^+ transport existed. Therefore, at Vp=0 we were able to observe changes in membrane current due to K^+ transport. On the other hand, at the hyperpolarized potentials, the driving force for K^- transport is reduced and the electrical driving force for Cl^- efflux or Na^+ influx enhanced.

Whole cell changes in ion conductance caused by intracellular Ca^{2+} accumulation were studied by adding the divalent cation ionophore, A23187 (1 μM) to the cell bath after the formation of the whole-cell patch and initiation of the pulse protocol described above. There was some variability in the whole cell current response indicated in the SUMMARY TABLE. As previously described, outward current at Vp=0 mV can be accounted for by outward K^+ transport and inward current at Vp=-60 mV accounted for by inward Na^+ or outward Cl^- transport. In 8 of 10 trials, addition of A23187 to the cell bath resulted in a marked transient stimulation in outward current at Vp=0 mV, corresponding to activation of K^+ channels with no significant change in current level observed at the hyperpolarizing potentials. Figure 1

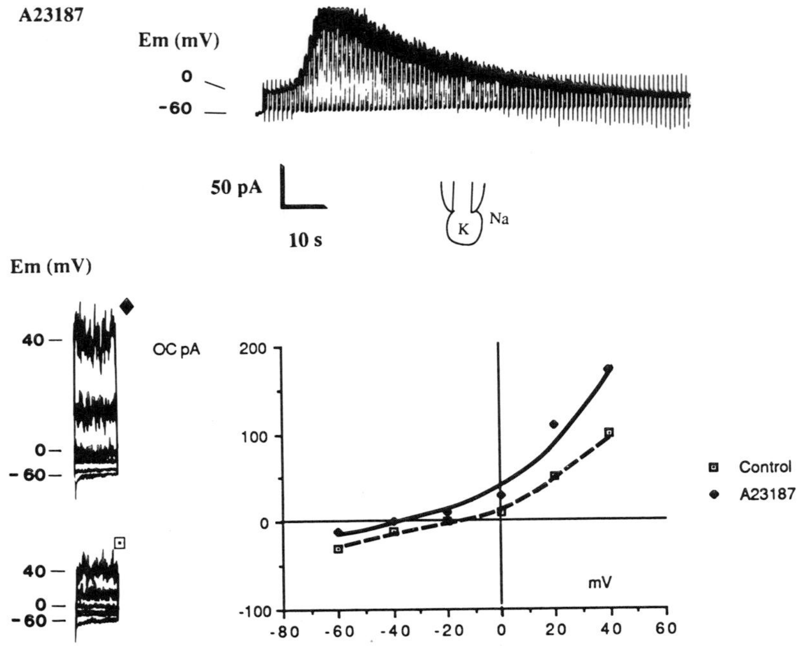

Fig. 1. Whole Cell Currents evoked with A23187 ($1\mu M$) (Upper Panel) Whole cell membrane currents were measured when the cell was alternatively voltage clamped at 0 and at -60 mV. The addition of A23187 ($1\ \mu M$) caused a transient increase in outward current at 0 mV, lasting between 30 to 60 s. (Lower Panel) Whole cell IV relationships have been shown for the above experiments prior to (■) and following the addition of the calcium ionophore (♦).

shows the time course of these changes in whole-cell current and the whole-cell current-voltage relationships prior to and after A23187 addition. The reversal potential of the I-V relationship shifted from -20 mV in the control to -40 mV, towards the equilibrium potential for K^+, confirming that A23187 evoked an increase in whole cell K^+ conductance.

A23187 Evoked K^+ Channel Activity

Figure 2 shows the effect of A23187 addition to the bath of a cell-attached membrane patch configuration. Approximately one minute after the addition of A23187 to the bath a transient burst of outward current steps was observed, corresponding to single channel openings in 4 of 6 trials. The I-V relationship of the activated channel in the cell-attached configuration has also been shown in Figure 2. Single channel conductance was determined in this and three similar experiments: 84.0±10.2 (SD), (Vp=0 to -100 mV). The reversal potential was estimated to occur

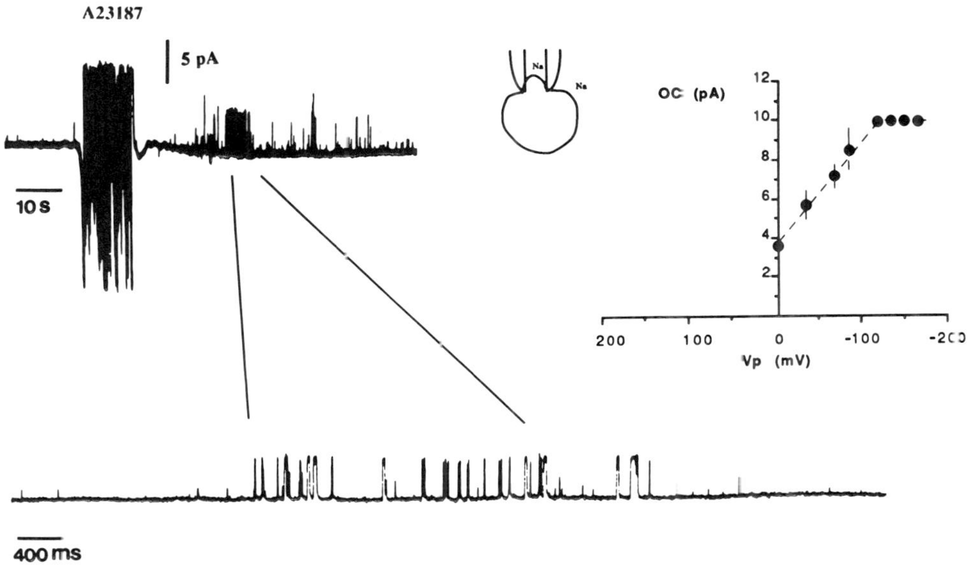

Fig. 2. Single Channels Evoked with A23187 (1 μM).
This figure shows the effect of A23187 addition (1mM) to the bath of a cell-attached membrane patch. The noise in the record was associated with interruption of Farraday shielding upon ionophore addition.

at Vp=50 mV, which corresponds to a negative membrane potential suggesting that this channel is selective for K^+.

This K^+ channel was studied in the excised, inside-out patch configuration in order to confirm that the stimulation observed following A23187 addition to cells was due to the direct effect of cytosolic Ca^{2+} on channel activity. The I-V relationship for the channel when exposed to asymmetric salt solutions, i.e. NaCl (140 mM) in the pipette and KCl in the bath or symmetrical solutions (with 140 mM KCl in both pipette and bath) has been shown in Figure 3. The unitary conductance calculated for this channel when exposed to quasi-physiological ion gradient was 153.2±18.6 (SD) (n=5). The conductance of the channel in symmetrical KCl solutions was calculated as 187.3±16.5 (SD) (n=5) . In Figure 3 the effect of membrane potential and cytosolic Ca^{2+} on channel activity has also been shown. When the free calcium concentration bathing the cytosolic surface was buffered at 10 nM, no single channel openings were observed regardless of the applied membrane potential, at 110nM Ca^{2+}. Channel activity was observed only at depolarized membrane potentials and finally in the presence of 720 nM free Ca^{2+} the channel was very active throughout a large range of membrane potentials. The effects of both cytosolic Ca^{2+} concentration and applied potential on channel open probability have been shown.

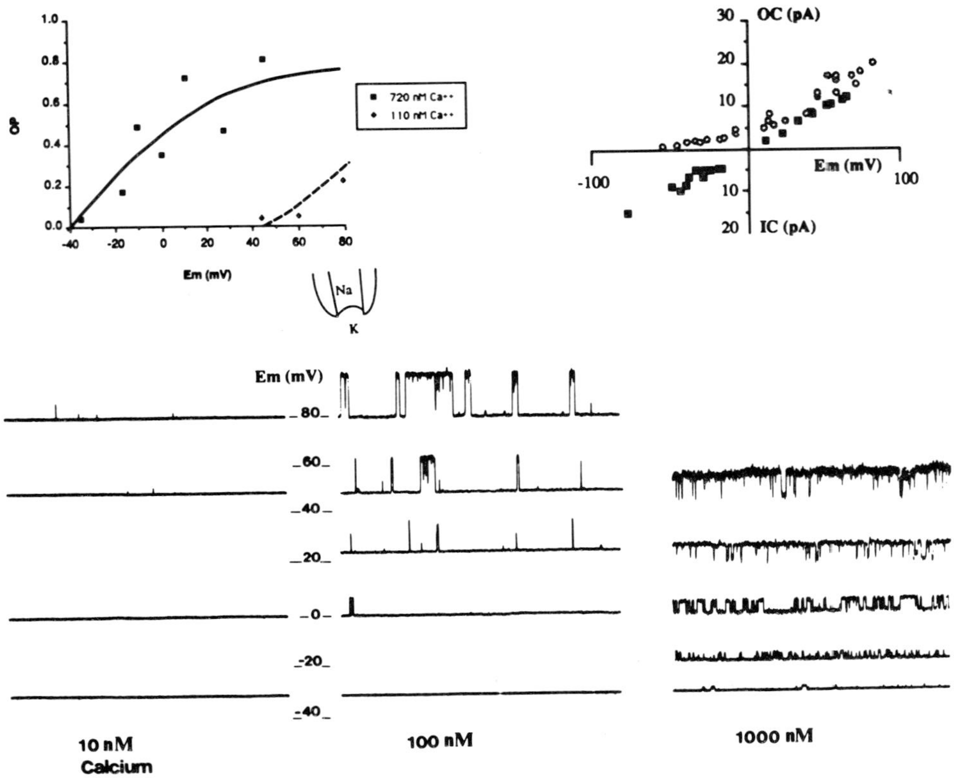

Fig. 3. Activation of K^+ Channel by Depolarization and Calcium Single channel I-V relationships have been shown for the channel in the inside-out patch configuration exposed to quasi-physiological salt solutions: (pipette=NaCl, bath=KCl) (○) or symmetric salt solutions: (pipette and bath=KCl) (▪) (upper right hand panel). The effect of membrane potential and cytosolic Ca^{2+} on single channel activity has also been shown (lower panel). Channel open probability was determined at varying membrane potentials and cytosolic $[Ca^{2+}]$ in an excised, inside-out patch exposed to quasi-physiological ion gradients (upper left hand corner).

Cyclic AMP-Mediated Whole Cell Ion Currents

The effect of intracellular cyclic AMP accumulation on whole cell membrane conductance was determined by applying the phosphodiesterase inhibitor, isobutyl-1-methylxanthine (IBMX) and the membrane permeant cAMP analogue, dibutyryl cAMP (dbcAMP). The effect of this treatment on whole cell membrane current was quite inconsistent as can be seen in the SUMMARY TABLE. However, 8 of 17 trials, a transient inward current was evoked within one minute after addition of both IBMX (100μM) and dbcAMP (100μM) (Figure 4). This increase in inward current occurred regardless

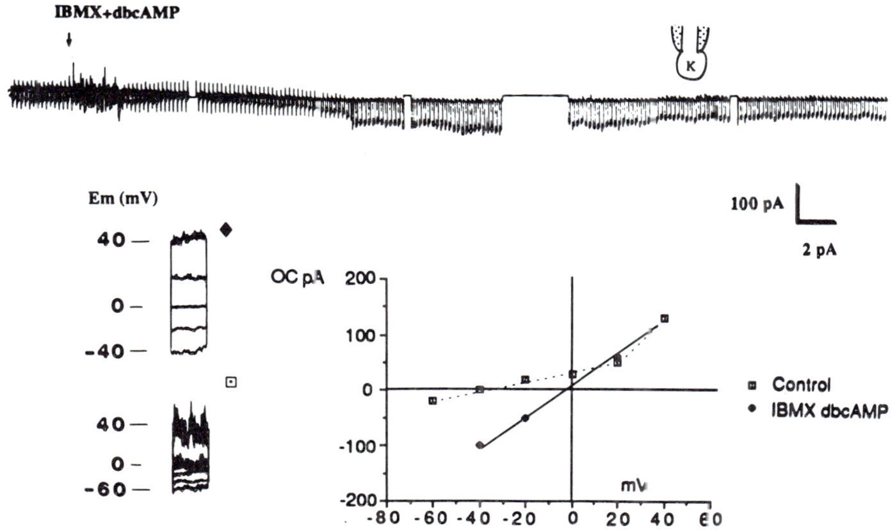

Fig. 4. IBMX and dbcAMP Activate Whole-Cell Current. (Upper panel) The addition of IBMX (100μM) and dbcAMP (100μM) to a fibroblast in the whole cell configuration evoked a transient increase in inward current when bathed in a choline chloride (140 mM)-containing bath. (lower panel) The whole cell I-V relationship and membrane potential were determined during breaks in record.

of whether the cell bath contained NaCl or choline chloride (n=8) suggesting that the cAMP-induced response was due to the outward conductance of chloride ion rather than inward Na^+ current. The whole cell I-V relationships prior to and following cAMP-activation in a typical experiment have been shown in Figure 4. As expected for the activation of chloride conductance the reversal potential shifted from -30 to 0 mV in both NaCl (140 mM) and choline Cl (140 mM) bath solutions.

Cyclic AMP Activates Single Ion Channels in Cell-Attached Membrane Patches

As in the whole-cell experiments, the effect of IBMX and dbcAMP on single ion channel activity in cell-attached patches was inconsistent. Addition of (100μM) of both IBMX and dbcAMP to the bath resulted in transient channel activation in 13 of 227 cell-attached membrane patches examined. Onset of channel activity was variable, ranging between 30 and 180 sec. Duration of channel activation was also variable but was generally short-lived, lasting between 30 and 60 seconds. The I-V relationship of the activated channel, shown in figure 5 was outwardly rectifying with a unitary conductance of 44.3±5.3 pS in the depolarizing voltage range and 15.0±2.6 pS in the hyperpolarizing range. This I-V relationship is similar to the outwardly-rectifying Cl^--selective channel previously described in fibroblasts in

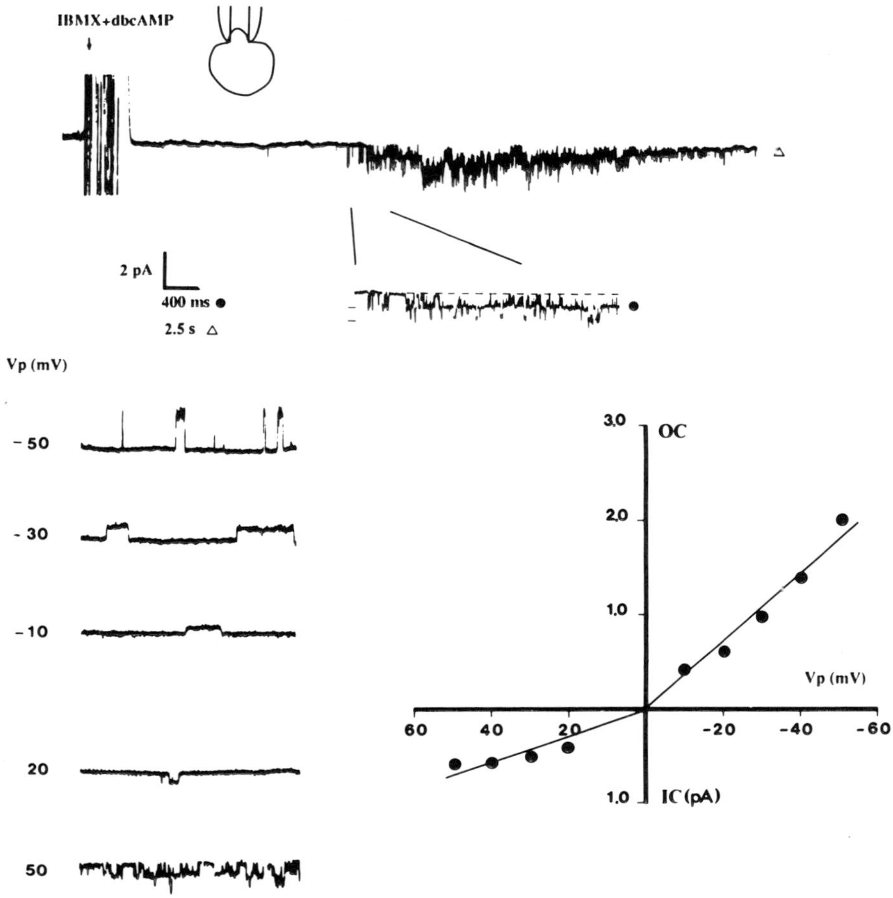

Fig. 5. IBMX and dbcAMP Stimulate Single Channel Activity. In the cell-attached configuration, transient single channel activity was evoked with the addition of IBMX and dbcAMP (100 μM),(Vp=58 mV). The I-V relationship for the channel activated by IBMX and dbcAMP (100μM) in the cell-attached configuration has been shown.

in their radio-isotopic flux studies, Chao et al.(9) did not observe changes in fibroblast membrane Cl^- permeability following this laboratory (7). The Cl^--selectivity of this channel was confirmed in the earlier report with ion substitution studies. As shown previously, this channel can be activated with patch excision followed with the application of membrane depolarization of +60 to +100 mV. Depolarization following patch excision evoked activity not only in the 13 patches stimulated with cAMP but also in an additional 17 inside-out patches. Therefore, intracellular cAMP accumulation lead to activation of 13 of 30 membrane patches containing chloride channels.

SUMMARY TABLE

Whole Cell Currents Evoked by A23187 and dbcAMP

Effect of A23187 Addition

no. of cells	none	↑OC	↑IC	↑OC and IC
10	**0**	**8**	**1**	**1**

Effect of IBMX and dbcAMP

no. of cells	none	↑OC	↑IC	↑OC and IC
17	**8**	**1**	**6**	**2**

DISCUSSION

The results of the present study showed that the predominant effect of intracellular Ca^{2+} accumulation was to increase K^+ conductance with no marked effect on Cl^- conductance observed. The results differ from other studies with fibroblasts, in which Ca^{2+}-mediated Cl^- conductance was reported. Both Lin et al. (8) and Chao et al. (9) reported that Ca^{2+} influx caused an increase in Cl^- conductance using different techniques, i.e. radioisotopic flux techniques and the chloride sensitive fluorophore 6-methoxy-N-[3-sulfopropyl quinolinium] (SPQ) respectively.The failure to observe Ca^{2+}-activated Cl^- conductance in the present studies could be related to dialysis of components for signal transduction in the whole-cell patch-clamp preparation. However, as whole cell K^+currents were activated in the present studies with the addition of A23187 it seems unlikely that Ca^{2+}-activated Cl^- channels would not be detected using the whole cell preparation.

It is clear that intracellular calcium accumulation stimulates chloride conductance in a number of epithelial tissues (10, 11). The mechanism through which Ca^{2+} activates Cl^- conductance remains unknown. Calcium may act either directly or indirectly on specific Cl^- channels. The failure of the Ca^{2+} ionophore, A23187 to evoke whole cell Cl^- currents may prove to be a distinctive feature of Cl^- transport regulation in fibroblasts.

On the basis of the experiments shown in this report and previous studies performed in this laboratory the effects of intracellular cAMP accumulation on ion conductance in skin fibroblasts appear inconsistent. In approximately half of the cells examined in the present study, a transient increase in Cl^- conductance was evoked with addition of IBMX and dibutyryl cAMP. Transient whole cell Cl^- currents were due possibly to the activation of the outwardly rectifying Cl^- channel previously described in fibroblasts (7). Although, cyclic AMP-mediated changes in Cl^- conductance have also been reported by Lin et al.(8)

dbcAMP treatment using the fluorescent dye SPQ. The reason for this variability and even the inconsistency observed in the present patch-clamp studies is not clear. The cAMP-evoked Cl^- conductance changes observed in the present patch-clamp studies were transient, lasting approximately one minute. Perhaps this short time course made detection difficult in the fluorescent dye studies. A physiological role for transient cyclic AMP evoked membrane conductance has not yet been determined in fibroblasts. It is unlikely that this conductance pathway is involved in cell volume regulation as it is not evoked with cell swelling (12).

In the case of several types of epithelial cell, i.e. airway and intestinal cells, cAMP activation of Cl^- conductance through a 30 pS outwardly-rectifying Cl^- channel has been well described and a role for this pathway in secretion implicated (1,2,3,11). The failure of B-agonists to evoke normal levels of epithelial water and salt secretion in CF patients suggests that the outwardly-rectifying Cl^- channel is altered in this disease and that CFTR may be important in the regulation of this channel. However our experiments show that fibroblasts, a cell type with low levels of CFTR expression, can exhibit cAMP-evoked Cl^- conductance, possibly through the outwardly-rectifying Cl^- channel. These findings appear to contradict the above hypothesis regarding CFTR function and suggest that CFTR may have an indirect effect on Cl^- conductance. There may be subtle but significant differences in the extent and duration of cAMP evoked Cl^- conductance between nonepithelial and epithelial cells which we will investigate in future comparative studies.

Acknowledgements

I would like to acknowledge the excellent assistance of Tamara Bond who worked as a summer student funded through a scholarship awarded by the Canadian Cystic Fibrosis Foundation. Her studies formed a basis for the experiments reported in this present study. I would also like to acknowledge the excellent technical assistance of Evangelica Reyes who maintained cell cultures and performed computer analysis of patch-clamp data.

This project was funded through grants awarded to C.E.B from the MRC (Canada) and NIH (Cystic Fibrosis Centre Grant).

References

1. Li, M., J.D. McCann, C.M. Leidtke, A.C.Nairn, P.Greenwood, and M.J. Welsh, 1988, Cyclic AMP-dependent protein kinase opens chloride channels in normal but not in airway epithelium, **Nature (Lond.)**, 331:358.
2. Shoumacher, R.A., R.L. Shoemaker, C.R. Halm, E.A. Tallant, R.W. Wallace, and R.A. Frizzell, 1987, Phosphorylation fails

to activate chloride channels from cystic fibrosis airway cells, **Nature, Lond.**, 330:752.
3. Hwang, T.-C., L. Lu, P.L. Zeitlin, D.C. Gruenert, R. Huganir and W. Guggino, 1989, Cl^- channels in CF: Lack of activation by protein kinase C and cAMP-dependent protein kinase, **Science**, 244:1351.
4. Cotton, C.U., M.J. Stutts, M.R. Knowles, J.T. Gatzy, and R.C. Boucher, 1987, Abnormal apical cell membrane in cystic fibrosis respiratory epithelium, **J.Clin.Invest.**,79:80.
5. Hamill,O., P.,A. Marty, E. Neher, B. Saknan and F.J. Sigworth, 1981, Improved patch-clamp techniques for high resolution current recording from cells and cell-free membrane patches, **Pflugers Arch. Eur. J. Physiol.** 391:85.
6. Findlay, I., 1984, A Patch-clamp study of potassium channels and whole-cell currents in acinar cells of the mouse lacrimal gland, **J.Physiol.**, 350:179.
7. Bear, C.E., 198 , Phosphorylation-activated chloride channels in human skin fibroblasts, **FEBS Lett.**, 237:145.
8. Lin, P.M., Ahluwalia and E. Gruenstein, 1988, Regulation of conductive Cl^- transport in hunan skin fibroblasts, **Am. J. of Physiol.**, 255:C552.
9. Chao, A.C., J.A. Dix, M.C. Sellers and A.S. Verkman, 1989, Fluorescence measurements of chloride transport in monolayer cultured cells, **Biophys. J.**, 56:1071.
10. Dharmasathaphorn, K., J.A. McRoberts, K.G. Mandel, L.D. Tisdale and H. Masui, 1984, A human colonic tumour cell line that maintains vectorial electrolyte transport, **Am. J. of Physiol.**, 246:G294.
11. Frizzell, R.A., D.R. Haln, G. Reckhemmer and R.L. Shoemaker, 1986, Chloride channel regulation in secretory epithelia, **Fed. Proc.**, 45:2727.
12. Rothstein, A.and C.E. Bear, 1989, Cell volume changes and the activity of the chloride conductance path, **Ann. N.Y. Acad. Sci.**, 574:294.

DISCUSSION

(Unidentified): Is the dibutiryl-cAMP simply added to the bath and allowed to cross the membrane?

BEAR: It was added to the bath, yes.

(same): The reason I ask the question is that the KD for the dibutiryl cAMP with a regulatory subunit of protein kinase A is about 1000-fold more than for cAMP itself. So I wondered if you'd studied higher concentrations of dibutiryl cAMP.

BEAR: No, that was the concentration that I stuck with, since I did see somewhat of a response, but now, since you raise that point, I'll check higher concentrations and also receptor-mediated changes in cAMP. Those are studies I have to perform as well.

GALIETTA: I just want to ask you, in which conditions do you perform the whole cell recordings? I mean, do you record from resting cells or from trypsinized cells?

BEAR: All of these studies were performed using cells that were in confluent monolayers. I trypsinized the cells and bathed them 5 to 6 times before I performed the studies. After that treatment, the cells were still stuck to the plate, but they were rounded. Perhaps that is a contaminating factor to consider.

GALIETTA: The confluent condition in your experiments could explain probably the problems we found to obtain cAMP-mediated activation of Cl^- channels.

BEAR: You mentioned in your paper that you couldn't see a marked calcium-sensitivity to your large potassium channel because you were looking at mytotic cells. I've studied those as well and I still saw sensitivity to physiological calcium concentrations. So there is possibly another factor.

GALIETTA: We tried in cell-attached patches to apply cAMP analogues or forskolin and we were not able to see cAMP mediated activation of chloride channels. As you know, we were very ineffective in finding the rectifying chloride channels. We found only the large conductance chloride channel in hundreds of experiments. Maybe there is some difference in our experimental conditions.

BEAR: Yes, I'm familiar with your work and I don't have a good explanation to describe that other than to use the catch-all that it is a culture phenomenon.

RUGOLO: I only wanted to make a comment on the fact that we have presented a poster in which we have studied the effect of hypotonic shock on skin fibroblast chloride transport, and chloride transport is increased. We have some evidence that a channel is involved, so I think that our data really are in good agreement just to say that there is RVD in skin fibroblasts.

BEAR: I haven't shown any single-channel data for the chloride channel which may be responsible, although in preliminary experiments it appeared as though there was a very small chloride selective channel, 3 to 5 pS which could be evoked. But the whole cell changes reflected predominant changes in total membrane current, and perhaps masked the activity of that smaller channel.

ROMEO: I am worried about the small percentage of fibroblasts in which you find this activation of the channel, as it is known that fibroblasts in culture are very heterogeneous, that is, there are different types of cells. The fibroblasts may be different in origin and this might be one type of explanation. I wonder whether you have done any subcloning of the fibroblasts.

BEAR: No, I haven't. I haven't tried to synchronize cell cycle either. Those are all important factors.

CHLORIDE TRANSPORT IN THE CYSTIC FIBROSIS ENTEROCYTE

Jan Bijman (1), Henk Veeze (3), Maarten Kansen (1), Ben Tilly (2), Bob Scholte (1), Andre Hoogeveen (1), Dicky Halley (1), Maarten Sinaasappel (3) and Hugo de Jonge (2)

(1)Dept of Cell Biology & Genetics
(2)Dept of Biochemistry I
(3)Dept of Pediatrics
Erasmus University, Rotterdam

ABSTRACT

Molecular mechanisms of intestinal chloride channel regulation and potential abnormalities in electrogenic chloride secretion in intestinal epithelium from cystic fibrosis (CF) patients were investigated by a combination of Ussing chamber, vesicle transport and off-cell patch-clamp analysis.

Short circuit current (I_{sc}) measurements in normal and CF rectal biopsies provided evidence for i) a defect in the cAMP-provoked activation of chloride secretion and a (hyper)expression of cAMP-dependent potassium secretion in all CF patients examined (n=11); ii) a defect in the carbachol-provoked chloride secretion and a (hyper)expression of carbachol-induced potassium secretion in 6/11 patients; iii) a residual (but still impaired) carbachol-induced chloride secretion in 5/11 CF patients (including 2 sibs). The latter class of CF patients appeared to consist genetically of compound heterozygotes for the major delta-F_{508} deletion, suggesting a correlation between the nature of the mutation in the CF gene and the severity of the chloride secretory defect in CF intestine.

In our search for a regulatory function of GTP-binding (G-) proteins detected previously in the luminal membrane of rat and human intestinal epithelial cells, evidence was found for the presence of a GTP[S]-activatable- and GDP[S]-inhibitable chloride conductance in the apical membrane of rat enterocytes and human colonocytes. In excised patches of human colonocyte membranes,this G-proteine-sensitive chloride conductance was identified further as a novel type of chloride channel (20pS; inwardly rectifying) that was different from the 33pS outwardly rectifying chloride channel activatable by cAMP-dependent proteinkinase (PK-A) and voltage depolarization. The possible role of this channel in intestinal chloride secretion and its functioning in CF intestine is under present investigation.

INTRODUCTION

Electrophysiological studies of the CF condition carried out in the past decade have been focused on the reduced chloride permeability of the epithelial tissues affected by the disease. These tissues comprise the sweat duct[1,2], airways[3,4,5,6] and intestine[7,8,9]. The abnormalities found in the regulatory characteristics of apical chloride channels in CF can be summarized as follows: (1) A defect in the activation of a chloride channel at a site distal to cAMP-dependent protein kinase in trachea, sweat gland, enterocytes and lymphocytes[1-7,9,10]. (2) A defect in the Ca^{2+}-activation distal to Ca^{2+}-mobilization in sweat duct and enterocytes but not in sweat coil and trachea[6,7,9,10]. (3) A defect in the activation of the channel at a site distal to cGMP-dependent protein kinase in the enterocyte[2,3]. This summary shows that the insensitivity of chloride channels to β-adrenergic (cAMP-related) activating signals is universally found in all CF epithelia examined, while cholinergic (Ca^{2+}-related) activation of the chloride channels is only partially affected (airways) or remains unaffected (sweat coil) in some CF epithelia. The molecular basis of the tissue-specific differences in Ca^{2+}-dependent responses is presently unknown but may be related to differences in the coupling mechanism linking the Ca^{2+}-signal to the chloride channel, or to the expression of a subset of Ca^{2+}-sensitive but CF-insensitive chloride channels in airway and sweat coil which are lacking in the sweat duct and intestinal epithelium. In the CF enterocytes the defect in Ca^{2+}-linked chloride secretion could be attributed either to the malfunctioning of a single (Ca^{2+}-and cAMP-sensitive) apical chloride channel or to the defective regulation of a separate Ca^{2+}-sensitive channel that is distinct from the cAMP-activated channel present in the same membrane. In our search for the possible expression of multiple types of apical chloride channels in the intestinal epithelium, evidence was obtained for the presence of a G-protein-regulated chloride channel that is clearly different from the cAMP-sensitive channel on the basis of biochemical and electrophysiological criteria. It is important to find out whether or not the G-protein-, Ca^{2+}- and cAMP-regulated chloride channels represent distinct protein molecules and whether each channel species contributes to intestinal chloride secretion and is similarly affected by the CF mutation. Such a finding would imply that the CF-gene-encoded- protein, CFTR[8], is unlikely to function as a chloride channel itself but would instead act indirectly as a regulator of multiple channels, e.g. by adjusting the level of a cytosolic inhibitor[11]. In the present report, focusing on the intestinal epithelial cell as a target tissue in CF, further evidence is provided for a defect in intestinal chloride secretion in rectal epithelium and for the existence of a possible relationship between the nature of the mutations in the CF gene and the severity of the physiological defect. Furthermore, several characteristics of the G-protein regulated chloride channel found so far exclusively in rat enterocytes and human colonocytes are presented in some detail.

RESULTS

Chloride Currents in Rectal Biopsies from Control and CF Patients Mounted in the Ussing-Chamber

In previous experiments it was shown that in ileal resection preparations and small intestinal biopsies the chloride secretory response to cAMP, cGMP and calcium-linked secretagogues was absent in CF tissue[2]. To further investigate the manifestation of the CF defect in the intestine we have studied the chloride secretory response of normal and CF rectal suction biopsies mounted in the Ussing chamber. As shown in Table 1 no difference beween normal and CF was found in the amiloride inhibitable part of I_{sc} which reflects sodium reabsorbtion.

TABLE 1

Changes in I_{sc} in response to several secretagogues (mol/l) in normal and CF rectal biopsies

Secretagogue	Concentration	delta-I_{sc} ($\mu A/cm^2$ ±SEM) Normal positive I_{sc}	negative I_{sc}	CF
Amiloride	(10^{-4})	-7.1±3.2 (10)	-7.0±1.5 (9)	
Carbachol	(10^{-4})	+26.7±3.6 (11)	-12.5±4.6 (6)	
"			-9.8±2.2 (3) ->	2.9±0.3 (3)
"				2.6±0.1 (2)
Indomethacin	(10^{-5})	-2.3±1.3 (10)	0.3±0.2 (11)	
8-Bromo-cAMP/ forskolin	(10^{-3})	+2.3±1.3 (10)	-0.6±0.5 (9)	
Histamine	(5.10^{-4})	+12.6±3.7 (5)	3.9±2.0 (2)	

Apparently, the increase in sodium permeability of CF airway tissue is not reproduced at the level of CF colon.

In CF rectal biopsies the short circuit current (I_{sc}) in response to cholinergic and cAMP-stimulation, is inversed when compared to the I_{sc} response of normal tissue (Table 1). The inversed I_{sc} in response to these secretagogues that was observed earlier in CF ileal preparations[2] most probably reflects potassium secretion in the absence of chloride secretion[12] unmasked or induced by the CF condition. As indicated in Table 1, the carbachol induced I_{sc} response in CF tissue was variable. An inversed response was observed in 9/11 biopsies. This inversed response is present in ileal mucosa also and can be blocked with barium (5 mmol/l) at the apical side which suggests the espression of potassium channels in this membrane[3]. In 3/11 patients the cholinergic response was biphasic, an initial inversed I_{sc} response followed by a transient increase in I_{sc}. A normal but reduced cholinergic response is seen in 2/11 biopsies (Table 1). Apparently, the observed net increase in I_{sc} in 5/11 patients reflects a capacity of residual chloride secretion. We have correlated these mixed carbachol I_{sc} responses of CF biopsies with the presence of the deltaF$_{508}$ mutation in the genome of the patients (Table 2). The results show that 5 out of 6 patients with absence of residual chloride secretion are homozygous for the deltaF$_{508}$ deletion. More importantly, all patients showing residual chloride secretion were compound heterozygotes. These results suggest that the manifestation of the disease may vary with the nature of the CF mutation.

TABLE 2

Presence of carbachol-induced chloride secretion in rectal biopsies of cystic fibrosis patients in relation to the deltaF$_{508}$ mutation

Residual Chloride Secretion	DeltaF/N	DeltaF/DeltaF
-	1	5
+	5[i]	0

[i]including 3 sibs

G-Protein Mediated Chloride Conductance in the Colon Carcinoma Cell Line HT-29cl.19a

A crucial role for G-proteins in the activation of cation conducting channels, like cardiac Ca^{2+}- and neuronal K^{+}-channels, has been demonstrated[13,14,15,16]. Recently, a number of GTP-binding proteins - including G_i, G_s, G_p and p21 ras - were detected in apical membranes of intestinal epithelial cells[16], despite the absence of evidence for G-protein-coupled hormone receptors in this membrane. The apical membrane of the enterocyte is a preferential domain for chloride channels which are active in salt secretion. However, the occurrence of G-protein activation of anion channels in general has not been established earlier. Therefore, we studied the regulation of conductive chloride efflux from apical HT-29cl.19a vesicles, obtained by differential Mg^{2+}-precipitation, by quantitating the potential-driven uptake of ^{125}I into vesicles[3,17] loaded with 150mmol/l KCl and suspended in 150mmol/l KGluconate (Table 1). A 2-3 fold increase in ^{125}I uptake was measured in vesicles preloaded with non-hydrolysable GTP[S]. The GTP[S] provoked increase in ^{125}I-uptake was completely abolished in the presence of potassium ionophore valinomycin

(10μmol/l) which short circuits the chloride diffusion potential across the vesicle membrane.

In another series of experiments, the quenching of the chloride sensitive fluorescent dye 6-methoxy-N-(3-sulfopropyl)-quinolium (SPQ) trapped into rat small intestinal brush-border membrane vesicles[18] was taken as a parameter to measure electrogenic chloride transport (Table 3). The addition of 25mmol/l KCl to the medium of chloride-free vesicles leads to rapid SPQ quenching, that was accelerated in the presence of valinomycin (10 μmol/l)[3].

TABLE 3

Effects of guanine nucleotides on k_1 of SPQ quenching (min^{-1}). Equimolar concentrations (100 μm) of various guanine nucleotides were trapped into rat brush-border membrane vesicles.

Control	100%	(10)	GTP	169 ± 15	(3)
GDP	102 ± 7	(3)	GTP[S]	238 ± 13	(10)
GDP[S]	141 ± 5	(3)	GMP-PNP	307	(1)

The valinomycin-sensitive component of the chloride influx was increased approximately 2-fold by entrapment of 10^{-5}mmol/l GTP[S] inside the vesicles. The effect of GTP[S] could not be explained by activation of potassium channels, since no difference in the rate of SPQ quenching was observed between control- and GTP[S]-containing vesicles, in the absence of valinomycin. Activation of the G-protein pathway was seen also with other non-hydrolysable analogues like GMP-PNP, or with GTP in the presence of GTPase inhibitor vanadate (1 mmol/l). GDP (100 μmol/l) or GDP[S] were effective antagonists of the GTP[S]-activated chloride flux into the vesicles. No additional effects on the chloride flux were observed with ATP (1 mmol/l), ATP[S] (1 mmol/l) or combinations of these nucleotides in GTP[S] loaded vesicles, indicating that the GTP response is not modulated by ATP or ATP analogues.

Channel Studies in Apical Membranes of HT-29cl.19a Cells

In a next series of experiments we studied the effects of GTP[S] on channel activation in excised apical membrane patches of HT-29cl.19a cells. Channel openings occurred within 2-20 min upon exposure of the membrane patch to 10μmol/l GTP[S] (Fig. 1). The channel displays inward voltage rectification like the cAMP-sensitive chloride channel in the oxyntic cell of amphibians[19].

Furthermore, the conductance of the channel is 20pS at OmV clamp potential and the channel activity is antagonized by GDP[S] (2 mmol/l). The channel was never observed in the absence of GTP[S] or in patches exposed to catalytic subunit of protein

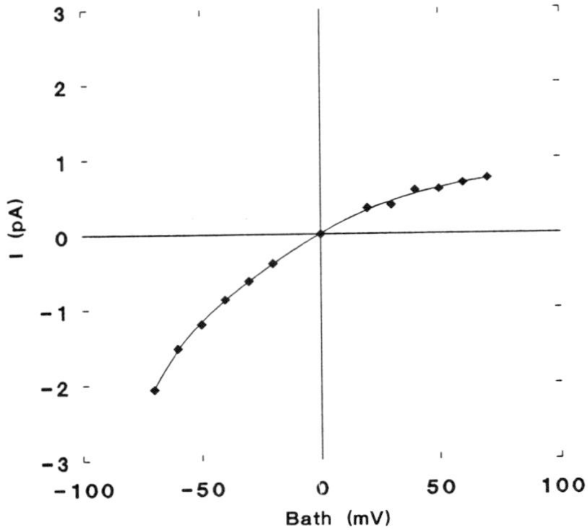

Fig. 1. Current to voltage relationship of a 20 pS chloride channel in the apical membrane of HT-29cl.19a cells.

kinase A and ATP. The latter condition resulted in the activation of a 35pS outwardly rectifying chloride channel in HT-29cl.19a apical membrane patches[3]. Instead, in some patches, the coexistence of both types of chloride channel could be demonstrated by voltage activation of the 35 pS channel followed by GTP[S] addition.

DISCUSSION

In this chapter an overview was given of chloride transport measurements in rectal biopsies from normal and CF individuals, and the results of chloride permeability measurements in apical membrane intestinal epithlial cells. The enhancement of electrogenic chloride secretion in response to cholinergic- or cAMP-linked adrenergic stimuli was either impaired or abolished in CF biopsies. Of notable interest is the variation in the carbachol induced chloride secretory response (Table 1) which showed correlation with the site of the CF mutation in the gene of CF patients (Table 2). The data suggest that the biopsies of patients heterozygous for this deletion have retained a capacity to secrete residual amounts of chloride in response to Ca^{2+}-linked secretagogues. Whether these patients also express a milder form of the disease is presently under investigation.

The discovery of a G-protein- in addition to a previously described cAMP-regulated chloride conductance pathway in the apical membrane of rat enterocytes and human colonocytes (Table 3) is of considerable interest to a further understanding of CF. It may imply that in intestinal tissue several separate pathways exist to mediate chloride secretion. Although our data suggest that in the digestive tract multiple pathways are affected by the disease, the data do not permit us to discriminate between a single channel defect or a defect affecting multiple channels. Further studies of channel characteristics by on-cell and whole-cell patch-clamp analysis in normal and CF epithelial cells are needed to gain a better insight into the effect of the CF mutation at the level of each chloride channel species.

References

1. J.H. Chen, H. Schulman and P. Gardner, 1989, A cAMP-regulated chloride channel in lymphocytes that is affected in cystic fibrosis, **Science**, 243:657.
2. H.R. de Jonge, J. Bijman and M. Sinaasappel, 1989, Relation of regulatory enzyme levels to chloride transport in intestinal epithelial cells, **Ped. Pulmonology Suppl.**, 1:54.
3. H.R. de Jonge, J. Bijman and M. Sinaasappel, 1988, **Ped. Pulmonology Suppl.**, 2:93.
4. R.A. Frizzell, 1987, Cystic Fibrosis: A disease of ion channels **Trends in Neurosci.**, 10:190.
5. M. Li, J.D. McCann, C.M. Liedtke, A.C. Nairn, P. Greengard, M.J. Welsh, Cyclic AMP-dependent protein kinase opens chloride channels in normal but not cystic fibrosis airway epithelium, 1988, **Nature**, 331:358.
6. P.M. Quinton and J. Bijman, 1983, Higher bioelectric potentials due to decreased chloride absorption in the sweat glands of patients with cystic fibrosis, **New Engl. J. Med.**, 308:185.
7. P.M. Quinton, 1983, Chloride impermeability in cystic fibrosis, **Nature**, 301:421.
8. J.R. Riordan, J.M. Rommens, B.-S. Kerem, N. Alon, R. Rozmahel, Z. Grzelczak, J. Zielenski, S. Lok, N. Plavsic, J.-L. Chou, M.L. Drumm, M.C. Iannuzzi, F.S. Collins and L.-C. Tsui, 1989, Identification of the CF gene: cloning and characterization of complementary DNA, **Science**, 245:1066.
9. R.L. Shoemaker, R.A. Frizzell, T.M. Dwyer and J.M. Farley, 1986, Single-channel currents from canine tracheal epithelial cells, **Biochem. Biophys. Acta**, 858:235.
10. M.J. Welsh and C.M. Liedtke, 1986, Chloride and potassium channels in cystic fibrosis epithelia, **Nature**, 322:467.
11. H.R. de Je Jonge, N. van den Berghe, B.C. Tilly, M. Kansen and J. Bijman, 1989, (Dys)regulation of epithelial chloride channels, **J. Biochem. soc. Trans.**, 17:816.
12. J.L. Goldstein, N.T. Nash, F. Al-Bazzaz, T.J. Layden, M.C. Rao, 1988, Rectum has abnormal ion transport but normal cAMP-binding proteins in cystic fibrosis, **Am. J. Physiol.**, 254:C719.
13. A. Yatani, J. Codina, A.M. Brown and L. Birnbaumer, 1987, A G protein directly regulates mammalian cardiac calcium channels, **Science**, 238:1288.
14. A. Yatani, J. Codina, A.M. Brown and L. Birnbaumer, 1987, Direct activation of mammalian atrial muscarinic potassium channels by GTP regulatory protein Gk, **Science**, 235:442.
15. M. Mohrmann, H.F. Cantiello and D.A. Ausiello, 1987, Inhibition of epithelial Na^+ transport by atriopeptin, protein kinase C and pertussis toxin, **Am. J. Physiol.**, 253:F372.

16. D.B. Light, D.A. Ausiello and B.A. Stanton, 1989, **J. Clin. Invest.**, 84:352.
17. B.C. Tilly, M. Kansen, P.G.M. van Gageldonk, N. van den Berghe, J. Bijman, H.R. de Jonge, 1990, Activation of intestinal chloride channels by GTP-binding regulatory proteins, submitted.
18. N.P. Illsey and A.S. Verkman, 1987, Membrane chloride transport measured using a chloride-sensitive fluorescent probe, **Biochemistry**, 26:1215.
19. J.R. Demarest, D.D.F. Loo and G. Sachs, 1989, Activation of apical chloride channels in the gastric oxyntic cell, **Science**, 245:402.

DISCUSSION

S.GUGGINO: Do the keratinocytes have MDR? I'm interested in the effects of vinblastine and the other cations, because they are transported by P-glycoprotein.

BIJMAN: In the lecture I showed that activity of the outward rectifying 30-50pS Cl^--channel is present in excised patches of cultured normal and CF keratinocytes. In 10 minute periods at physiological holding potential the normal channel is not activated by 500 nM PK-A and PK-C, in the presence of 1 nmol ATP and 10^{-8}mol/l Ca^{2+}. With bath 10^{-6}mol/l Ca^{2+} the activity of the voltage activated channel is not inhibited by PK-C. These experiments may suggest that stimulatory/inhibitory phosphorylation sites may be absent on the outward rectifier, and that the channel is part of a channel family with similar electrical properties but different regulation sites. In addition, I have shown that the channel activity is inhibited by chemically unrelated compounds like probenicid, NPPB, DIDS and substrates of the MDR pump like colchicin, vinblastine and adriamycin. Therefore, the apparent sensitivity of the anion transport system to unrelated cat- and anions may suggest that a factor (c.f. Kunzelman et al., this volume) may reside in the CF cell that prevents expression of channel activity. Such a factor could potentially also be a substrate of CFTR as suggested by its homology with MDR-protein, the latter I am not sure is expressed in the keratinocyte.

(Unidentified): Just a question for understanding: the last experiments were all done on isolated patches, excised inside-out?

BIJMAN: Yes.

WINE: I just wanted to follow up on Dr. Guggino's question. I wonder if the MDR has been identified in the cells that we all know and love, T84, HT29, and so on. Are there any data on that?

RIORDAN: Specifically in T84 cells P-glycoprotein-1 gene is expressed at an intermediate to low level. MDR1 is expressed in epithelial cells in general. It's in the apical membrane. Keratinocytes have not been looked at rigorously. It's not in skin fibroblasts. I'm not really sure whether it's in keratinocytes or not. But it's in the cells that we all know and love, quite substantially: that's its primary site of localization in normal tissues, in bipolar cells. We have looked at T84 specifically side by side with P-glycoprotein and CFTR probes and they are together.

S.GUGGINO: I have another question. When you're looking for the inward rectifying channel on the HT29 cells and you isolate vesicles, what is the marker that you use for apical membrane vesicles and, secondly, are you isolating them in a certain state, for example non confluent cells, so that you're getting this kind of channel as opposed to the outward-rectifying channel?

BIJMAN: The Cl^--channels we study are in patches obtained from apical membranes of cells in the middle of confluent monolayers. With respect tothe markers I believe Dr. de Jonge can answer your question best.

DE JONGE: Membrane vesicles used for characterization of the GTP-gamma-S activated chloride conductance were obtained from both plastic-grown monolayers of the HT-29.19A subclone and from freshly isolated rat small intestinal villus cells by the well-known differential Mg^{2+}-precipitation technique. The rat intestinal brush border membrane vesicles were 15-20 fold enriched in the apical membrane marker sucrase and virtually depleted of the basolateral membrane marker Na^+, K^+-ATPase. Unfortunately, the Cl^--secreting HT-29.19A subclone resembles membrane marker enzymes. The sole criterium for suggesting an apical origin of the vesicles was therefore the relatively low activity of Na^+m K^+-ATPase (enrichment factor 1.3-1.5) in the vesicle preparation.

FRIZZELL: Jan, I was wondering what you know about the macroscopic conductance properties of the keratinocytes with respect to activation by receptor-mediated events that work through cAMP or calcium.

BIJMAN: You mean in the Ussing chamber or in the patches?

FRIZZELL: Any macroscopic measurements of transport, other than single channel.

BIJMAN: In the Ussing chamber the normal and CF keratinocytes respond electrically to carbachol, A23187 and isoproterenol but not to cAMP and forskolin. Also, these responses are similar in the absence of extracellular Cl^-. This suggests that cAMP regulated Cl^--secretion is not an intrinsic property of the cells. The responses probably are related to K-channel activation which we observe in the on-cell patch. Spontaneous Cl^--channel activity is sparsely observed in cell-attached patches, but channel activity is not significantly induced by any of the aforementioned secretagogues.

KUNZELMANN: You saw the activation of chloride channels in excised patches of HT29 cells by GTP-gamma-S. Could you reverse this application also in these experiments by GDP-beta-S?

BIJMAN: Yes

KUNZELMANN: My second question is the explanation for the effects of trypsin in keratinocytes, in the activation of chloride channels.

BIJMAN: Indeed, at physiological holding potential all Cl^--channels present in excised apical patches of keratinocytes are within 1 minute activated by 0.05% trypsin. The precise mechanism of trypsin channel activation is presently not clear.

AUSIELLO: Could you comment on the toxin sensitivity of your GTP-gamma-S sensitive inwardly rectifying chloride channel?

BIJMAN: In our laboratory Dr. de Jonge is characterizing the action of these compounds in Ussing-chamber experiments.

DE JONGE: Pretreatment of monolayers of the HT-29 colonocytes with pertussis toxin did not affect the GTP or GTP-gamma S-activation of the chloride conductance in the apical membrane vesicles and did not modulate Cl^- secretion in the Ussing chamber either in the absence or presence of secretagogues. In contrast, pretreatment with choleratoxin led to a 3-4 fold increase of the basal chloride conductance and a partial loss of the GTP-gamma S response in the membrane vesicles and to a dramatic stimulation of Cl^- currents in the Ussing chamber. At present we try to dissect the classical action of choleratoxin, that is G_s activation, adenylate cyclase activation, and cAMP-activation of the outwardly rectifying chloride channel, from a possible direct effect of activated G_s on the inwardly rectifying Cl^- channel, for example by inhibiting the classical pathway by adenylate cyclase and protein kinase inhibitors. In conclusion, we did not find any evidence for a role of a pertussis toxin-sensitive G_i protein as a modulator of apical chloride channels but we cannot yet exclude the possibility that choleratoxin-sensitive G_s proteins play a role as activators of the GTP gamma S-sensitive Cl^- channel.

CHLORIDE TRANSPORT PATHWAYS IN HUMAN KERATINOCYTES

Michela Rugolo, *Michele De Luca and
Teresa Mastrocola

Dip. di Biologia E.S.
Università di Bologna, Bologna
and *Ist. Nazionale per la Ricerca sul Cancro, Genova

INTRODUCTION

The transport of Cl^- across cell membrane plays an important role in a number of biological functions, including the maintenance of cell volume and the regulation of intracellular pH. A variety of systems capable of transporting Cl^- across the plasma membrane have been described in mammalian cells. They include the electroneutral Cl^- /anion exchange and Cl^-/cation cotransport, and the electrically conductive Cl^- channels. The Cl^-/ anion exchange has been shown to be involved in regulation of intracellular pH in hamster fibroblasts (L'Allemain et al., 1985) and in Vero cells (Olsnes et al., 1986). In some cell types, it has been reported that the Cl^-/cation cotransport can be activated by cell shrinkage induced by hypertonic media in a process named regulatory volume increase (for a review see Hoffmann, 1986). Conversely, in human lymphocytes (Grinstein et al., 1982), Ehrlich ascites tumor cells (Hoffmann, 1986), and human fibroblasts (Rugolo et al., 1989), cell swelling induced by hypotonic shock has been clearly demonstrated to activate electrically conductive Cl^- channels, in a process named regulatory volume decrease.

In several epithelial cells, Cl^- channels have been shown to be activated by an increase in the cytoplasmic concentration of Ca^{2+} and cyclic AMP (Welsh, 1986; Frizzell et al., 1986). In the epithelial cells of patients with cystic fibrosis (CF), evidence has been presented that cyclic AMP fails to activate Cl^- channels, whereas the Ca^{2+} dependent activation remains intact (Welsh and Liedtke, 1986; Schoumacher et al., 1986).

In the present study we report a preliminary investigation on Cl^- transport in epidermal keratinocyte lines from healthy individuals. We have identified the nature of the pathways by which Cl^- moves across the plasma membrane. Furthermore, we have

The Identification of the CF (Cystic Fibrosis) Gene
Edited by L.-C. Tsui *et al.*, Plenum Press, New York, 1991

determined the influence of some cAMP and calcium dependent effectors on Cl^- transport, to assess the possibility to use keratinocytes as a cell model system for the study of the molecular defect of CF.

MATERIALS AND METHODS

Keratinocyte culture

Human epidermal keratinocytes were obtained from skin biopsies and cultured on feeder layer of lethally irradiated 3T3-2J cells, according to the method described by Rheinwald and Green, 1975. Briefly, $2cm^2$ skin biopsies were minced and trypsinized at 37°C for 3h. Cells were collected every 30min, plated ($2x10^5$ cells/75-cm^2 flask) on lethally irradiated 3T3-2J cells and cultured in 5% CO_2 in keratinocyte growth medium: Dulbecco-Vogt Eagle's and Ham's F12 media (3:1 mixture) containing 10% fetal calf serum, insulin (5μg/ml), transferrin (5μg/ml), adenine (0.18mM), hydrocortisone (0.4μg/ml), cholera toxin (0.1nM), triiodothyronine (20pM), EGF (10ng/ml) and penicillin/streptomycin (50 IU/ml).

For the experiments, cells were trypsinized and seeded with the 3T3-2J cells feeder layer at a density of $2x10^5$ cells into multiwell dishes containing 6x35-mm diameter wells and were fed for 7 days. Experiments were performed using monolayers of keratinocytes immediately after confluence was reached.

Chloride efflux

Cells in 35-mm diameter 6 well culture dish were washed twice with 10 ml of 0.9% NaCl and 10mM Na-HEPES, pH 7.4, and incubated for 90 min in a water bath at 37°C and ambient CO_2 with 1ml of the incubation medium containing 135mM NaCl, 3mM KCl, 1.3mM $CaCl_2$, 1.5mM $MgCl_2$, 10mM D-glucose, 20mM Na- HEPES, pH 7.4 and 1.5μCi/ml of $Na^{36}Cl$. Efflux was determined as described by Lin and Gruenstein, 1988, with minor modifications. In brief, efflux was initiated by aspiring the radioactive medium from the well and rapidly washing the cells three times with 10ml of ice-cold 0.3M sucrose and 2.5mM Na-HEPES, pH 7.4. The entire wash procedure took ca. 15s. At the end of the wash, the cells were incubated with the efflux medium (0.8ml) at 37°C in a water bath. At 15s after the start of efflux, the efflux medium was removed from the well and delivered to a scintillation vial for counting. 0.8ml of the same efflux medium was added at the same well within 2-3s. The procedure of removing efflux medium and add fresh efflux medium was repeated at the indicated time points. At the end of the experiment, the cells were dissolved in 1ml 1M NaOH (15min at 55°C). An aliquot of the cell lysate was collected for determination of protein content (Lowry et al., 1951) and the remainder was transferred in a scintillation vial for counting. The total radioactivity in the cells present at the beginning of efflux (at zero time) was calculated by summing the radioactivity determined in the efflux medium at the times indicated plus the radioactivity remaining in the cells. The results are expressed as nmoles remaining in the cells/mg protein.

Materials

$Na^{36}Cl$ was obtained from Amersham; NPPB was kindly provided by Prof. R.Greger, Albert Ludwig Universität, Freiburg. Bumetanide was a gift of Prodotti Roche, Milano.

RESULTS AND DISCUSSION

The mammalian epidermis is composed principally by a single cell type, the keratinocyte, in various stages of differentiation. In vivo, the basal layer, separated by a basement membrane from the fibroblasts of the underlying dermis, contains the basal or proliferating keratinocytes.In cell culture conditions, keratinocytes depend on the support of fibroblasts to initiate proliferation. When human dermal keratinocytes from skin biopsies, are inoculated together with lethally irradiated 3T3 cells (Rheinwald and Green, 1975), the keratinocytes grow from single cell to macroscopic colonies. Each colony grows at the perifery and starts the differentiation process in the center and eventually gives rise to a stratified squamous epithelium, which has been shown to be suitable for grafting on burned patients (Green et al., 1979).

Living human fibroblasts, which were present in the originary tissue and whose moltiplication in culture should be greatly stimulated by the presence of EGF in the growth medium, were indeed almost absent, due to the inhibitory influence of irradiated 3T3 cells (Green et al., 1979).

The first step of our study has been the characterization of the pathways involved in Cl^- transport in human keratinocytes. To this purpose, we have determined $^{36}Cl^-$ efflux under steady-state conditions. In the semilogarithmic plot shown in fig.1, the time course of Cl^- efflux from human keratinocytes is described. The measured efflux exhibited first order exponential decay. The value of the rate constant of Cl^- efflux (k), determined from the experimental points by a computer program based on standard nonlinear least square minimization procedure (Bevington, 1969) was 0.145 min^{-1}. The amount of Cl^- released by the cells during the first minute of efflux, which accounts for approximately 15-18% of total Cl^- content, was not taken into consideration for determination of the k of efflux. It is likely that the efflux observed within the first min is due to cells which come off the plate during the early efflux periods, as it has been reported to occur in other cellular systems (Lin and Gruenstein, 1988).

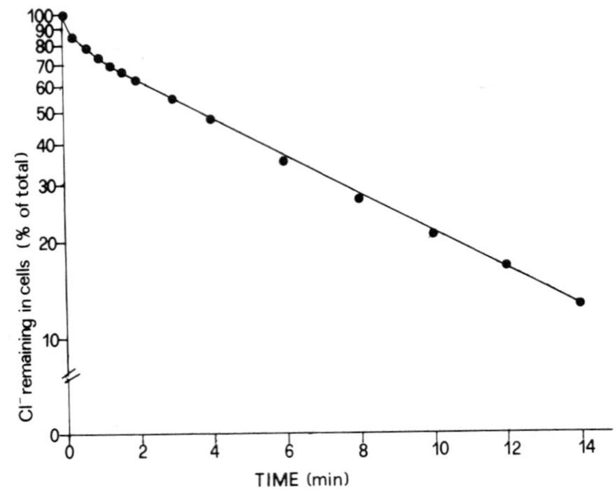

Fig. 1. Steady-state chloride efflux from human keratinocytes. Cells were loaded with $^{36}Cl^-$ for 90 min and efflux was determined as described in Materials and Methods.

The Cl^- efflux shown in fig.1 may be thought to be due to the contribution of various transport pathways, among which the anion exchange, the anion/cation cotransport and the electrically conductive pathway. To evaluate the relative contribution of these pathways, we have analyzed the effect of some specific inhibitors on the k of Cl^- efflux from keratinocytes and the results obtained have been summarized in fig.2.

The sensitivity to DIDS was used to reveal the activity of the anion exchange. 0.1mM DIDS induced a maximal 40% inhibition of Cl^- efflux (fig.2), indicating that approximately 40% of total efflux is mediated by this anion transport mechanism. To verify that the observed effect was due to inhibition of the anion exchange, Cl^- was removed from the extracellular medium and substituted by the impermeant anion gluconate, as an alternative means of inhibiting anion exchange. Under this condition, Cl^- efflux was 40% inhibited (not shown), suggesting that DIDS specifically inhibits the Cl^-/anion exchange component.

The addition of 10μM bumetanide, a well known inhibitor of the anion/cation cotransport, to the efflux medium was without effect on Cl^- efflux, however a weak (ca. 10%), but significant inhibition was observed when this inhibitor was added together with DIDS, as shown in fig.2. It seems therefore that the small inhibition determined by bumetanide can be determined only under conditions when the operation of the anion exchange is abolished. This finding is confirmed by the fact that bumetanide was able to produce the same 10% inhibition of Cl^- efflux in Cl^--free gluconate containing efflux medium (result not shown).

An anion selective channel has been identified in the plasma membrane of human keratinocytes by means of patch-clamp studies

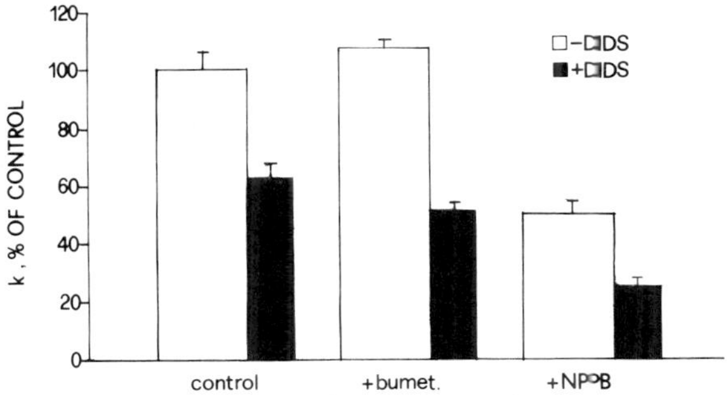

Fig. 2. Effect of DIDS, bumetanide and NPPB on the rate constant (k) of Cl^- efflux from keratinocytes. Cells were loaded with $^{36}Cl^-$ as described in Material and Methods. When indicated, 0.1mM DIDS, 10μM bumetanide and 0.1mM NPPB were added in the efflux medium. Results are expressed as % of total Cl^- efflux rate constant (k). Values represent means ± SD of 4 determinations.

(Galietta et al., 1989; see also this book, Bijman et al.). This channel is permeable to Cl^- ions and can be activated by membrane depolarization in excised patches. Interestingly, the biophysical behaviour of this channel is similar to that of the Cl^- channel of the respiratory epithelium. To assess the contribution of this pathway to overall Cl^- efflux, we have examined the effect of NPPB, a blocker of the Cl^- channel in several epithelial cells. Fig. 2 shows that the addition of 0.1mM NPPB induced a 50% inhibition of the k of Cl^- efflux. It is noteworthy that, as shown in fig. 2, the effects of DIDS and NPPB are additive, indicating that the anion exchange and the Cl^- channel are independent pathways in human keratinocytes.

From the data obtained with the inhibitors, it can be concluded that total Cl^- efflux from keratinocytes consists of ca. 40-50% of Cl^-/anion exchange, 10% Cl^-/cation cotransport and 50% electrically conductive pathway, accounting for approximately 100% of total Cl^- efflux.

A preliminary analysis of the regulatory mechanisms that might activate Cl^- transport in keratinocytes has been therefore undertaken. We have determined whether elevation in the cytoplasmic cAMP- and calcium-dependent concentrations was able to increase the k of Cl^- efflux from keratinocytes. As reported in table 1, Cl^- efflux has been determined both in the absence and in the presence of DIDS, to abolish the operation of the Cl^-/anion exchange. The cAMP analog 8-Br-cAMP, the beta-agonist isoproterenol, the inhibitor of the phosphodiesterase IBMX, the activator of the catalytic subunit of the adenylate cyclase forskolin, failed to induce any significant activation of Cl^- efflux. It has to be noticed that to test the effect of these cAMP-dependent effectors on Cl^- transport, cholera toxin was omitted from the keratinocytes growth medium.

In table 1 it is also reported that both the Ca^{2+} ionophore A23187 and the vasoactive peptide bradykinin, which was shown to induce a rapid transient rise in the intracellular free Ca^{2+} concentration in keratinocytes (Wheeler and Sachs, 1989), were unable to induce any increase in the k of Cl^- efflux.

TABLE 1

EFFECT OF CYCLIC AMP AND CALCIUM DEPENDENT EFFECTORS ON CL^- EFFLUX FROM KERATINOCYTES

RATE CONSTANT OF CL^- EFFLUX (min^{-1})

Additions	NaCl medium	NaCl medium + 0.1mM DIDS
none	0.151 ± 0.006 (6)	0.111 ± 0.003 (6)
isoproterenol 5μM	0.145 ± 0.004 (3)	0.108 ± 0.004 (3)
IBMX 0.2mM	0.160 ± 0.009 (3)	0.115 ± 0.006 (3)
IBMX+isoproterenol	0.161 ± 0.007 (3)	0.115 ± 0.006 (3)
forskolin 1μM	0.154 ± 0.002 (3)	------
IBMX+forskolin	0.158 ± 0.005 (3)	------
8 Br-cAMP 0.1mM	0.147 ± 0.004 (3)	0.013 ± 0.006 (3)
A23187 3uM	0.153 ± 0.005 (6)	------
Bradykinin 0.1μg/ml	0.157 ± 0.006 (5)	------

The experiments with cAMP effectors were performed with keratinocytes grown in the absence of cholera toxin. The cAMP effectors were preincubated with the cells for 90 min and added also to the efflux media. A23187 was added only to the efflux medium. Bradykinin was preincubated with the cells for 1 min and added also to the efflux medium.

These data taken together clearly indicate that both cAMP and Ca^{2+} ions do not seem to play any direct role in the regulation of Cl^- transport in keratinocytes. It has however to be noticed that the contribution of the conductive pathway to overall Cl^- transport is quite high in keratinocytes (40- 50%), in comparison with 20% reported in human fibroblasts and 6% in ascite cells. It is possible to suppose that, in keratinocytes under resting conditions, the Cl^- channel might be already activated by some intracellular mechanism. If this should be the case, the lack of the effect by any cAMP- and Ca^{2+}-dependent effectors on Cl^- transport could be explained. Further studies are required to verify this hypothesis.

In conclusion, if the cAMP dependent activation of the Cl^- channel is in some way directly linked to the expression of the CF gene, it follows that our negative results obtained with cAMP-dependent effectors, might be taken as evidence that keratinocytes are not a suitable cell model system for the study of the molecular defect responsible for CF. At present, no data are available on the expression of the CF gene in keratinocytes. In preliminary experiments, a very low level of expression of CF gene has been determined in keratinocytes either by means of Northern blot analysis (J.Riordan, personal communication) and with amplification of DNA by polymerase chain reaction (P. Ronchetto and G.Romeo, personal communication). The availability of definitive results on CF gene expression will provide the direct and conclusive evidence for the possible use of keratinocytes in CF studies.

ACKNOWLEDGMENTS: We wish to thank Prof. G. Lenaz, University of Bologna, Prof. G. Romeo and Dr. L.J.V. Galietta, Ist. G. Gaslini Genova, for helpful discussion and suggestions. We appreciate the expert technical assistance by Mrs. M. Guastella, Cellife, Genova, in growing keratinocyte lines.

ABBREVIATIONS

HEPES: N-2-hydroxyethylpiperazine-N'-2-ethane sulfonic acid
DIDS: 4,4'-diisothiocyanatostilbene-2,2'-disulfonic acid
NPPB: 5-nitro-2-(3-phenylpropylamino) benzoate
IBMX: 3-iso-butyl 1-methyl xantine
8Br-cAMP: 8 bromo cyclic AMP

References

Bevington, P. R., 1969, "Data reduction and error analysis for the physical sciences", McGraw-Hill, New York.

Frizzell, R. A., Rechkemmer, G. and Shoemaker, R. L., 1986, Altered regulation of airway epithelial cell chloride channels in cystic fibrosis, **Science**, 233:558.

Galietta, L. J. V., Mastrocola T., De Luca M., Bondanza S., Cancedda R. and Romeo G., 1989, Uso dei cheratinociti umani in coltura come modello per lo studio elettrofisiologico del difetto molecolare della fibrosi cistica, Abstract "IV Congresso Nazionale F.I.S.M.E.", Milano.

Green, H., Kehinde, O. and Thomas, J., 1979, Growth of cultured human epidermal cells into multiple epithelia suitable for grafting, **Proc. Nat. Acad. Sci. USA**, 76:5665.

Grinstein, S. A., Rothstein, A., Sarkadi, B. and Gelfand, E. W., 1982, Responses of lymphocytes to anisotic media: volume regulating behaviour, **Am. J. Physiol.**, 246:C204.

Hoffmann, E. K., 1987, Volume regulation in cultured cells, in: "Current Topics in Membranes and Transport", R. Gilles, A. Kleinzeller and L. Bolis, eds., Academic Press, 30:125.

L'Allemain, G., Paris, S. and Pouyssegur, J., 1985, Role of a Na^+ dependent Cl^--HCO_3 exchange in regulation of intracellular pH in fibroblasts, **J. Biol. Chem.**, 260:4877.

Lin, P. and Gruenstein, E., 1988, Pathways of Cl^- transport in human fibroblasts, **Am. J. Physiol.**, 255:C112.

Lowry, O. H., Rosenbrough, N. J., Farr, A. L. and Randall, R.

J., 1951, Protein measurements with the Folin phenol reagent, **J. Biol. Chem.**, 193:265.
Olsnes, S., Tonnessen, T. I. and Sandvig, K., 1986, pH- regulated antiport in nucleated mammalian cells, **J. Cell. Biol.**, 102:967.
Rheinwald, J. G. and Green, H., 1975, Serial cultivation of strains of human epidermal keratinocytes: the formation of keratinyzing colonies from single cells, **Cell**, 6:331.
Rugolo, M., Mastrocola, T. Flamigni, A. and Lenaz, G., 1989, Chloride transport in human fibroblasts is activated by hypotonic shock, **Biochem. Biophys. Res. Comm.**, 160:1330.
Schoumacher, R. A., Shoemaker, R. L., Halm, D. R., Tallant, E. A., Wallace, R. W. and Frizzell, R. A., 1987, Phosphorylation fails to activate chloride channels from cystic fibrosis airway cells, **Nature**, 330:752.
Welsh, M. J., 1986, An apical membrane chloride channel in human tracheal epithelium, **Science**, 232:1648.
Welsh, M. J. and Liedtke, C. H., 1986, Chloride and potassium channels in cystic fibrosis airway cells, **Nature**, 330:752.

DISCUSSION

ROMEO: Since we are partly responsible for the notion that CFTR is present in keratinocytes, I would like to just describe briefly the experiment which was run in our laboratory by Dr. Patrizia Ronchetto. The PCR, which showed the presence of the messenger in keratinocytes, had to be pushed to at least 3 different rounds of 30 cycles. In this way you can really demonstrate that almost everything is expressed, and with this small amount of messenger nobody would conclude that this represents expression. What I would like to do is confront our data with those of Francis Collins, who is not now here, because we have to be careful in the implications which are drawn from such experiments.

RIORDAN: I'd like to continue with that point, Giovanni. We had looked in a number of keratinocyte RNAs, both lines and primary cultures, mainly with Northern blots. So the discrepancies in order of magnitude are two, in terms of sensitivity. But at that level you do not detect CFTR. I think that is a relevant point, given that this afternoon a lot of things have been based on correlations or lack of correlations between the amount of messenger RNA and some very elaborate transport measurement which I have some reservations about in general. But I think it is fair to say that in our hands there is at least, probably approaching two orders of magnitude less CFTR RNA in a keratinocyte than in T84 cells, and certainly at least one order of magnitude less than in some other cells, more normal cells like sweat glands and things. I wouldn't despair about the fact, if you'd like to continue making a correlation positively rather than negatively, I think you're justified in doing that.

GREGER: I would like to comment that you are in a happy family, in a way, because we listened yesterday to Dr. Quinton, and he was showing us convincingly that even though he has the preparation for CF, the sweat duct with chloride channels, he had serious problems in stimulating it via the cyclic pathway. So I think cAMP regulation and the CF defect are really not necessarily linked, and that's the take-home message.

CORCELLI: Michela, I would like to ask you, does it make sense to try the effect of 8-bromo-AMP under hypotonic conditions? Because if I remember well, I saw a paper in the literature in which they test the role of cAMP as second messenger in the osmotic response?

RUGOLO: No, I don't think there is evidence of the role of cAMP on cell volume regulation, at least in the response to hypotonic shock.

FRIZZELL: Keratinocytes do seem to have a conductive pathway for efflux, even though you can't find a regulator of it. That reminded me of the experiments that Kevin Kirk at the University of Alabama had done using sweat duct cells in primary culture, where he was able to show that the spontaneously active conductance of those duct cells could be reduced by inhibiting prostaglandin formation using indomethacin, and could then be restored in the presence of indomethacin by adding exogenous prostaglandins back. So I wonder if there is a possibility that even though you can't find regulation, it might simply be due to some autacoid production or something like that.

RUGOLO: Maybe. I have no answer.

BOUCHER: Have we heard if CF keratinocytes are distinctly different?

RUGOLO: We have not tested them.

BOUCHER: Does anybody know that?

SATO: I'm a dermatologist, but I know of no skin disease that is specific to CF, and I think the epidermis is quite happy in CF. So I don't see any reason why defects in keratinocytes should be expected.

A CHLORIDE CONDUCTANCE EVOKED BY HYPOTONIC SHOCK IN EPITHELIAL CELLS

L.J.V. Galietta (*), V. Barone (*),
D.C. Gruenert (+) and G. Romeo (*)

(*) Lab. di Genetica Molecolare
Ist. G. Gaslini, Genova, Italy
(+)Cardiovascular Research Institute
University of California, San Francisco, CA

Single channel recordings have shown that airway epithelial cells possess apical rectifying Cl^- channels whose activity is regulated by the cAMP pathway. Several reports (Frizzell et al., 1986; Welsh and Liedtke 1986; Schoumacher et al., 1987; Li et al., 1988) indicate an altered regulation of these chloride channels as the possible functional defect causing cystic fibrosis (CF). We asked whether the defective regulation of Cl^- channels in CF can also be detected in the whole-cell configuration of the patch-clamp technique. This approach, which allows to record macroscopic membrane currents, gives a more representative view of the whole cell membrane conductance. The possible existence of different types of Cl^- currents can be more easily detected. Furthermore, in polarized cells, the whole-cell technique allows to record currents flowing through the basolateral membrane which is not accessible to the patch pipette. Recent reports using the whole-cell approach have shown that colonic carcinoma T84 cells (Worrell et al., 1989) and primary culture respiratory cells (Schoppa et al., 1989; McCann et al., 1989) have a Cl^- conductance characterized by inactivation at positive membrane potentials. Two of these papers (Worrell et al., 1989; McCann et al., 1989) report that the inactivating Cl^- current is sensitive to cell volume changes. It has also been shown that this current is activated by forskolin (Schoppa et al., 1989) or by the cAMP dependent protein kinase (McCann et al., 1989). Our experiments on T84 and transformed respiratory cells confirm that the inactivating Cl^- conductance can be activated by hypotonic shocks. On the other hand, we failed to see a cAMP-dependent activation. This suggests that the volume-sensitive Cl^- conductance is different from the secretory Cl^- conductance

The Identification of the CF (Cystic Fibrosis) Gene
Edited by L.-C. Tsui *et al.*, Plenum Press, New York, 1991

which is regulated by cAMP. This conclusion seems supported by the fact that the activation of the volume-sensitive Cl^- current is unaltered in CF cells.

METHODS

Cell Culture

Colon carcinoma T84 cells were a gift from J.R. Riordan. Transformed airway epithelium cell lines were established as reported by Gruenert et al. (1988). Two lines (9HTEo- and 4AHTE2o-) were obtained from the tracheal epithelium of normal individuals whereas a third line (CFNPE9o-) was from the nasal polyp epithelium of a cystic fibrosis patient. The culture medium was MEM or RPMI 1640 supplemented with 10% fetal calf serum, L-glutamine, and antibiotics. Transformed respiratory cells were plated onto dishes coated with a solution containing 1 mg of human fibronectin, 10 mg of bovine serum albumin, and 1 ml of Vitrogen 100 (Collagen Corp.).

Patch-Clamp Recordings

Outside-out patch and whole-cell recordings were performed as described elsewhere (Hamill et al., 1981; Marty and Neher, 1983). The pipette solution contained (in mM): 140 NaCl, 0.18 $CaCl_2$, 2 EGTA, 1 $MgCl_2$, 1 Na_2ATP, 10 Na-Hepes (pH=7.3; free $Ca^{2+} = 1 \cdot 10^{-8}$ M). The osmolality was adjusted to 290 mosmol/Kg by addition of mannitol. The extracellular isotonic medium contained (in mM): 130 NaCl, 3 KCl, 2 $CaCl_2$, 2 $MgCl_2$, 10 glucose, 10 Na-Hepes (pH=7.3) and mannitol to adjust the osmolality to 290 mosmol/Kg. The extracellular hypertonic medium had the same composition but a higher mannitol concentration in order to reach a value of 340 mosmol/Kg. The extracellular hypotonic solution (210 mosmol/Kg) contained no mannitol and a lower NaCl concentration (90 mM).

The cell capacitance was cancelled using the analogical circuit of the EPC 7 (List-Medical) amplifier. In part of the experiments, the series resistance was analogically compensated using the appropriate circuitry of the patch-clamp amplifier. In the rest of the experiments, in which this was not possible, the voltage drop caused by the current flowing through the series resistance was calculated and used to correct the applied voltage. Data were stored on a conventional tape of a video-cassette recorder after analogical to digital conversion using a pulse code modulator (Sony 501 ES). Voltage stimulation and data analysis were performed with an IBM-AT computer equipped with a Data Translation 2801A board and a software package programmed by J. Dempster (1987). Membrane potentials are indicated taking the outside of the cell as ground. Outward currents represent anions flowing into the cell.

RESULTS

Whole-cell experiments performed with isotonic bath and pipette solutions showed that a chloride conductance usually appeared in a few minutes after the access to the cell was established. As depicted in FIG. 1, the cells initially displayed a low membrane conductance. The input resistance was in the range 1-2 GOhm and in some cases it was possible even to detect single channel activity. After 3-5 minutes there was a spontaneous increase of membrane currents which reached a value 5- to 15-fold larger than the initial value. FIG. 2A and 2B show samples of currents obtained from the same experiment of FIG. 1. The traces of FIG. 2B were recorded 10 minutes after establishing the whole-cell configuration. Compared to those obtained at the fourth minute (FIG. 2A), the currents are much larger and show a time dependent inactivation at the most positive potentials (+60, +80 mV). The replacement of external chloride by gluconate caused a shift of the zero-current potential towards positive values suggesting a Cl^- selectivity (see FIG. 2C).

The spontaneous activation of chloride currents was a common finding in our experiments. It was observed in 11/16 experiments on normal respiratory cells and 13/16 experiments on CF respiratory cells. It was also observed in T84 cells (9/10 experiments) confirming the results reported by Worrell et al (1989). These authors have explained this phenomenon postulating that the equilibration between the cytosol and the pipette solution produces a cell interior hypertonic with respect to the extracellular medium causing cell swelling and chloride current activation. This interpretation is confirmed by experiments of the type shown in FIG.3. In fact, the perfusion with the hypertonic extracellular medium (340 mosmol/Kg) since the beginning of the experiment prevented the appearance of Cl^- currents in 22/26 experiments on normal respiratory cells and 8/3 experiments on CF cells. As also shown in FIG. 3, the lowering of the external osmolality to 210 mosmol/Kg caused the activation of large

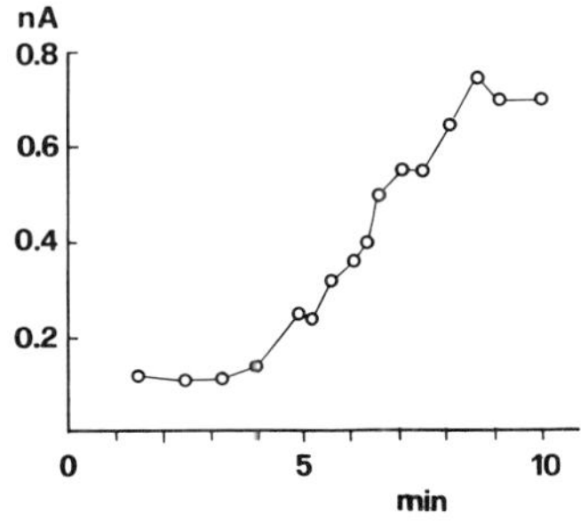

FIG. 1. Spontaneous increase of whole-cell current in a transformed respiratory cell (CFNPE9o-) with isotonic (290 mosmol/Kg) bath and pipette solutions. Peak current values obtained by voltage steps to +80 mV are plotted versus the recording time. The holding potential was -50 mV.

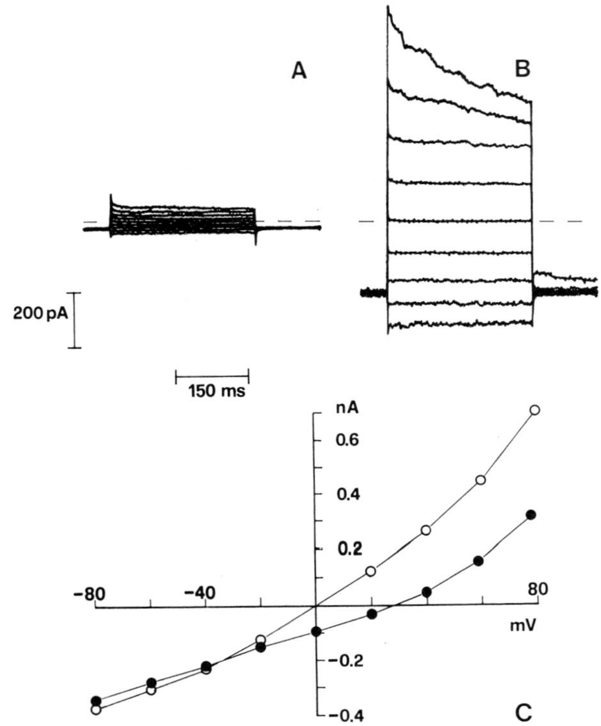

FIG. 2. Whole-cell currents from a transformed respiratory cell. The experiment is the same of FIG. 1. The currents of FIG. 2A and FIG. 2B were respectively obtained 4 and 10 minutes from the beginning of the experiment. They were elicited by voltage pulses to membrane potentials between -80 and +80 mV starting from a holding potential of -50 mV. The dashed line indicates the zero-current level. FIG. 2C shows the current-voltage relationships with symmetrical NaCl (open circles) or after partial replacement (93%) of external Cl^- by gluconate (closed circles).

membrane currents. Anion replacement experiments indicated that these currents were chloride selective.

The hypotonic shock was able to activate Cl^- currents in all the experiments carried out on normal (n=28) and CF (n=16) respiratory cells as well as in 4/5 experiments on T84 cells. The delay between the beginning of the hypotonic perfusion and the onset of currents was relatively short (1-2 min.) and no significant differences between normal and CF cells were found. The time-course of the Cl^- conductance during the hypotonic shock was found to be dependent on the voltage stimulation protocol. Using the protocol indicated in FIG.3, i.e. a holding potential

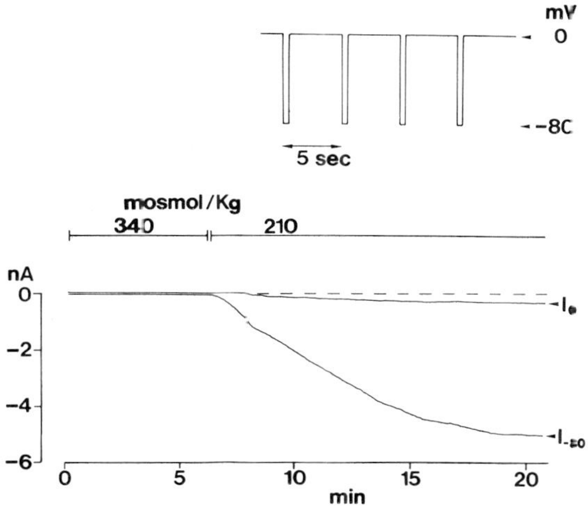

FIG. 3. Activation of Cl^- currents by hypotonic shock. The continuous lines represent the time-course of the currents at 0 mV (I_0) and -80 mV (I_{-80}) during a whole-cell experiment on a normal respiratory cell (9HTEo-). The dashed line indicates the zero-current level. The stimulation protocol is represented in the upper part of the figure. As evident from the plot, the membrane conductance remained low during perfusion with the hypertonic medium, but a large increase of membrane currents was obtained with the hypotonic medium.

close to the equilibrium potential of Cl^-, a continuous increase of Cl^- currents was observed for several minutes during the perfusion with the hypotonic medium. Steady-state values were difficult to achieve because they required very long times and were so large that the voltage drop along the series resistance became a significant fraction of the applied membrane potential. The increase of the external osmolality to the hypertonic value (340 mosmol/Kg) caused the membrane conductance to slowly decrease. Full recovery of the initial membrane resistance required times longer than 10-15 min.

FIG. 4A shows an example of Cl^- currents activated by the hypotonic shock. Note the typical inactivation at positive voltages similar to that shown by spontaneously activated currents. Note also the reduction of the tail current absolute value following inactivating voltage pulses. As shown in FIG. 4B, the replacement of the external chloride by gluconate produced a shift of the reversal potential to +40 mV. A value of +64 mV is expected for a perfectly chloride selective membrane in our experimental conditions. The difference between the observed reversal potential and the Nernstian potential for chloride could indicate that chloride channels underlying the macroscopic current are not completely impermeable to gluconate. The presence of a relevant Na^+ or cation conductance was excluded because the

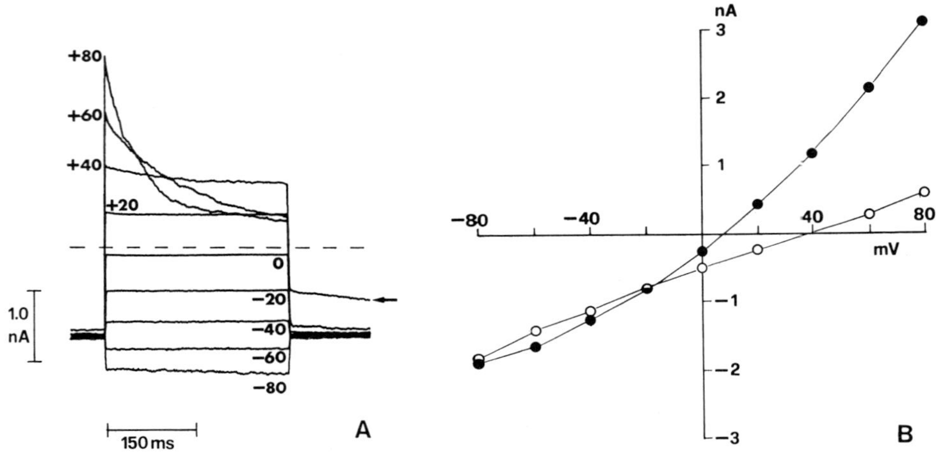

FIG. 4. Whole-cell currents activated by the hypotonic medium. The plot of FIG. 4A represents superimposed membrane currents obtained by voltage steps to potentials between -80 and +80 mV starting from a holding potential of -50 mV. The arrow marks the tail current after the step to +80 mV. FIG. 4B shows the effect of external Cl^- replacement on membrane currents. Closed circles indicate the current-voltage relationship obtained with the hypotonic medium (100 mM Cl^- outside/ 140 mM Cl^- inside); open circles instead represent the current-voltage relationship with 90% of the external Cl^- replaced by gluconate.

application of an external solution containing the large cation N-methyl-D-glucamine instead of Na^+ did not change the current-voltage relationship.

As shown in FIG. 4A, the recovery of currents after the inactivating voltage pulse was relatively slow. We investigated this process by means of a two voltage steps protocol similar to that used by Shoppa et al. (1989). A test voltage step to +80 mV was applied at increasing intervals following a conditioning pulse to the same membrane potential. The ratio of the current obtained with the test pulse (I_1) to the current elicited by the conditioning pulse (I_2) was plotted in FIG. 5 versus the interval duration. This was performed for two holding potentials: -50 and -90 mV. As evident from FIG. 5, the reactivation process shows a double exponential time-course. Most of the current recovery is obtained in the first 100-200 msec, but the rest is fully reactivated only after long times. It is also evident that the reactivation process is faster at the more negative membrane potential (-90 mV). The existence of a long lived inactivated state forces to the use of long waiting times (5-6 sec.) between pulses in order to achieve full reactivation of currents.

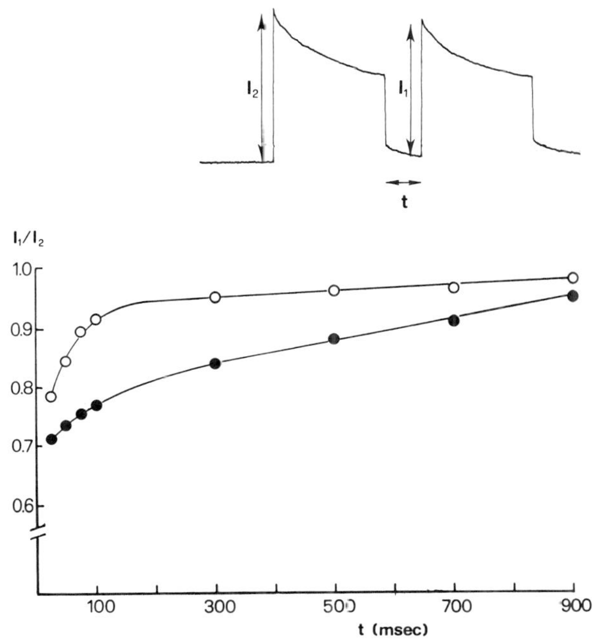

FIG. 5. Voltage- and time-dependence of current recovery from inactivation. Two voltage steps to +80 mV were applied at varying intervals. The ratio of the current elicited by the test pulse (I_1) to the current evoked by the conditioning pulse (I_2) is plotted versus the interval duration. Data shown in the figure are mean values from 3 different experiments. Holding potential was -50 mV (closed circles) or -90 mV (open circles).

In 40% of the experiments it was possible to obtain outside-out patches after the whole-cell recording. The apparent seal resistance after patch formation was in the range 5-20 GOhm. This demonstrates that seal currents in our whole-cell recordings were negligible. In almost all the patches, it was found the activity of channels whose current-voltage relationship was outwardly rectfying as shown in FIG. 6A. Anion replacement experiments indicated that these channels were chloride selective. Channels quickly inactivated when the membrane potential was stepped to large positive potentials (see FIG. 6B). Channel reactivation was obtained when the membrane potential returned to negative values. These kinetic features resemble those of the macroscopic Cl^- currents.

Attempts were made to find the regulatory pathway underlying osmotically activated Cl^-currents. Extracellular Ca^{2+} seems not required because a hypotonic solution containing very low Ca^{2+} (2 mM EGTA and nominal zero Ca^{2+}) was able to activate Cl^- currents (n=4).

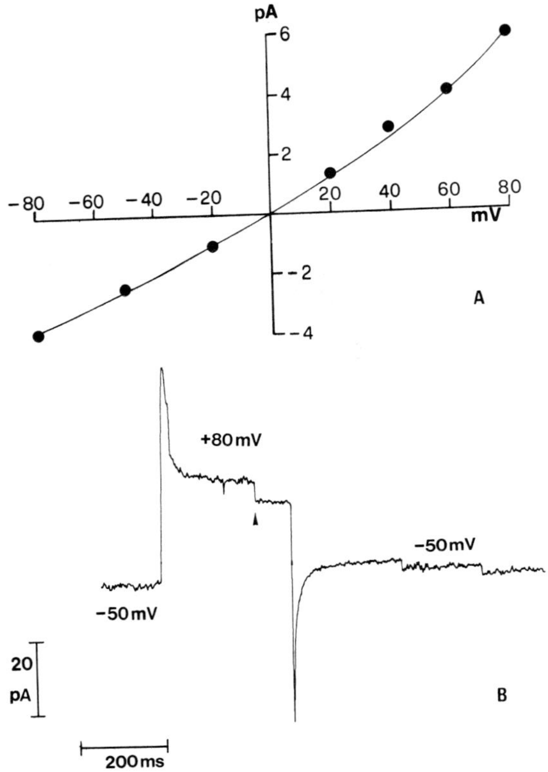

FIG. 6A. Current-voltage relationship of a Cl^- channel obtained from an outside-out patch. The FIG. 6B shows the voltage dependent inactivation of Cl^- channels observed in outside-out patch recordings. As depicted in the figure, the step to +80 mV causes channel inactivation (arrow). A fast closure of a second channel is probably masked by the capacitive current. The return of the membrane potential to -50 mV evokes the reactivation of two channels.

The role of cAMP was also investigated. These experiments were performed with the hypertonic medium as the external solution in order to inhibit spontaneous activation of currents. Application of external forskolin (50 μM; n=4), external dibutyryl-cAMP (0.5 mM; n=5), or internal cAMP (50 μM; n=5) was ineffective on cells in which the subsequent exposure to the hypotonic medium was able to activate large chloride currents.

DISCUSSION

Our experiments confirm the presence of a volume-sensitive Cl^- current in T84 cells. We have also shown that this current represents a considerable component of the membrane conductance in transformed cell lines obtained from airway epithelia. It is not possible to establish whether the volume-sensitive conductance has a role in the secretion of Cl^- in the epithelium. Our experiments suggest that this conductance is different from the secretory Cl^- conductance which is activated by cAMP elevating agents. This conclusion is also supported by the normal activation of Cl^- currents found in our experiments on CF cells. Furthermore, the fast inactivation of Cl^- channels produced by positive membrane potentials has not been described for apical Cl^- channels activated by cAMP elevating agents or by protein kinases (Frizzell et al., 1986; Welsh and Liedtke, 1986; Schoumacher et al., 1987; Li et al., 1988). Therefore, the volume-sensitive chloride conductance, as argued by Worrell et al. (1989), may represent a non-specialized, housekeeping function involved in cell volume regulation and widespread among different epithelial cell types. The lack of a cAMP-activated Cl^- conductance in our experiments could be due to a dedifferentiation process occurred in our cell culture conditions. It is possible that the expression of this conductance is under the control of exogenous regulatory factors. For example, Zeitlin et al. (1989) have shown that steroid hormones increase the response of the Cl^- transport to cAMP in tracheal epithelial cells. In future studies we will investigate the correlation between the expression of the CFTR and the presence of the cAMP activated Cl^- conductance. This will help to understand the function of the protein and the molecular basis of cystic fibrosis.

ACKNOWLEDGEMENTS

We are indebted to Dr. J. Dempster for kindly giving us his programs for patch-clamp and whole-cell data analysis.

References

Dempster, J., 1987, A range of computer programs for electrophysiological signal analysis using the IBM PC AT, **J. Physiol.**, 392:20P.

Frizzell, R.A., G.R. Rechkemmer, and R.L. Shoemaker, 1986, Altered regulation of airway epithelial cell chloride channels in cystic fibrosis, **Science**, 233:558.

Gruenert, D.C., C.B. Basbaum, M.J. Welsh, M. Li, W.E. Finkbeiner, and J.A. Nadel, 1988, Characterization of human tracheal epithelial cells transformed by an origin-defective simian virus 40, **Proc. Natl. Acad. Sci. USA**, 85:5951.

Hamill, O.P., A. Marty, E. Neher, B. Sakmann, and F.J. Sigworth, 1981, Improved patch-clamp techniques for high-resolution current recordings from cells and cell-free membrane patches, **Pflügers Arch.**, 391:85.

Li, M., J.D. McCann, C.M. Liedtke, A.C. Nairn, P. Greengard, and M.J. Welsh, 1988, Cyclic AMP-dependent protein kinase opens chloride channels in normal but not cystic fibrosis airway epithelium, **Nature**, 331:358.

Marty, A., and E. Neher, 1983, Tight-seal whole-cell recording, in: "Single-Channel Recording," B. Sakmann and E. Neher, eds., Plenum Press, New York.

McCann, J.D., M. Li, and M.J. Welsh, 1989, Identification and regulation of whole-cell chloride currents in airway epithelium, **J. Gen. Physiol.**, 94:1015.

Schoppa, N., S.R. Shorofsky, F. Jow, and D.J. Nelson, 1989, Voltage-gated chloride currents in cultured canine tracheal epithelial cells. **J. Membr. Biol.**, 108:73.

Schoumacher, R.A., R.L. Shoemaker, D.R. Halm, E.A. Tallant, R.W. Wallace, and R.A. Frizzell, 1987, Phosphorilation fails to activate chloride channels from cystic fibrosis airway cells, **Nature**, 330:752.

Welsh, M.J., and C.M. Liedtke, 1986, Chloride and potassium channels in cystic fibrosis airway epithelia, **Nature**, 322:467.

Worrell, R.T., A.G. Butt, W.H. Cliff, and R.A. Frizzell, 1989, A volume-sensitive chloride conductance in human colonic cell line T84, **Am. J. Physiol.** 256:1111.

Zeitlin, P.L., M. Wagner, D. Markakis, G.M. Loughlin, and W.B. Guggino, 1989, Steroid hormones: modulators of Na^+ absorption and Cl^- secretion in cultured tracheal epithelia, **Proc. Natl. Acad. Sci. USA**, 86:2502.

DISCUSSION

SCHLATTER: We have also what is really just a confirmation in a different cell type. We've been looking at nasal epithelial cells, primary cultures of brushed turbinates, at the different whole cell currents. Among the others that we see, under conditions where we have swelling but no calcium, no cAMP stimulation, we see a similar current that inactivates with time. During the inactivation we can record single channels which are outward rectifying, about 75 pS. When we reshrink the cells we lose that channel activity. We can block with DIDS or IAA, so one question is, has anybody (you or Ray Frizzell) looked at the IAA sensitivity inside-outside of these swelling-induced currents?

GALIETTA: No, not yet.

WINE: I just want to comment on the remarkable consistency with this swelling-induced chloride current. We've got one thing nailed down thanks to this, and it really looks great. That's the sort of consistency we would like to see now with those cystic fibrosis relevant chloride currents.

CONTI: I didn't see any picture of the inactivation of these channels in the steady state. I didn't see any inactivation curve. Has anybody done this? I mean, if you keep the polarization for a long time, how far does that current go down.

GALIETTA: We didn't perform long depolarizing pulses for technical reasons and furthermore because the cell membrane can be disrupted and damaged by these long pulses. We saw also that in some way the channels can be irreversibly inactivated by these long depolarizing pulses.

FRIZZELL: We had some luck in T84 cells maintaining fairly long depolarizing voltages. It seemed to take somewhere on the order of 15 to 20 seconds before one gets into a sort of quasi steady state of current.

CONTI: Is there any sort of inactivation curve? I mean, any activation parameter as a function of voltage? Does it go to zero?

FRIZZELL: No, I can't remember what the minimum of the rate constant is, but it's in the American Journal of Physiology, 1989.

GUGGINO: About how we're seeing consistency here, I thought we saw consistency before, and now we're not seeing it again. So I think that maybe you should wait with your statement about consistency. That's just a comment. I wanted to ask, do you think that the stimulation that some people see of chloride secretion in CF epithelial cells with A23187 is due to cell swelling?

GALIETTA: We didn't try the ionophore yet.

WINE: With regard to the comment on consistency, have you ever tried to publish a negative result?

SCHLATTER: With regard to ionophore stimulation and seeing swelling, in the brushed nasal cells with ionomycin and looking at a video camera at the same time, we don't see any obvious swelling. But they're fairly confluent.

GARDNER: There is an ionomycin stimulated chloride pathway in CF cells, so the question is whether that pathway is the same one that is being activated by volume regulation, however that pathway is activated. So I think that would make a nice correlation, to look at either drug sensitivity or whatever you want to do to see whether those are both the same pathways, and if those are different from the pathway that has a common lesion in CF cells, which is the PKA-PKC activated pathway, which appears, maybe, to be a separate chloride channel.

REGULATION OF LYMPHOCYTE CHLORIDE CHANNELS

Phyllis Gardner, Tom McDonald, Ikuo Nishimoto,
John Wagner, Muhammad Schumann, Jennifer Chen,
and Howard Schulman

INTRODUCTION

The chloride permeability defect which characterizes the apical membrane of secretory epithelial cells in cystic fibrosis has been difficult to study in part because of lack of accessibility of tissue for study. This problem has fostered attempts by investigators both to immortalize epithelial cell lines and to search for other cells which may serve as a model to study chloride conductances. As we describe here, transformed lymphoblasts have whole cell chloride currents under conditions of stimulation by cAMP agonists, calcium ionophore, or hypotonicity which are quite similar to the aggregate chloride currents measured under similar conditions in epithelial cells. In addition, transformed B and malignant T cells have an outwardly rectifying, 40pS, chloride selective channel that is activated during patch clamp recording by one of at least three ways: 1) patch excision and depolarization; 2) phosphorylation by cAMP-dependent protein kinase (PKA); 3) by exposure of the lymphoblast to Ca^{2+} ionophore during cell-attached recording. This channel is closely similar or identical to the outwardly rectifying chloride channel identified by patch clamp recording from the apical membrane in secretory epithelial cells (Frizzell et al., 1986; Welsh, 1986; Welsh and Liedtke, 1986; Schoumacher et al., 1987; Li et al., 1988). Thus lymphocytes are an easily accessible and abundant tissue which can sometimes serve as a model for the study of the chloride permeability which is affected in cystic fibrosis. Accordingly we have performed some preliminary experiments to further delineate the Ca^{2+}-dependent pathway of Cl^- channel activation.

METHODOLOGY

Where indicated, experiments were performed on normal (GM3299) EBV-transformed human B lymphoblasts (NIGMS Human Mutant Genetic Cell Repository), cells of the malignant T-lympho-

The Identification of the CF (Cystic Fibrosis) Gene
Edited by L.-C. Tsui *et al.*, Plenum Press, New York, 1991

cyte cell line Jurkat E6-1 (ATCC), cells of the malignant T84 human colonic adenocarcinoma cell line, and in one case (Fig. 2H), an origin-deficient SV40 virus-transformed CF mutant human airway cell (CFNPEO-). Before recording cells were washed twice and resuspended in the bathing solutions indicated. All experiments were performed at 28-30°C. Currents were recorded using the cell-attached, inside-out and whole cell recording configurations (Hamill et al., 1981) as indicated.

RESULTS

Comparison of Whole Cell Chloride Currents in EBV-Transformed B Lymphoblasts to T84 Epithelial Cells

Using standard whole cell patch clamp recording methodology, we studied whole cell chloride currents in both B lymphoblasts and T84 colonic adenocarcinoma epithelial cells with solutions designed to minimize the contribution from other ionic species. Under these conditions baseline current of both the B cells and the T84 cells is small (Fig. 1, first traces) without apparent voltage- or time-dependence. When either the lymphoblasts or the T84 cells are exposed to the membrane permeable cAMP analog, cpt cAMP (400mM), a prolonged increase in Cl^- conductance is seen after a delay of 30 sec to approximately 3 minutes (Fig. 1, second traces). This cAMP-induced Cl^- conductance in either case is ohmic in character, with little or no voltage- or time-dependence. Alternatively, when either lymphoblasts or T84 cells are exposed to the calcium ionophore ionomycin (500nM), increases in Cl^- conductance are seen after a delay of 20 seconds to several minutes (Fig. 1, third traces). In the cases of the ionomycin-induced Cl^- currents, voltage- and time-dependence can be demonstrated. Slowly activating Cl^- current is seen at large depolarizations (≥+80mV). Finally, whole cell Cl^- currents in both lymphoblasts and T84 cells can be induced when the extracellular solution is made hypotonic relative to the intracellular solution (Fig.1, fourth traces). In the case of the hypotonicity-induced Cl^- currents, voltage- and time-dependence is also seen. At large depolarizations a characteristic inactivating Cl^- current component is seen in both cases. These hypotonicity-induced Cl^- currents have been previously described in T84 cells (Worrell et al., 1989).

Thus, comparable Cl^- currents can be induced in lymphocyte and epithelial cells after stimulation by cpt cAMP, ionomycin and Ca^{2+}, or hypotonicity.

Single Channel Recordings of the Outwardly Rectifying Chloride-Selective Channel from Jurkat T Leukemia Cells and EVB-Transformed B Lymphoblasts

One prevailing view is that an unusual Cl^- selective channel in the apical membrane of secretory epithelial cells plays an essential role in the pathophysiology of CF. By use of single

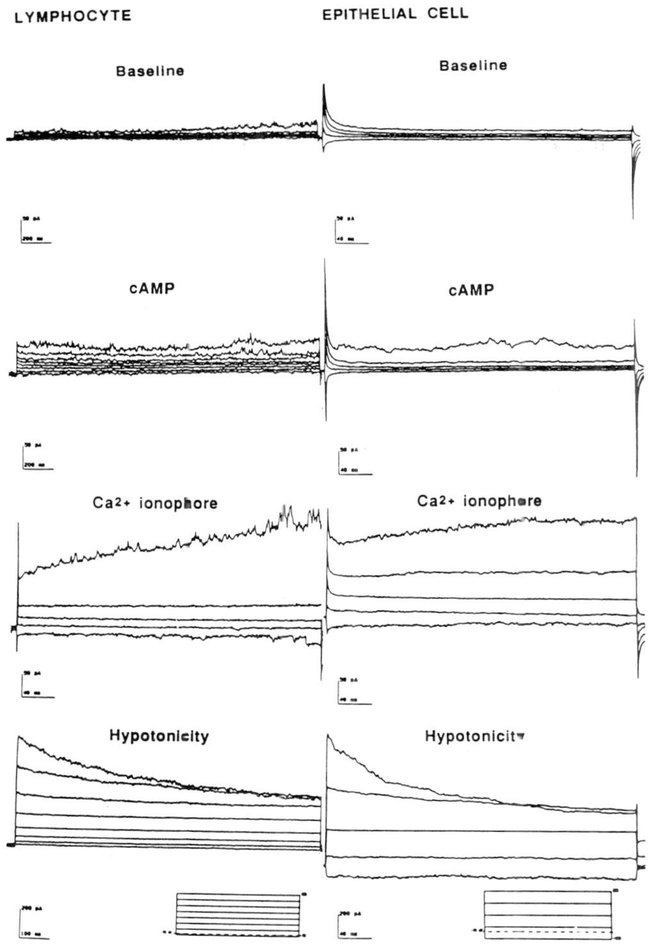

Fig. 1. **Comparison of whole cell Cl: currents in transformed B lymphoblasts and malignant or transformed epithelial cells.**
Cells were suspended in a solution of (in mM): 170 TrisCl, 2.5 $CaCl_2$, 1 $MgCl_2$, 10 glucose, 5 HEPES, pH 7.4, 325mosm. The pipette recording solution contained (in mM): 140 CsCl, 1 $MgCl_2$, 2 Mg ATP, 0.5 EGTA, 5 HEPES, pH 7.4, 300 mosm. The holding potential was -70mV, and the lower insert diagrams the voltage pulse protocol. All records are from separate cells and are representative of the indicated response. First line, baseline current recorded in B lymphoblast and in T84 epithelial cells. Second line, current in B lymphoblast and T84 cell stimulated by bath addition of 400mM cpt cAMP. Third line, current in B lymphoblast and T84 cell stimulated by bath addition of ionomycin (500nM). Fourth line, current stimulated in B lymphoblast and in CFNPEO-cell (transformed CF human airway cell) when the bath solution is made hypotonic relative to the pipette solution.

channel patch clamp recording, we have shown that a chloride channel with conductance and regulatory properties similar to those described for epithelial chloride channels is present in both T and B lymphocyte cell lines (Chen et al., 1989). Characteristic single channel Cl^- currents are recorded under one of several circumstances. First, activity of the Cl^- channel is identified by excising membrane patches into a HEPES-buffered Ringers solution and imposing large membrane potentials for variable periods from 10 seconds to more than 10 minutes (Fig. 2). The Cl-channel currents are characterized by a nonlinear current-voltage (I-V) relation in symmetrical solutions, with slope conductance (approximately 40pS at zero mV) higher at depolarized voltages. Anion selectivity of the channel can be demonstrated by the appropriate reversal potential shift in an imposed 1:4 Cl^- gradient (Fig. 2).

Similar Cl^- channel activity in B and T cells can be also seen under a second circumstance. If the cells are exposed to membrane permeable cAMP analogs or to cAMP-dependent agonists (e.g., 10^{-6}M prostaglandin E_1), outwardly rectifying Cl-selective currents open during cell-attached recording in a fraction of patches. In the case of cell-attached recording the channel open probability appears to be quite low, with infrequent channel openings demonstrated. The underlying mechanism for cAMP-dependent channel activation appears to be phosphorylation by PKA. Exposure of the cytoplasmic surface of excised patches to catalytic subunit of PKA plus ATP (1mM) activates a similar outwardly rectifying, Cl^- selective channel (Fig. 3).

Finally, there is a third way in which the outwardly rectifying Cl^- channel can be activated in lymphocytes. Exposure of Jurkat T lymphocytes to calcium ionophore A23187 (10^{-7}M) activates the outwardly rectifying Cl^- channel in cell-attached patches to a similar extent as the excision-depolarization method. The channel open probability appears to be lower during cell-attached recording, as seen with cAMP agonists, than with excised patches. Ca^{2+} does not appear to be acting as a direct ligand, since subsequent patch excision and perfusion with low internal Ca^{2+} solutions does not cause channel closure.

The outwardly rectifying lymphocyte Cl^-channel, induced by excision depolarization, by cAMP-dependent phosphorylation, or by an uncharacterized Ca^{2+}-dependent pathway, appears closely similar to the outwardly rectifying Cl^- channel in the apical membrane of epithelial cells (Frizzell et al., 1986; Welsh, 1986; Welsh and Liedtke, 1986; Schoumacher et al., 1987; Li et al., 1988).

Preliminary experiments to delineate the Ca^{2+}-dependent activation pathway

Since the Cl^- channel is activated by a Ca^{2+}-dependent pathway, in which Ca^{2+} acts not as a direct ligand but by a Ca^{2+}-dependent effector, we performed whole cell and single channel

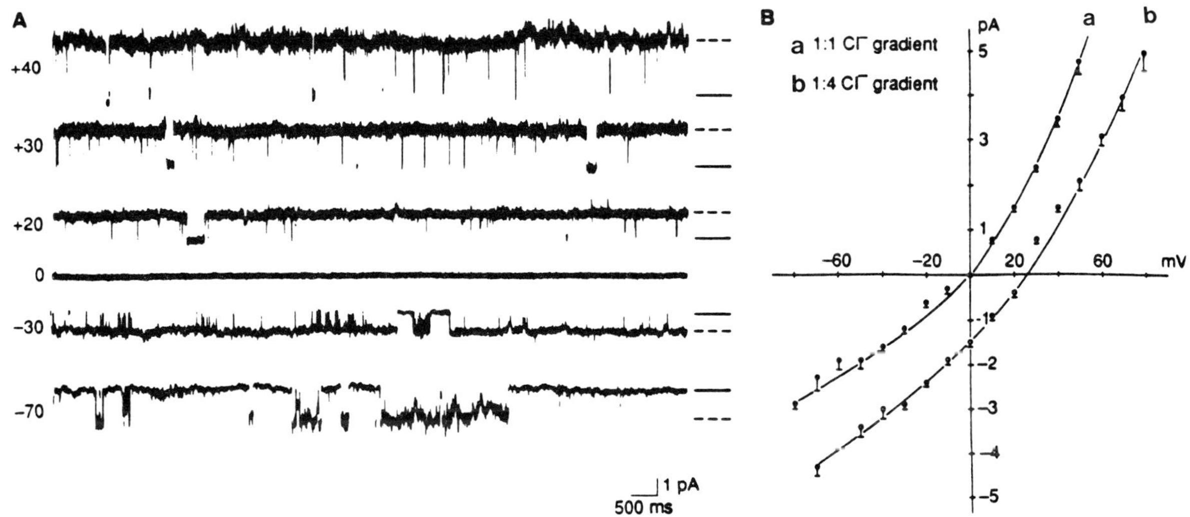

Fig. 2. **Tracings and I-V relation of Cl^- channels in Turkat E6-1 T lymphocytes.** (A) Single Cl^- channel currents recorded from an excised, inside-out patch at holding voltages. The polarity of transmembrane voltages is expressed with reference to the pipette solution. Outward currents are upward deflections. Baseline current level is indicated by the solid line and open-channel current level by the dashed line at the right of each trace. Bath and pipette solutions were identical, consisting of 154mM NaCl, 5.5 mM KCl, 2 mM $CaCl_2$, 1 mM $MgCl_2$, and 10 mM HEPES-NaOH (pH7.3). Records were filtered at 1 kHz. (B) Current-voltage relation obtained from the channel shown in (A) and a channel recorded from a patch of the same bath solution, but with a pipette containing 37.5 mM NaCl and 5 mM HEPES-NaOH (pH 7.4). Imposition of the approximate 1:4 Cl^- gradient shifted reversal potential by +28 mV (Nernst potential predicted shift is +36 mV for a perfectly Cl^- selective channel). Error bars represent based on n = 50 amplitude measured per potential. This figure is reprinted with permission from Chen et al., **Science**, 243:658; copyright 1989 by AAAS.

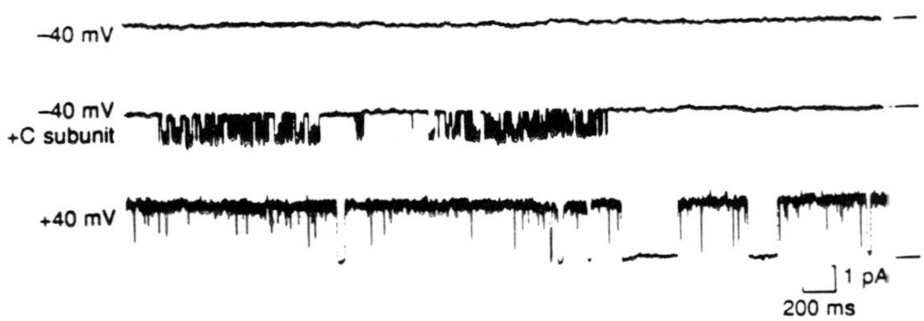

Fig. 3. **Cyclic AMP-dependent activation of Cl: channels in Turkat T lymphocytes.**
Effect of catalytic subunit of cAMP-dependent protein kinase on a Cl^- channel. Tracings are from an excised, inside-out patch. Baseline current level is indicated by a solid line at the right of each, and holding potential is indicated at the left of each trace. During the first 3 min after excision of the patch into a bath containing 1 mM ATP, no channels opened (trace 1). Subsequent addition of 150 nM of purified catalytic subunit activated the channel at -40 mV (trace 2). Channel activity continued when the patch was depolarized to +40 mV (trace 3). The bath solution consisted of 140 mM NaCl, 2 mM $MgCl_2$, 0.1 mM $CaCl_2$, 1.1 mM EGTA (calculated free Ca^{2+} 10^{-8}M), and 10 mM HEPES-NaOH (pH 7.2). The pipette solution consisted of 154 mM NaCl, 5.5 mM KCl, 2 mM $CaCl_2$, 1 mM MgCl, 10 mM Hepes-NaOH (pH 7.3). The catalytic subunit of the cAMP-dependent protein kinase was purified to homogeneity from bovine heart essentially as described (Beavo et al., 1975). The cAMP kinase stock (10mM) is in 25 mM KPi (pH 6.7), containing 10 mM dithiothreitol and 50% glycerol (w/v). The specific activity is 1.2mmol/min per milligram with histone as substrate. Addition of an equal volume of vehicle without the catalytic subunit had no effect on eight tested patches. Addition of 1 mM ATP alone induced channel activation in 2/20 patches, but only after sustained depolarization by +80 mV. The latter finding was attributed either to endogenous kinase or to the effect of sustained depolarization. This figure is reprinted with permission from Chen et al., **Science**, 243:658; copyright 1989 by AAAS.

recording to determine this effector in lymphocytes. Bradykinin (BK; 1-20μM) causes a dose-dependent increase in intracellular Ca^{2+} concentration. In the same concentration range, BK causes an increase in a DIDS-blockable, Cl^- selective current. as measured by whole cell current clamp. BK appears to be generating the Cl^- current via the second messenger Ca^{2+}, since this BK effect on Cl^- currents is mimicked by A23187 (10^{-7}M) and eliminated by prior perfusion of the cells with the membrane permeable calcium chelator, BAPTA AM (500μM). The BK effect on Cl^- currents was significantly inhibited by intracellular perfusion with 10μM CaMK (273-302), (HRSTVASCMHRQETVDCLKKFNARRKLKGA), a peptide which is based on the inhibitory domain of multifunctional calcium/calmodulin dependent protein kinase (CaM kinase) and which inhibits CaM kinase with an IC_{50} of 1.0μM and PKC with an IC_{50}>200μM (Malinow et al., 1989). The BK effect on Cl^- currents was also significantly inhibited by intracellular perfusion with 1.0μM CaMK (296-311) (CRKLKGAILTTMLATR), a peptide which is based on the calmodulin binding domain of CaM kinase and which inhibits CaM kinase with an IC_{50} of 0.1μM and PKC with an IC_{50} of 5.0μM (Malinow et al., 1989). On the other hand, BK-induced Cl^- currents were unaffected by cellular perfusion with CaMK (284-302) (QETVDCLKKFNARRKLKGA), a truncated version of CaMK(273-302) with an IC_{50} of >80μM for CaM kinase and 25μM for PKC (Malinow et al., 1989). These results are compatible with the interpretation that CaM kinase is the Ca^{2+}-dependent effector molecule involved in Cl^- channel activation.

To test this hypothesis further we performed single channel recording with various combinations of $[Ca^{2+}]_i$ (100μM), ATP (1mM), and calmodulin 5μg/ml. Perfusion with any combination of the above activated channels in less than or equal to 10% of patches (percentage of patches with Cl^- channels was 30-40% as confirmed by the excision-depolarization method). In contrast, when CaM kinase (purified as described in Schulman, 1984) was added to the Ca^{2+}, ATP, and calmodulin perfusion, characteristic outwardly rectifying Cl^- channels were activated in 36% of tested patches. Thus it appears very likely that in the Ca^{2+}-dependent activation pathway lymphocytes, outwardly rectifying Cl^- channels are activated by phosphorylation by CaM kinase.

CONCLUSION

Both lymphocyte and epithelial cell lines have cAMP, Ca^{2+}, and hypotonicity induced whole cell Cl^- currents which appear similar in character. In addition both cell types have an outwardly rectifying Cl^--selective channel (30-40pS at zero mV) which can be activated in several ways, including: 1) in excised patches after sustained depolarization; 2) in excised patches after phosphorylation by PKA; 3) in cell-attached patches after exposure of the cell to calcium ionophore. The lymphocyte appears to be a reasonable model for some questions concerning these Cl^- conductances. Accordingly, preliminary experiments were performed to further delineate the Ca^{2+}-dependent activation pathway of the Cl^- channel. Ca^{2+} agonists stimulate whole cell Cl^- conductances in lymphocytes. This effect is inhibited by synthetic peptides which are inhibitory for CaM kinase with a reasonable specificity. In addition, purified CaM kinase, in combination with ATP, calmodulin and Ca^{2+}, activated the outwardly rectifying Cl^- channel in excised patches. These preliminary data strongly suggest that phosphorylation by CaM kinase mediates the Ca^{2+}-dependent activation pathway of the outwardly rectifying Cl^- channel in lymphocytes.

References

Beavo, J.A., Bechtel, P.J. and Krebs, E.G., 1975, Activation of protein kinase by physiological concentrations of cyclic AMP, **Methods Enzymol.**, 38:299.

Chen, J.H., Schulman, H. and Gardner P., 1989, A c-AMP regulated chloride channel in lymphocytes that is affected in cystic fibrosis, **Science**, 243:657.

Frizzell, R.A., Rechkemmer, G. and Shoemaker, R.L., 1986, Altered regulation of airway epithelial cell chloride channels in cystic fibrosis, **Science**, 232:558.

Hamill, O.P., Marty, A., Neher, E., Sakmann, B. and Sigworth, F.J., 1981, Improved patch-clamp techniques for high-resolution current recordings from cells and cell-free membrane patches, **Pfluegers Arch.**, 391:85.

Li, M., McCann, J.D., Liedtke, C.M., Nairn, A.C., Greengard, P. and Welsh, M.J., 1988, Cyclic AMP-dependent protein kinase opens chloride channels in normal but not in cystic fibrosis airway epithelium, **Nature**, 331:358.

Li, M., McCann, J.D., Anderson, M.P., Clancy, J.P., Liedtke, C.M., Nairn, A.C., Greengard, P. and Welsh, M.J., 1989,Regulation of chloride channels by protein kinase C in normal and cystic fibrosis airway epithelia, **Science**, 244:1353.

Malinow, R., Schulman, H. and Tsien, R.W., 1989, Inhibition of postsynaptic PKC or CamMKII blocks induction but not expression of LTP, **Science**, 245:862.

Schoumacher, R.A., Shoemaker, R.L., Halm, D.R., Tallant, E.A., Wallace, R.W. and Frizzell, R.A., 1987, Phosphorylation fails to activate chloride channels from cystic fibrosis airway cells, **Nature**, 330:752.

Schulman, H. J., 1984, Phosphorylation of microtube-associated proteins by a Ca2+/calmodulin dependent protein kinase, **Cell Biol.**, 99:19.

Welsh, M.J., 1986, An apical-membrane chloride channel in human tracheal epithelium, **Science**, 232:1648.

Welsh, M.J. and Liedtkec C.M., 1986, Chloride and potassium channels in cystic fibrosis epithelia, **Nature**, 322:467.

DISCUSSION

FRIZZELL: Phyllis, I was wondering what you meant by "once it started working" and what was going on. You said that initially in the experiments, they were negative with the CAM kinase.

GARDNER: These experiments were done by several post-doctoral fellows, one of whom was Dr. Ikuo Nishimoto. We don't have as much CAM-kinase as we did of the catalytic subunit of protein-kinase-A. Because we had it in low concentrations we had to perfuse it on, we couldn't just put it into a bath. So Dr. Nishimoto started out by trying to hand inject it and he tried to perfuse it slowly. For a variety of reasons, he just couldn't get it to work until he got a pump-driven system with a micropipette injector and had it in close proximity to the patch. He thinks that there is a very important need to keep that perfusion going for between 30 sec. and up to 2 min. Mostly about a minute. That if you keep perfusing the patch you see this. Now I'm not sure. These are all done at room temperature, by the way, and he just thinks it's very critical that you perfuse the patch very carefully that entire time. Now, of course, if you aliquote it in, it's diffusing out in the bath, you don't get the proper concentration. So it probably is a concentration effect.

GUGGINO: That's a nice study. I think you can show that these experiments can be done in a proper way and have proper controls to assure yourself that you are getting kinase activation of the channel. And one way to do it is to show that you need every single component to get activation. And another is to show that if you have a specific inhibitor of that kinase, then you don't get activation. Have you done that in excised patches?

GARDNER: We haven't done that yet. The two things that people have asked me about would be putting on, of course, the CAM-kinase peptide inhibitor once you've started this, and the second thing would be to put on a phosphatase. And I think those are good experiments.

GUGGINO: Because the PKA activation of the channel is inhibited by the Walsh peptide. And staurosporine will block PKC activation of the channel.

GARDNER: Is that after you activate it with PKA, you then add the Walsh inhibitor?

GUGGINO: No. You can add Walsh inhibitor, PKA, ATP, and then you don't see activation of the channel and then you can voltage activate to show that there is a channel there.

GARDNER: Right. Because in this particular case we'd have to perfuse it on and I would think it would be hard to activate and then not activate with the peptide inhibitor. I think to do what you said you would have to use the phosphatase experiment and you'd have to do both sets of studies.

GUGGINO: Right. And then with PKC you need diusolglycerol to activate it so you perfuse PKC, ATP, and then put the diusolglycerol on later and show that it's only after you have all the components that you get activation.

GARDNER: Yes. In this particular case you definitely...that table didn't quite include everything, because we also did CAM-kinase plus ATP, CAM-kinase plus calmodulin, CAM-kinase plus ATP plus calmodulin. And what we found was that ATP is an absolute requirement, that CAM-kinase doesn't work if you don't have ATP. On the other hand, in two circumstances out of 20 (or maybe it was even three), without calmodulin we did get activity. But as we know, the cytoplasm may contain calmodulin, and also sometimes purified CAM-kinase is constituitively active without calmodulin.

KUNZELMANN: Did you observe additive effects of cAMP and ionomycin in the whole cell recording?

GARDNER: That, I think, is a really hard experiment to assess. In some cases, they do not appear additive, and this is in both epithelial cells and lymphocytes. In some cases they do appear additive. I've never been able to assess it because, first of all, I think that cAMP doesn't get in as well as ionomycin or A23187 and I'm not sure how effectively we're activating protein kinase A vs. other kinases. Secondly, both of these responses can be transient in both epithelial cells and lymphocytes and it's hard to know where you are on that curve, because they have time-dependent components. I could show you results where they look additive and I could show you results where they don't look additive. They are the same order of magnitude. But sometimes the ionomycin ones look bigger.

SCHLATTER: I'd like to just take a minute to talk about normal lymphocytes, because we've been studying the normal peripheral T-cell at the resting phase (I'll talk about the resting state). What we find (we've been studying all the chloride channels that we can find) are big differences between cell-attached and excised patches in the properties of the chloride channel and in the prevalence of seeing it. When we excise, in 50% of patches (this is over 150 patches) we see a multiconductance large chloride channel. If we are going to see that channel, we're going to see it within minutes, so we have the other 50% that are silent for 30 minutes to an hour, and this is important for what Dr. Greger and everyone is thinking about: spontaneous activation and how you test kinases. So in some of those patches in which they were quiescent for 30 minutes to an hour, we could add protein kinase A catalytic subunit, ATP etc., and get activation of this channel. And we could reverse the activation by washing out the ATP, or washing out the kinase. We could also, in some of those silent patches, get activation 100% of the times we tried, which is nine out of nine, with staurosporine. We can also get inhibition of active channels with TPA. So we think chloride channels are activated by kinase A, inactivated by kinase C. One of the other interesting points is that with regard to the lymphocyte channels, tiny ones have been seen, middle-sized rectifying ones, of course (that Phyllis is describing), and large conductance. And we think they are all related, through coordination of gating. So we have been studying this quite a bit. We see little tiny channels, about 2-3 pS, that are not very voltage sensitive and have linear IV relations, as nearly as we can tell for a small channel. We see them in the same patch, starting to coordinate their gating, grow up to larger ones (6 to 10 pS rectifying outward) then growing larger (20 to 40, 20 to 60 pS rectifying), then growing into the maxi. And this spontaneous event seems to be always in the direction of increasing conductance. Then we start seeing the subconductance states, from a large conductance. We can get that kind of activation with kinase A. We can get it spontaneously. We can get it with staurosporine and we can get it in cell-attached patches when we activate. We are fairly sure that this is a coordinated gating and not just a loss of small ones or masking by large, because the voltage dependence also changes. So you've got changes in rectification, changes in single channel conductance, changes in voltage dependence, such that at previously permissive voltages for the tiny ones, we no longer see them after the coordination. So one point I'm trying to get across is that tiny, medium, large, linear rectifying, multiconductance - they've all been seen in epithelial as well of course - and I think that the lymphocyte, from our results may suggest they are all interrelated through coordination of gating that may be under control by at least a couple of kinases. We believe there are other controls too.

GARDNER: Thanks. I think it goes along with what Dennis said, that a channel can be very dynamic and can change over time. I wanted to say one thing. Our epithelial cells, the results that I showed were mostly from T84 cells but also we are working with Dieter Gruenert's transformed cells and actually one of the results was from Dieter's cells and they look very similar. I didn't mention that during my talk.

MEHTA: Do you see a difference between the time of activation from protein-kinase A vs. CAM-kinase? Is there absolutely no difference?

GARDNER: I can't say that absolutely. We did those experiments two years ago, and there are different people doing them. But it seems to be. Our experience with protein-kinase A was about some time between 30 sec. and the far-out side was 3 minutes. But most of them came about between 1 and 2 minutes and that was at room temperature, and that is the same experience we have with the CAM-kinase.

MEHTA: Are the calmodulin kinase inhibitor peptides actually available?

GARDNER: The sequence is published and actually I think one of the interesting things is that CAM-kinase, if you follow the literature, was cloned from neuronal cells. It's about 2% of the protein in neurons and it has a lot of in vitro phosphorylation substrates. It has very few in vivo functions that have been identified with it, despite its high level. Now, it is not only present in neurons, of course; it's ubiquitous to tissues and actually Paul Neim in our lab is now cloning both lymphocyte and epithelial cell CAM-kinase. It probably has some sequence differences. But the peptide inhibitors were designed in collaboration - Howard Schulman did them in collaboration with Dick Tsien, who then used them to show that CAM-kinase has a role in long term potentiation. In their paper in Nature, the sequence of these peptides, which you can make quite easily, is published and that was this year sometime. He may also have them for giving out. I'm sure he does. I'm sure if you wrote Howard he would try to provide them to you. They have a problem, actually. You have to be very careful. For some reason, even though they're very straightforward peptides, they seem to lose their activity and they're not sure exactly what that has to do with, and so Howard keeps testing the activity of these peptides at all times in his lab.

BUCHWALD: You reported initially that you could detect differences between normal and CF lymphoblasts (or lymphocytes, I can't remember). These latest that you've discussed today, were done, you say, with both leukemic cells and with lymphoblasts. Interchangeably?

GARDNER: No. The whole-cell results that we have, yes. They have been done both on Jurkat-T leukemic cells and EBV transformed B cells. You're bringing up the difference between normal and CF derived ones. If you like, I can tell you what our follow-up on that is, because it's one of my problems: trying to put together the knowledge about CFTR and what some of the experimental results are. I think Ray Frizzell has some experimental results which are very provocative, because if CFTR is not expressed in lymphocytes, I cannot see any way in which you should have a deficit or a difference between normal and CF derived ones. And yet there is some evidence that there is a difference. In our personal case, what we initially published was that protein kinase A in the normals activated what we thought was a channel about 50% of the time. In the CF-derived ones, in 37 patches we never saw it. But in retrospect, that could easily have been flawed, as false negatives, because the channel density varies, and very soon after we did those studies, our channels disappeared for several months. So I don't know if we were on the tail-end of that. We have since done whole-cell clamp studies. In the normal lymphocytes, we've done about 80 of those using both caged cAMP and CDT cAMP. With the caged, you have the cyclical compound inside and you flash it and it becomes active and with the other, obviously, you add it in the bath. About 60-70% of normals respond. And they respond in that quite characteristic fashion that I showed you. In the CF-derived lymphocytes (we've only done 16 and we have not done caged, and we want to do all those things - it just takes a lot of time) I can tell you right now 7 were completely negative - no response - 6 had an inward calcium current so we knew they were responding to cAMP, and they had this very beautiful inward rectifying calcium current which, as Jim Bubien has described, is felt to be due to CD20. This has been cloned and it is a calcium channel which appears to be activated by cAMP in the B-lymphocyte. Then there were two - one had a very transient outward current that was activated, and the second one looked (if I were reading it blind, which I was) positive. So one out of 16. The problem, of course, with whole-cell clamp studies is that you can also get an increase in the linear ohmic type current just from breakdown. So, at this point, that could have been just a spurious result, or it could be that what we generally see is that the CF cells don't seem to respond to the same degree but that they do respond. And, finally, we did cAMP-induced flux. There is a cAMP-induced flux in the normal cell and there is one in the CF-derived lymphocyte. In the epithelial cell there is a cAMP-induced normal flux and it is absent in the CF cell. What we didn't do was put a calcium chelator in our cells, and as you know these are calcium-current stimulated and we could have gone through the calcium pathway. Dr. Frizzell, maybe you want to wait on your results, but you see a difference between normal and CF-derived lymphocytes and if CFTR is not expressed in lymphocytes, I can't explain that.

BUCKWALD: What about these latest calmodulin-kinase experiments? Did you compare normal and CF?

GARDNER: No, we haven't had a chance.

GUANINE NUCLEOTIDE BINDING PROTEINS REGULATE EPITHELIAL Na^+ CHANNELS

Dennis A. Ausiello, Jennifer L. Stow, Charles R. Patenaude, and Horacio F. Cantiello

Renal Unit, Massachusetts General Hospital and Departments of Medicine and Pathology Harvard Medical School, Boston, Massachusetts, U.S.A.

INTRODUCTION

Regulation of ion movements via transport proteins is of particular importance in absorbing and secreting epithelia where the intracellular ionic concentrations must be carefully balanced to enable the cells to perform transepithelial electrolytic movements. The ability of epithelia to respond in an acute fashion to environmental signals relies, in part, on their capacity to induce a set of secondary messengers, including the intracellular signals Ca^{2+}, pH and cAMP, generated by interaction of hormones with receptors at the basolateral membrane (1). The resultant physiological response at the apical membrane (e.g. increased Na^+ transport) is assumed to develop from the transmission of these second messenger systems through, for example, protein kinases in biochemical pathways which result in the phosphorylation of one or more apical membrane proteins, including the Na^+ channel itself (2). Recent evidence from studies of partially purified Na^+ channels from the apical membrane of A6 cells suggest that one of the proteins in this complex (of at least 7 proteins) is phosphorylated by cAMP-dependent protein kinase (3). Changes in the intracellular cAMP concentration upon activation of the V2 receptor by vasopressin have been reported to increase Na^+ transport in tight epithelia, such as toad urinary bladder and the A6 toad kidney cell line (4,5).

Recently pertussis toxin-sensitive guanine nucleotide regulatory (G_i) proteins have been implicated in the direct regulation of receptor-activated channels from various excitable tissues. Patch-clamp studies of muscarinic K^+ channels from myocytes and Ca^{2+} channels from excitable tissue indicate that the α_i and β-gamma subunits of these G_i proteins serve as intermedia-

The Identification of the CF (Cystic Fibrosis) Gene
Edited by L.-C. Tsui *et al.*, Plenum Press, New York, 1991

ries of channel activity by coupling reactions between hormone receptors and the ion channels themselves (6-15). It should be noted that these studies were conducted in nonpolar cells where hormone receptors, G_i proteins, and ion channels are uniformly distributed on the plasma membrane.

Although G_i proteins may directly couple and activate channels, there is also evidence that they may activate phospholipases (16,17) and lipoxygenases (15) as an intermediate step to channel regulation. For example, one of the G_i subunits, β-gamma, has been shown, when used in nanomolar concentrations, to be linked to the activation of phospholipase, the products of which in turn activate a K^+ channel (15). This effect of β-gamma was mimicked by addition of arachidonic acid, and, at least partially, by the addition of 5-lipoxygenase products such as leukotrienes B4 and C4. The fact that antibodies against PLA_2 or the addition of nordehydroguayaretic acid, a 5-lipoxygenase inhibitor, prevented channel activity, implicates both phospholipase products and leukotriene products in channel regulation.

We have previously demonstrated that in the cultured pig kidney epithelia, LLC-PK_1, pertussis toxin can inhibit amiloride-sensitive Na^+ channel mediated $^{22}Na^+$ influx across the apical membranes of intact cells (18). In order to ascertain whether G_i proteins directly modulate Na^+ channel activity on apical membranes of renal epithelia, we applied the patch clamp technique to A6 cells to monitor single Na^+ channel activity. We report that activation of endogenous pertussis toxin-sensitive proteins in apical membranes or the addition of the activated $G\alpha_{i-3}$ subunit stimulates Na^+ channel activity. These data provide evidence that the apical plasma membrane Na^+ channels can be modulated by G_i proteins. As this event occurs distal to signal transduction pathways in the basolateral membrane of A6 cells, this raises the possibility of a topologically distinct signal transducing pathway colocalized with the Na^+ channel.

MATERIALS AND METHODS

A6 Cell Culture. A6 cells (ATCC # CCL102) derived from *Xenopus laevis* renal tubular cells were purchased from the American Type Culture Collection (Rockville, MD) and used between passages 20-36. Cells were cultured on glass coverslips in tissue culture media as previously described (17). Briefly, cells were grown in a Coon's modification of Ham's F-12 and Liebovitz's F-15 media modified to contain 105 mM NaCl and 25 mM $NaHCO_3$. The mixture was supplemented with 10% fetal bovine serum (Gibco). Cells were maintained at 27°C in an atmosphere of humidified air plus 5% CO_2. Cells were passaged and used between days 3 and 4 when cells were 60-70% confluent. Cells grown under these conditions express high conductance Na^+ channels (19).

Patch Clamp Measurement of Na^+ Channel Activity. Patch-clamp pipettes were made of KG-33 glass capillaries (Garner Glass Co.

Claremont, CA), pulled on a PB-7 pipette puller and heat-polished on an MF-9 pipette polisher (Narishige, Tokyo, Japan). All pipette shanks were coated with Silgard 184 prior to polishing. Patch pipettes were prepared with an internal diameter of 0.1-0.8 mm and a resistance between 10 and 40 megaohms in symmetric NaCl solutions. Actual currents and command voltages were obtained and driven with a PC-501 patchclamp amplifier (Warner Instruments, Handen, CT) using a head stage with a feedback resistance of 10 gigaohms and mounted onto a MN-2 micromanipulator (Narashige, Tokyo, Japan). The resultant signal was filtered at 500 Hz with an 8pole Bessel filter (Frequency Devices, Haverhill, MA). Data were acquired, digitized and stored in a hard disk of a personal computer (Unitech, Boston, MA) through a TL-1 interface (Tecmar, Solon, OH) until further analysis with PClamp 5.03 (Axon Instruments, Burlingame, CA). Patch-pipette solution was in mM: 115 NaCl, 5 KCl, 0.8 Mg SO_4, 1.2 $CaCl_2$ and 10 Hepes, pH 7.4. The perfusion solution was a modification of the patch pipette solution where Na^+ was replaced equimolarly with K^+ as to contain 115 mM KCl and 5 mM NaCl, all other solutes remaining the same. Both solutions were filtered (Millipore, 0.22 mm pore size) before each experiment. Data were obtained solely from excised inside-out patches and are presented as % of total open time, i.e. 100 x P_o (P_o = open probability). Only single-channel tracings were analyzed. Data are expressed as mean ± S.E. where n equals number of channels analyzed. % open time in spontaneous channels ranged between 7.1 and 75% of total time analyzed (> 20 sec). Open probability did not statistically change at a given holding potential for up to 15 minutes. Channels for different experimental protocols were chosen depending on the spontaneous activity, i.e. channels that displayed lower % open time were chosen for GTP or $\alpha_{i\text{-}3}$ activation and channels with higher activity (> 40%) were exposed to either GDPβS or pertussis toxin.

Pertussis toxin addition. Pertussis toxin was obtained from Peninsula Laboratories, Belmont, CA and stored at -20°C. The toxin was activated in a stock solution containing 100 mM NaCl, 20 mM Na_2HPO_4, pH 7.0,10 mM dithiothreitol and 1 mM NAD^+ as previously described (6). Stock solutions were kept at 0°C until the time of the experiment. Activated pertussis toxin (2 ml) was added to the chamber (0.4 ml) perfusing the cytosolic side of the excised membrane patches to a final concentration of 100 ng/ml.

GTP and analog additions. GTP, GTP-gamma-S, and GDPβS were added to the perfusion chamber bathing the cytosolic side of the membrane patch from a stock solution containing 10 mM Hepes, pH 7.6, at 0°C to a final concentration of nucleotide of 0.1 mM.

Activated $\alpha_{i\text{-}3}$ addition. Human $\alpha_{i\text{-}3}$ activated with GTP-gamma-S and purified by ion-exchange chromatography (6) was the kind gift of Dr. Lutz Birnbaumer. Activated $\alpha_{i\text{-}3}$ was stored in a solution containing 10 mM Tris-HCl, pH 8.0,1 mM EDTA, 1 mM dithiothrietol, 7.5 mM $MgCl_2$ and 20 mM KCl in bovine serum albumin-coated tubes at -70°C. 2 μl of the activated $\alpha_{i\text{-}3}$ stock solution at 20°C was added to the perfusion chamber at the cytosolic surface of the patch with an albumin coated pipette dispenser to a final concentration of $\alpha_{i\text{-}3}$ of 20 pM in most experiments.

RESULTS

Immunolocalization of Na^+ channel in A6 cells. Immunocytochemistry was carried out with a specific polyclonal antibody (20) to localize the Na^+ channel on cultured A6 toad kidney

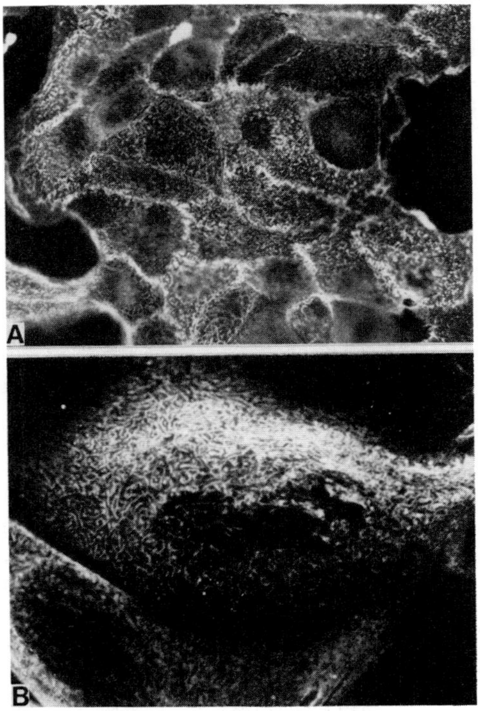

Fig. 1. **Immunofluorescence localization of the Na^+ channel on A6 toad kidney cells.** A6 cells were fixed in paraformaldehyde and incubated with a polyclonal antibody which reacts specifically with the Na^+ channel and then with a goat anti-rabbit IgG-FITC. A, shows a monolayer of cells viewed by conventional epifluorescence. The Na^+ channel staining is seen in a fibrillar pattern associated with the microvilli on the apical surfaces of these polarized cells. B, the apical cell surface is shown in greater detail when the cells are viewed by confocal microscopy. The long microvilli can be seen covering the cell surface in ridge-like arrays and the darkened nucleus can be seen projecting up toward the apical surface in the center of this cell. The Na^+ channel staining is localized specifically to the projecting microvilli and is not seen on any other domains of the cell surface.

cells. These studies showed that Na^+ channels are localized on the microvilli at the apical cell surface of these polarized cells as predicted by their function on transepithelial transport.

Basal Na^+ channel activity. We used the patch-clamp technique to obtain single-channel currents from the apical surface of A6 epithelial cells grown to partial confluence. Spontaneous channel activity was present in approximately 60% of successful patches. The Na^+ channel displayed no detectable rectifying properties between ± 100 mV, was selective for $Na^+:K^+$, 5:1, and had a conductance of 8.87 ± 0.34 pS (n = 78) in symmetrical Na^+ solutions (Na^+ = 115 mM) (Fig. 2). Single-channel activity was amiloride-sensitive with an apparent $K_i < 10^{-7}$ M. These data are in agreement with those previously reported by Hamilton and Eaton (19).

Effect of guanine nucleotides and pertussis toxin on single Na^+ channel activity. The effect of GTP, its analogs, and pertussis toxin was assessed on control channel activity as well as patches that did not show any spontaneous channel activity. The % open time of spontaneous channels in 4/4 patches increased from 6 to 50% (Fig. 3, A and B) following addition of 0.1 mM

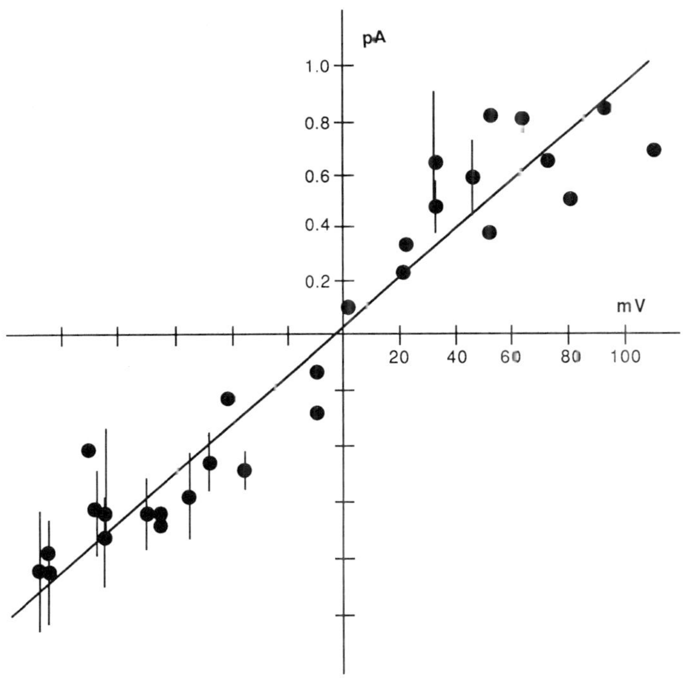

Fig. 2. **Current-voltage relationship for high conductance amiloride-sensitive Na^+ channel in A6 cells.** Na^+ conductance was determined in symmetrical Na^+ solutions (Na^+ = 115 mM). Data are mean ± S.E., n = 78.

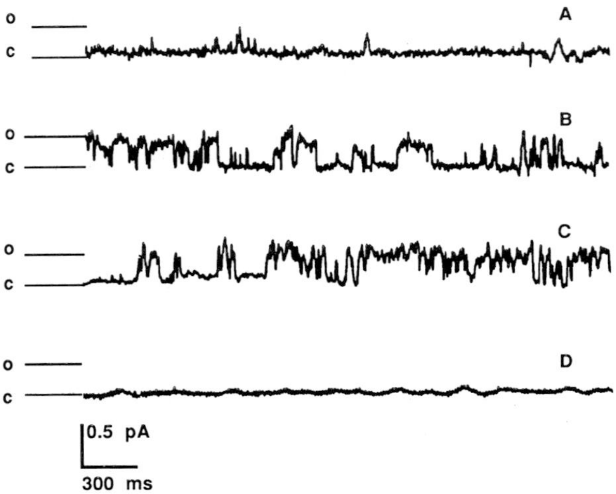

Fig. 3. **Guanine nucleotide regulation of single epithelial Na^+ channels.** Single channel currents are shown from excised inside-out patches from the apical membrane of partially confluent A6 cells. Upward deflection indicates open state, i.e. an inward Na^+ current. Holding potential was + 60 mV. Tracing A, patch with minimal spontaneous Na^+ channel activity (% open time = 2.5,10 events). Tracing B, same patch as A, 1 min after addition of 0.1 mM GTP to the perfusion chamber (% open time = 30.6, 56 events)). Tracing C, patch with high spontaneous Na^+ channel activity (% open time = 60.5, 75 events). Tracing D, same patch as C, 2 min after addition of 0.1 mM GDPS (% open time = 0, 0 events).

GTP. Na^+ channel activity was also observed on addition of 0.1 mM GTP to the bathing solution in 2/5 patches that had not displayed spontaneous Na^+ channel activity (data not shown). In contrast, GDPβS inhibited Na^+ channel activity from spontaneously active channels (Fig. 3, C and D) by approximately 84% without an effect on the single channel conductance. Addition of preactivated pertussis toxin (100 ng/ml) to the cytosolic bathing solution inhibited spontaneous Na^+ channel activity within 1 min by 85.2% from 46.5 ± 9.3 to 6.87 ± 2.49 ($p < 0.01$) % open time (see Fig. 4) and reduced the % open time to zero in all 11 patches treated after approximately 3 min. Pertussis toxin was also without effect on the single channel conductance, 9.35 ± 2.35 vs. 8.84 ± 2.19 pS. Removal of pertussis toxin did not reverse its effect. In contrast, addition of 0.1 mM GTP-gamma-S restored Na^+ channel activity in all four patches previously exposed to pertussis toxin where % open time increased from 0 to 35.4 ± 10.6 ($p < 0.01$) (Fig. 4). The addition of 0.1 mM GTP was not able to reverse pertussis toxin inhibition of Na^+ channel activity (data not shown). The addition to the perfusion buffer of ATP (0.1 mM) alone or with cAMP (0.1 mM) or phorbol

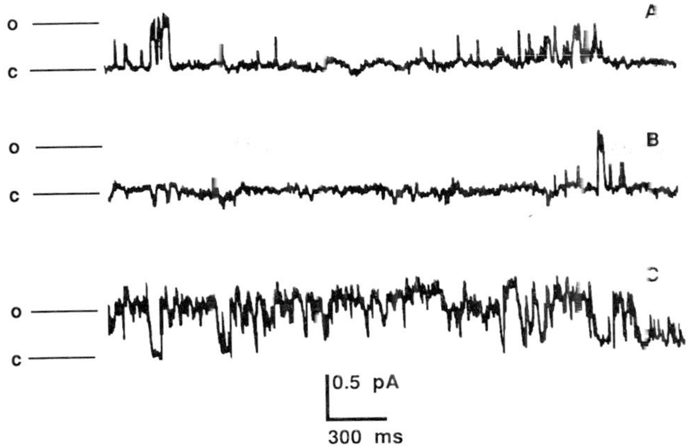

Fig. 4. **Pertussis toxin inactivation of single epithelial Na^+ channels.** Tracings A and B illustrate the effect of pertussis toxin (100 ng/ml) after 1 and 2 min of addition, respectively, to the perfusion chamber. Tracing C, reversal of pertussis toxin effect by 0.1 mM GTP-gamma-S. Upward deflection indicates open state. Holding potential was + 80 mV.

esters (0.1 mM) did not stimulate Na^+ channel activity in the presence or absence of pertussis toxin.

Effect of purified α_{i-3} subunit on single Na^+ channel activity. To demonstrate that the effect of pertussis toxin was mediated through the action of a G_i-like protein, purified activated human α_{i-3} subunit (Fig. 5 and Table I) was added to the bathing solution of the cytosolic side of the patch. α_{i-3} (20 pM) reversed the effect of pertussis toxin on Na^+ channel activity in 100% of the patches studied. The effect of α_{i-3} was dose dependent, with as little as 0.2 pM of the purified subunit activating the Na^+ channel (data not shown). The addition of GTP-gamma-S alone in a 5000 fold excess of its equimolar concentration with the chromatographically purified α_{i-3} (6) was without effect, making it unlikely that stimulation of the Na^+ channel by activated α_{i-3} was due to contaminating GTP-gamma-S. To rule out

TABLE I

Effect of activated human purified α_{i-3} subunit on single Na^+ channel activity from A6 cells

	Control	Pertussis Toxin	α_{i-3} (20 pM)
% open time	46.5±9.29 (4)	0 (4)	32.4±13.1 (4) $p < 0.05$
g (pS)	9.6±2.1 (3)		12.1±4.4 (3) NS*

*Not significant

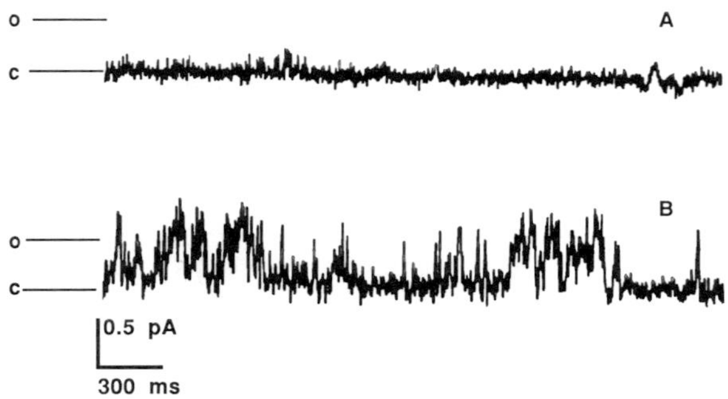

Fig. 5. **Effect of activated human purified α_{i-3} subunit on Na^+ channel activity in pertussis toxin-treated membranes.** Addition of activated purified human α_{i-3} at a final con centration of 20 pM immediately overcame pertussis toxin-inhibition of Na^+ channel activity. Tracing A, pertussis toxin-treated patch after 5 min. Tracing B, same patch as A after 2 min of addition of α_{i-3} subunit. Upward deflection indicates open state. Holding potential was + 80 mV.

this possibility, GDPβS (100 mM) was added to several patches prior to the addition of activated α_{i-3} (20 pM). The α_{i-3} subunit stimulated Na^+ channel activity to the same extent as α_{i-3} in the absence of GDPβS (data not shown).

DISCUSSION

The regulation of apical membrane ion channels in epithelia has traditionally been thought to occur through hormone receptor coupled events at the basolateral membrane. Vasopressin stimulation of Na^+ channel activity, for example, can be mediated at the basolateral membranes of renal epithelia by signal transduction pathways coupling hormone receptors with G proteins to stimulate adenylyl cyclase activity and cAMP production (1,5). The steps subsequent to cAMP generation that are responsible for eliciting the increase in Na^+ transport at the apical membrane remain unknown. We have previously demonstrated that in the cultured pig kidney epithelia, LLC-PK_1, pertussis toxin could inhibit amiloride-sensitive Na^+ channel activity (18). It thus seemed probable that the apical membrane itself contained regulatory components that modulated the Na^+ channel. In the present report, we have summarized the studies from our laboratory (21) that were undertaken to ascertain whether G_i proteins directly modulate Na^+ channel activity in apical membranes of the renal epithelia, A6, as demonstrated with the application of the patch clamp technique. These data demonstrate that a pertussis toxin-sensitive G_i protein is present as a modulator of the Na^+ channels in the apical membrane of these cells. This G_i protein is topographically distinct from those G proteins, both G_s and G_i, which are known to be functionally coupled to hormone receptors at the basolateral membrane and which can alter cAMP production and in turn apical Na^+ channel activity (1,5). These data expand the role of G_i proteins in modulating epithelial Na^+ transport. It thus appears likely that epithelia use G_i proteins both at proximal and distal steps of signal transduction. G proteins at the basolateral membrane mediate changes in second messengers that activate intracellular metabolic pathways (protein kinases) which possibly modify the activity of G_i proteins coupled to ion channels at the apical membrane. Epithelia may also use G_i proteins localized in apical membranes as independent signal transduction pathways capable of directly modifying ion channels. The inability of protein kinase activators to stimulate Na^+ transport in our present study is further evidence for a distinct apical membrane signal transduction pathway. Whether this G_i-mediated pathway is coupled to an apical membrane receptor remains to be defined. The present data are consistent with the presence of an apical membrane receptor coupled to Gi proteins, although this is not obligatory for their activation (see discussion in reference 21).

Irrespective of the signal responsible for the activation of G_i proteins, it is now clear from our recent data that the α_{i-3} subunit does not directly activate the Na^+ channel. Rather, its action is mediated by the regulatory control of phospholipase A_2 and lipoxygenase products. We have been able to demonstrate that the phospholipase inhibitor, mepacrine, decreases both spontaneous and α_{i-3} stimulated Na^+ channel activity. The addition of arachidonic acid overcomes this inhibition suggesting that fatty acid metabolism is responsible for Na^+ channel activity. Furthermore, by blocking the lipoxygenase pathway, we can demonstrate that arachidonic acid is no longer effective in modulating the Na^+ channel, but that this block can be overcome by the addition of leukotriene D4. From these data we can conclude that the G_i protein activation of apical Na^+ channels in A6 cells is probably mediated through the activation of PLA_2 and the generation of arachidonic acid and lipoxygenase products. G_i protein coupling to PLA_2 activity has been observed in other cell types (16,17) and G_i protein activation of channels has been proposed to be the consequence of the ability of the β-gamma subunit to activate phospholipase (15). This effect of β-gamma, however, required nanomolar concentrations of the subunit. Our data suggest that picomolar concentrations of the α_{i-3} subunit are capable of activating the phospholipase A_2 (unpublished observations). It is thus possible that data obtained with partially purified β-gamma is actually the result of contamination with the α_i subunit. There is clearly still much more we need to learn about this intriguing G_i protein and phospholipid signalling pathway. However, our data demonstrate that the regulation of Na^+ channel activity in renal epithelia must now include a role for apically localized pertussis toxin-sensitive G_i proteins.

Acknowledgements

We thank Dr. Lutz Birnbaumer for the gift of α_{i-3}. We also thank Ms. Lisa Firicano for secretarial work. This work was funded by N.I.H. grant DK 19406.

References

1. Palmer, L. G., and H. Sackin, 1988, Regulation of renal ion channels, **FASEB J.**, 2:3061.
2. Ling, B. N., and D. C. Eaton, 1989, Effects of luminal Na^+ on single Na^+ channels in A6 cells, a regulatory role for protein kinase C, **Am. J. Physiol.**, 256: F1094.
3. Sariban-Sohraby, S., E. J. Sorscher, B. M. Brenner, and D. J. Benos, 1988, Phosphorylation of a single subunit of the epithelial Na^+ channel protein following vasopressin treat ment of A6 cells, **J. Biol. Chem.**, 263:13875.
4. Leaf, A., 1966, On the functional structure of the transport system in the toad bladder, in: "Proceedings of the Third International Congress of Nephrology," Karger, New York.
5. Perkins, F.M., and J.S. Handler, 1981, Transport properties of toad kidney epithelia in culture, **Am. J. Physiol.**, 241: C154.
6. Codina, J., A. Yatani, D.Grenet, A.M. Brown, and L. Birnbaumer, 1987, Specificity of action of guanine nucleotide-binding regulatory protein subunits on the cardiac muscarinic K^+ channel, **Science**, 236:442.
7. Logothetis, D. E., Y. Kurachi, J. Galper, E. J. Neer, and D. E. Clapham, 1987, The β-gamma subunits of GTP-binding proteins activate the muscarinic K^+ channel in the heart, **Nature**, 325:321.
8. Yatani, A., R. Mattera, J. Codina, R. Graf, K. Okabe, E. Padrell, R. Iyengar, A.M. Brown, and L. Birnbaumer, 1988, The G protein-gated atrial K^+ channel is stimulated by three distinct $G_i\alpha$-subunits, **Nature**, 336:680.
9. VanDongen, A.M.J., J. Codina, J. Olate, R. Mattera, R. Joho, L. Birnbaumer, and A.M. Brown, 1988, Newly identifled brain potassium channels gated by the guanine nucleotide binding protein G_o, **Science**, 242:1433.
10. Yatani, A., H. Hamm, J. Codina, M.R. Mazzoni, L. Birnbaumer, and A.M. Brown, 1988, A monoclonal antibody to the α subunit of G_k blocks muscarinic activation of atrial K^+ channels, **Science**, 241:828.
11. Mattera, R., A. Yatani, G.E. Kirsch, R. Graf, K. Okabe, J. Olate, J. Codina, A.M. Brown, and L. Birnbaumer, 1989, Recombinant α_{i-3} subunit of G protein activates G_K-gated K^+ channels, **J. Biol. Chem.**, 264:465.
12. Mattera, R., M.P. Graziano, A. Yatani, Z. Zhou, R. Graf, J. Codina, L. Birnbaumer, A.G. Gilman, and A.M. Brown, 1989, Splice variants of the α subunit of the G protein G_s

activate both adenylyl cyclase and calcium channels, **Science**, 243:804.
13. Tsunoo, A., M. Yoshi, and T. Narahashi, 1986, Block of calcium channels by enkephalin and somatostatin in neuroblastoma x glioma hybrid, **Proc. Natl. Acad. Sci. USA**, 83:9832.
14. Yatani, A., J. Codina, and Y. Imoto, 1987, A G protein directly regulates mammalian cardiac calcium channels, **Science**, 238:1288.
15. Kim, D., D. L. Lewis, L. Graziadei, E. J. Neer, D. Bar-Sagi, and E.E. Clapham, 1989, G-protein β-gamma-subunits activate the cardiac muscarinic K^+-channel via phospholipase A_2, **Nature**, 337:557.
16. Nakamura, T., and M. Ui, 1985, Simultaneous inhibitions of inositol phospholipid breakdown, arachidonic acid release, and histamine secretion in mast cells by islet-activating protein, pertussis toxin. A possible involvement of the toxin-specific substrate in the Ca^{2+}-mobilizing receptor-mediated biosignaling system, **J. Biol. Chem.**, 260:3584.
17. Burch, R.M., A. Luini, and J. Axelrod, 1986, Phospholipase A2 and phospholipase C are activated by distinct GTP-binding proteins in response to α1-adrenergic stimulation in FRTL5 thyroid cells, **Proc. Natl. Acad. Sci. USA**, 83:7201.
18. Mohrmann, M., H. F. Cantiello, and D. A. Ausiello, 1987, Inhibition of epithelial Na^+ transport by atriopeptin, protein kinase C and pertussis toxin, **Am. J. Physiol.**, 253: F372.
19. Hamilton, K. L., and D. C. Eaton, 1985, Single-channel recordings from amiloride-sensitive epithelial sodium channel, **Am. J. Physiol.**, 249:C200.
20. Brown, D., E.J. Sorscher, D.A. Ausiello, and D.J. Benos, 1989, Immunocytochemical localization of Na^+ channels in rat kidney epithelia, **Am. J. Physiol.**, 256:F366.
21. Cantiello, H.F., C.R. Patenaude, and D.A. Ausiello, 1989, G protein subunit, α_{i-3}, activates a pertussis toxin-sensitive Na^+ channel from the epithelial cell line, A6, **J. Biol. Chem.**, 264:20867.

DISCUSSION

GREGER: May I just start off with one question regarding patch clamping that you were referring to at the end. When you use something to denude the patch, can you then still show the leukotriene effects and NDGA effects? So are you sure that is the most distal component?

AUSIELLO: Yes. It is a little bit more complex, because denuding the patch itself has a direct effect on the channels. If one takes a patch with relatively low spontaneous sodium channel activity and adds cytochalasin-D, the sodium channels are activated. However, when one accounts for that component, there is still the G-protein/lipase/lipoxygenase pathway intact. On the other hand, if you inhibit the lipoxygenase pathway, the effect of either denuding or breaking up the cytoskeleton is modified, suggesting that the cytoskeleton might actually be involved with the lipoxygenase. These are very preliminary data, but I think it's possible that one way to activate a lipoxygenase is to manipulate the cytoskeleton.

GREGER: Would you like to comment on the role of this kind of channel in volume regulation? It does not work as a sodium channel then.

AUSIELLO: Yes. Of interest is that the changes that we've seen, in preliminary experiments with chloride, are opposite to what we see with sodium. When we activate the lipoxygenase pathway, we inhibit the chloride channel. When we induce solation of microfilaments, we inhibit the chloride channel. Now, with regard to volume regulation, of course many people have proposed that the cytoskeleton may play a dynamic role, rather than just a passive role, in ion channel modulation. I believe we have to pay close attention, not only to the structural relationship of the cytoskeleton with channels, but to the structural relationship of the cytoskeleton with regulatory proteins. The phospolipase/lipoxygenase pathway is one that could very well work on in such a manner.

GREGER: Could you just add some comments on the Cl^--channels you were referring to?

AUSIELLO: These are in the A6 cells that Doug Eaton and colleagues have studied extensively and published on in the last 6 months.

FRIZZELL: I was wondering, with respect to Rick's comments on the possibility of an apical purinergic receptor, whether you know what receptor the α_{i-3} is?

AUSIELLO: That's an excellent question Ray. Our study has been conducted in the opposite direction. Normally we start off with a ligand and identify a receptor activation pathway. Here, we've begun with the pathway. It leaves the question, does the G-protein couple to a receptor, either an intracellular or an extracellular receptor, or has the epithelial cell simply spatially dissociated this signal transduction pathway, which is only sensitive to classical second messages such as protein kinases. We don't have a definitive answer. We have begun to explore two possibilities. The fact that this channel is always modulated in the excised patch suggests to me that if there is a receptor coupled to the G-protein, then it's either a lipid or an autacoid. The most likely autacoid would be adenosine or some derivative. The most likely lipids, of course, would be similar to those lipid products that we have just shown modulate the sodium channel. We are examining these possibilities.

KUNZELMANN: I had the impression from your recordings that the effects of arachidonic acid or leukotrienes were transient. Would you agree?

AUSIELLO: These effects last for 30 or 40 minutes in our patch and are reversible.

ATTITUDES OF PARENTS OF CYSTIC FIBROSIS CHILDREN TOWARDS NEONATAL SCREENING AND ANTENATAL DIAGNOSIS

*L. N. Al-Jader, ^M. C. Goodchild, +H. C. Ryley and *P. S. Harper

*Institute of Medical Genetics
^Cystic Fibrosis Unit, Department of Child Health
+Department of Medical Microbiology
University Hospital of Wales, Cardiff

Information on parents' attitudes towards neonatal screening for cystic fibrosis (CF) and antenatal diagnosis by chorion villus biopsy (CVS) has been derived from a detailed questionnaire administered to parents of CF babies diagnosed early following newborn screening (18 babies, mean age at diagnosis 8 weeks, range 4-13 weeks) and later on account of clinical criteria (11 babies, mean age at diagnosis 29 weeks, range 8-116 weeks). Screening was by measurement of immunoreactive trypsin (IRT) on Guthrie card blood spots, which was the basis of the Wales/West Midlands IRT Screening Survey, 1985-1989 (Ryley et al.1988). Families questioned were from Wales.

Most parents supported screening: parents of 15/18 screened babies (83%) and 10/11 unscreened babies (91%). 52% of families (15/29) would abort a CF foetus following antenatal diagnosis.

Neither standard of education nor social class correlated with attitudes to screening or antenatal diagnosis, although these factors were related to the parents' knowledge of CF in general. Questions on parent-child bonding revealed that all parents experienced reactions and anxieties. 11 families, of whom 9 were screened, were most severely affected as they endured longer delays. Four mothers acknowledged temporary rejection of their babies during the period of uncertainty or following the procedures of diagnosis.

Several families emphasised the importance of minimal delay between the initial mention of the possibility of CF on IRT testing and confirmation (or otherwise) of the diagnosis.

These findings have implications for the widespread introduction of neonatal screening for CF and for programmes of antenatal detection associated with CF carrier screening.

Reference

Ryley, H.C., Deam, S.A., Williams, J., Alfaham, M., Weller, P.H., Goodchild, C., Carter, R.A., Bradley, D., Dodge, J.A., 1988, Neonatal screening for cystic fibrosis in Wales and West Midlands: 1. Evaluation of immunoreactive trypsin test, **J. Clin. Pathol.**, 41:726.

DNA ANALYSIS IN CF FAMILIES BY BIOTINYLATED PROBES AND POLYMERASE CHAIN REACTION TECHNIQUE

C. Camporese, L. Picci, N.A. Greggio,
M. Abdol Mohammady, A. Barbato, L. Chiandetti,
F. Anglani and F. Zacchello

Dept. of Pediatrics, University of Padova, Italy

INTRODUCTION

About 70% of the mutations in the cystic fibrosis (CF) gene in North American, British and Dutch patients are caused by a 3 bp deletion referred to as DeltaF508 (Kerem et al., 1989).

In South European population the mutation is present in only 46.2% of CF chromosomes confirming the heterogeneity of CF mutation in Southern Europe hypothesized by studies with closely linked markers near the CF locus (Estivill et al. 1989). This finding has important implications for prenatal diagnosis and carrier detection in this population since direct deletion analysis detects only 26% of CF patients and 46% of carriers. The purpose of our study is to analyse for the presence of DeltaF508 mutation seven CF families that have been typed with the closely linked markers METH, METD, J311, XV2c and KM19.

MATERIAL AND METHODS

Families

During the last two years 7 families with 1:4 risk of cystic fibrosis and with at least one living affected child have been typed to establish the informative probes for carrier detection and prenatal diagnosis.

Southern Blot Analysis

The DNA probes were labelled by enzymatic incorporation of biotinylated 11-dUTP by nick translation (Gibco-Brl). The protocols for hybridisation and signal detection have been reported elsewhere (Anglani et al., 1990).

PCR Amplification of F508 Mutation

50 bp fragment containing the 3 bp deletion of CF gene was amplified according to the method of Mathew et al., 1989. Briefly: primer sequences were 5'-GGCACCATTAAAGAAAATATC- 3' and 5' -CTATATTCATCATAGGAAAC-3'. 250ng of DNA was denatured for five minutes at 93°C and amplified by Taq polymerase for 30 cycles of 0.8 min., 93°C; 0.8 min, 53°C; 1.5 min, 72°C, with final extension of 5 min at 72°C. Products were analysed by vertical electrophoresis in 12% polyacrilamide gel containing ethidium bromide for 2 hours at 90mA.

RESULTS AND DISCUSSION

In each family a fully informative pattern has been established by RFLPs analysis and the carrier status of the normal siblings has been assigned. Mutation analysis has revealed that none of the 7 patients were homozygous for DeltaF508 deletion, but 5 were compound heterozygotes and 2 patients did not have the mutation. Correlation of the DeltaF508 deletion according to haplotypes for KM19 and Xv2c is given in table.

HAPLOTYPES ANALYSIS FOR XV2c AND KM19

Designation	Haplotype		CF chromosomes		Normal
	Km19	XV2c	DeltaF508	Others	chromosomes
A	1	1	1	2	5
B	2	1	2	2	4
C	1	2	-	-	4
D	2	2	2	5	1
Total number of chromosomes:			14		14

In North European populations the haplotype 1/2 for Xv2c/ KM19 is the almost exclusive haplotype background for the mutation (Kerem et al., 1989). In South European populations every case of DeltaF508 mutation is associated with the haplotype defined by KM19/Pst1 allele 2 and pM6PD/Msp1 allele 2 suggesting that an independent mutation could be present preferentially in Southern Europe CF families (Estivill et al., 1989). In our study, despite the preferential involvment of the KM19/Pst1 allele 2 on CF chromosomes, the DeltaF508 has occured in only 4/11 CF chromosomes bearing the KM19 2 allele. In one istance, it has occured on CF chromosome bearing the KM19 1 allele. Our cases illustrate that in our country the discovery of the DeltaF508 mutation has not had the same impact on prenatal diagnosis of cystic fibrosis as in the other countries of Northen Europe or North America. In fact in all our families the prenatal diagnosis of CF must still rely on RFLPs analysis. Until the second major mutation of CF gene predicted by linkage disequilibrium analysis in South European population (Estivill et al., 1989b) is discovered, direct analysis of the gene will be limited to a small number of families.

References

Anglani, F., Camporese, C., Greggio, N.A., Murgia, A., Zacchello, F., 1990, The use of biotinylated probes in the DNA analysis for diagnostic purpose, **CHIMICAoggi, Inter. J. of Chem. and Biotech.**, (in press).

Kerem, B.S., Rommens, J., Duchanan, H., Markiewicz, D., Cox, T., Chakravarti, A., Buchwald, M., et al., 1989, Identification of the cystic fibrosis gene: genetic analysis. **Science**, 245:1073.

Mathew, C.G., Roberts, R.G., Harris, A., Bentley, D.P., Bobrow, M., 1989, Rapid screening for DeltaF508 deletion in Cystic Fibrosis, **Lancet**, II:1346.

Estivill, X., Gasparini, F., Novelli, G., Benitez, E.F.J., Williamson, P., et al., 1989a, DeltaF508 gene deletion in cystic fibrosis in southern Europe, **Lancet**, ii:1404.

Estivill, X., Gasparini, P., Novelli G., Casals T., Nunes V., Gallano, P., Savoia, A., Ruzzo, A., Dallapiccola B., Pignatti, P.F., 1989b, Linkage disequilibrium for DNA haplotypes near the cystic fibrosis locus in two South European populations, **Hum. Genet.**, 83:175.

PREVALENCE OF THE MAJOR MUTATION OF THE CF GENE IN BELGIAN PATIENTS

P. Cochaux (a), R. Van Geffel (c), D. Baran (b) and G. Vassart (a)

(a) Service de Genetique and
(b) Service de Pneumologie
Hopital Erasme (Bat.C), 1070 Brussels, Belgium
(c) Service de Pediatrie, Hopital de Braine L'Alleud
1420 Braine L'Alleud, Belgium

We had previously examined the KM19-XV2C 2-1 haplotype frequency in a sample of Belgian cystic fibrosis patients (ref.1). The results obtained were similar to those published by Beaudet et al (ref.2) : this haplotype was present in 89% of the CF chromosomes.

We have now analysed the prevalence of the DF508 deletion, evidenced by Tsui et al. in 70% of the CF chromosomes in a Caucasian population (ref.3). We have studied 87 patients (from different families) and could demonstrate that 50 of them presented the deletion on both chromosomes, 31 on only one chromosome, and 6 didn't have the deletion at all. Thus the prevalence of the DF508 mutation as responsible of CF in our Belgian sample is 75 %.

The distribution of the KM19-XV2C 2-1 haplotype was correlated to the presence of the deletion (see table below).

On the other hand, the identification of this mutation allowed us to improve several genetic diagnoses that were only partially informative using RFLP studies (for instance, when no DNA of the index case was available), and also allowed us to confirm a clinical diagnosis (the supposed CF child died before the sweat test could be performed).

Finally, we have found 1 carrier of the DF508 deletion in a sample of 71 non-CF individuals.

TABLE 1

HAPLOTYPES		CF						N
KM19	XV2C	DF508	(%)	non DF508	(%)	Total		
A. 1	1	1	(1%)	2	(7%)	3	(2%)	36%
B. 2	1	110	(97%)	15	(56%)	125	(89%)	16%
C. 1	2	0	(0%)	7	(26%)	7	(5%)	36%
D. 2	2	2	(2%)	3	(11%)	5	(4%)	12%
Not characterized		23		14		37		

CF (or N) : cystic fibrosis (or normal) chromosomes
Polymorphisms : KM19/Pst I ; XV2C/Taq I
1 (or 2) : absence (or presence) of the restriction site
Detection of the DF508 mutation was performed by ASO-hybridization after dot blotting of amplified DNA (primers C16B and C16D, ref.3)

References

1. Devoto et al., 1989, **Genomics**, 5:894-898.
2. Beaudet et al., 1988, **The New Eng. J. of Med.**, 318-1:50-51.
3. Kerem et al., 1989, **Science**, 245:1073-1080.

CYSTIC FIBROSIS IN THE SICILIAN POPULATION: LINKAGE DISEQUILIBRIUM AND PRENATAL DIAGNOSIS BY POLIMERASE CHAIN REACTION

E. D'Alcamo, P. Sammarco, R. di Marzo, L. Iapichino, F. Pardo, A. Maggio, V. Balsamo and F. Caronia

Centro Regionale FC
Ospedale dei Bambini, Palermo, Italy

We have been studying 106 chromosomes from 21 Sicilian families with one or more subjects affected with cystic fibrosis. Using XV2C and KM19 probes we have shown a linkage disequilibrium between haplotype B and CF allele (68%). However this association was less frequent than previously described in other Caucasian CF families (Estivill et al., 1987; Vidaud et al., 1989; Weber et al., 1988). Moreover, the presence in our population of a higher frequency of haplotype A (15.7%) compared to European and North American population was found, suggesting some heterogeneity of CF gene defects in our country.

We have also been studying the feasibility of prenatal diagnosis in our population with XV2C, KM19 and CS7 polymorphism. Of these 21 families, 8 (33.3%) are 100% informative with KM19 and XV2C while the others are 50% or not informative.

Recently, the identification of the CF gene (Kerem et al., 1989; Rommens et al., 1989; Riordan et al., 1989) has made it possible to perform direct DNA studies by PCR and dot-blot analysis. A three bp deletion (DeltaF508 mutation) in the Cystic Fibrosis gene is associated with a lot of CF cases in North American and North European populations.

We have been able to observe the DeltaF508 deletion in 104 CF Sicilian families' chromosomes by PCR and dot-blot analysis and we showed that it accounts for 58% of these CF chromosomes.

Finally, we report data about prenatal diagnosis of CF performed by DeltaF508 deletion study: The parents of a CF child who had died were both heterozygotes for the DeltaF508 deletion. We performed prenatal diagnosis by CVS at the 11th week of gestation; the small amount of villus was amplified with oligonucleotide primers specific C16B and C16D. PCR amplified product was blotted and hybridized with 32P-gamma-dATP-labelled oligonucleotides CF-DeltaF508 and CF-N; fetal DNA analysis showed that fetus was heterozygote.

These data show a higher heterogeneity of CF Sicilian chromosomes than other European and North American populations, although the possibility of carrying out prenatal diagnosis by polymorphism or direct analysis is high as for other genetic diseases (thalassemia) (Di Marzo et al., 1988).

Acknowledgements

We thank the Regional Cystic Fibrosis Association for the support given to E. D'Alcamo for this work and the C.N.R. for the project support n°89.0030875.

References

Di Marzo, R., Douling, C.E., Wong, C., Maggio A., Kazazian, H.H., 1988, The spectrum of β-thalassemia mutations in Sicily, **Br. J. of Haematol.**, 69:393.

Estivill, X., Scambler, P.J., Wainwright, B.J., Hawley, K., Frederick, P., Schwartz, M., Baiget, M., Kere, J., Williamson, R., Farrall, M., 1987, Patterns of polymorphism and linkage disequilibrium for cystic fibrosis, **Genomics**, 1:257.

Kerem, B., Rommens, C.M., Buchanan, J.A., Markeiwicz, D., Cox, T.K., Chakravarti, A., Buchwald, M., Tsui, L.C., 1989, Identification of the cystic fibrosis gene: genetic analysis, **Science**, 245:1073.

Riordan, J.R., Rommens, J.M., Kerem, B., Alon, N., Rozmahel, R., Grzelczak, Z., Zielenski, J., Lok, S., Plavsic, N., Chou, J.L., Drumm, M.L., Iannuzzi, M.C., Collins, F.S., Tsui, L.C., 1989, Identification of the cystic fibrosis gene: cloning and characterization of complementary DNA, **Science**, 245:1066.

Rommens, J.M., Iannuzzi, M.C., Kerem, B., Drumm, M.L., Malmer, G., Dean, M., Rozmahel, R., Cole, J.L., Kennedy, D., Zsiga, N.H.M., Buchwald, M., Riordan, J.R., Tsui, L.C., Collins, F.S., 1989, Identification of the cystic fibrosis gene: chromosome walking and jumping, **Science**, 245:1059.

Vidaud, M., Kilzis, A., Ferec, C., Bozon, D., Dumur, V., Giraud, G., David, F., Pascal, O., Auvinet, M., Morel, Y., Andre, J., Chomel, J.C., Saleun, J.P., Farriaux, J.P., Roussel, P., Labbe, A., Dastugue, B., Locotte, G., Monnier, N., Foucaud, P., Goossens, M., Feingold, J., Kaplan, J.C., 1988, Confirmation of linkage disequilibrium between haplotype B (AV2C, allele 1; KM19 allele 2) and cystic fibrosis allele in the French population, **Hum. Genet.**, 81:183.

Weber, J., Aulehla-Scolz, C., Kaiser, R., Eigel, A., Neugebaur, M., Horst, J., Olek, K., 1988, Cystic fibrosis: typing 89 German families with linked DNA probes, **Hum. Genet.**, 81:54.

STUDY OF RECONSTITUTION OF THE RABBIT PAROTID $Na^+/K^+/2Cl^-$ COTRANSPORTER

Concetta Ferri, Monica Garnier and Angela Corcelli

Istituto di Fisiologia Generale
Universita' di Bari, Italy

The critical role of Cl^- transepithelial transport in eliciting a normal fluid secretion or absorption in various epithelia is strongly suggested from clinical and pathological studies. Modifications of epithelial function caused from bacterial toxins or genetically determined, as in cystic fibrosis, appear to be specifically dependent on some alteration in Cl^- transport across epithelia. The apical and basolateral membranes of epithelial cells are endowed with Cl^- transport systems of different specificities and characteristics. According to the current view on ion transepithelial transport, the entry of Cl^- into the cell can occur via cation coupled transport systems. In many absorbing and secreting epithelia a quaternary cotransport system of stoichiometry $1Na^+:1K^+:2Cl^-$ has been characterized. In order to understand the mechanism of transport and ion coupling of such complex transport system, it is important to have it in a functional state in reconstituted systems. In this type of study the availability of an high affinity inhibitor of the transporter can provide a useful tool to label the protein in various steps of solubilization and reconstitution. As regards the $Na^+/K^+/2Cl^-$ cotransporter the appropriate labelling agent to use for this purpose appears to be bumetanide. Recently, it has been shown that a high affinity bumetanide binding site is associated to the $Na^+/K^+/2Cl^-$ cotransporter in rabbit parotid acinar cell basolateral membrane (Turner et al., 1988). In a previous study we described a method to reconstitute the rabbit parotid bumetanide binding site in proteoliposomes (Corcelli et al., 1990). Here we report preliminary results of a further study of reconstitution in which, by modifying the previous method, we were able to incorporate the bumetanide binding site together with a Cl^- transporter in proteoliposomes.

The Identification of the CF (Cystic Fibrosis) Gene
Edited by L.-C. Tsui *et al.*, Plenum Press, New York, 1991

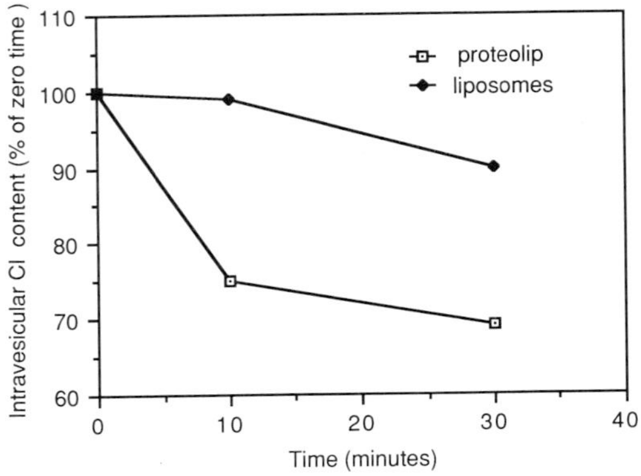

Fig. 1. CHLORIDE EFFLUX FROM RECONSTITUTED VESICLES
Reconstituted vesicles were prepared in 100mM mannitol, 10mM N-2- hydroxyethylpiperazine-N'-2-ethansulfonic acid(HEPES) buffered with tris(hydroxymethyl)aminomethane (Tris) to pH7.4 1mM EDTA plus 100mM sodium gluconate, 95mM potassium gluconate, 5mM KCl, 6mM Tris/$^{36}Cl^-$ (4μCi/ml). As control, parallel studies were carried out on liposomes (no protein incorporated). The efflux was started by dilution of an aliquot (60 μl) of $^{36}Cl^-$-loaded vesicles (proteolip. and liposomes) in 1.5 ml of buffer containing 300mM mannitol, 10mM Hepes/Tris pH 7.4, 1mM EDTA. After the appropriate time of incubation at room temperature, the efflux was terminated by fitration under suction over 0.45μm HAWP Millipore filters; filters were then washed with 6ml of stop solution. The stop solution contained buffer A plus 100mM Kgluconate. Data are means of triplicates, the S.D. was within 5% of the mean value. Data are expressed as % of the intravesicular $^{36}Cl^-$ content at zero time.

A protein extract was obtained from basolateral membrane vesicles (BLMV) of rabbit parotid acinar cells by solubilization with Triton X100 (0.3% w/v detergent at 1mg/ml membrane protein) and centrifugation at 100,000gx15min. This protein extract, supplemented with 0.185% soybean azolectin plus 0.085% pure egg phosphatidylcholine, was treated with Bio-Beads SM2 to remove the detergent. In control experiments the protein extract was replaced by a 0.3% detergent solution. The material obtained after bead removal was used for binding and transport studies.

Equilibrium ^3H bumetanide binding to BLMV, extract and reconstituted vesicles (PL) was measured using a nitrocellulose filtration assay. Bumetanide binding was determined in the presence of 100mM Na^+, 100mM K^+ and 5mM Cl^- or 100mM Cl^-.

Typically the bumetanide binding in rabbit parotid BLMV is stimulated by low (up to 5mM) chloride concentrations and is inhibited by high chloride concentrations (Turner et al., 1988). The difference between the binding found at low and high Cl^-

concentration (respectively 5 and 100mM) is here referred to as the Cl^- dependent bumetanide binding.

The table shows that at least 100% of BLMV Cl^- dependent bumetanide binding was recovered in the extract and, after the removal of detergent, in PL; no specific bumetanide binding was observed in control liposomes (not shown). The chloride efflux was measured from $^{36}Cl^-$ loaded PL and control liposomes. As illustrated in the figure, we found that the chloride efflux rate from PL was much higher than from control liposomes. These data indicate reconstitution of a Cl^- transporter together with specific bumetanide binding activity in PL. The dependence of this reconstituted Cl^--transport on cation transport and its inhibitor sensitivity is currently under investigation.

Table 1

RECONSTITUTION OF Cl^--DEPENDENT BUMETANIDE BINDING SITE

<u>Cl^-dependent ^3H bumetanide binding (cpm/40μl of sample)</u>

BLMV	1850
EXTRACT + LIPIDS	2000
PL	2050

Differences between bumetanide binding mean values (from triplicates) determined at 5 and 100mM chloride are reported. The S.D. of each determination was within 10% of mean value. The protein content of extract and PL was about 65% of BLMV content.

References

1. Corcelli, A., Turner, R.J., 1990, Reconstitution of the rabbit parotid basolateral membrane Na/K/Cl dependent bumetanide binding site, **Annals New York Academy of Sciences**, 574:106.
2. Turner, R.J., George, J.N., 1988, Ionic dependence of bumetanide binding to the rabbit parotid Na/K/Cl cotransporter, **J. Membrane Biol.**, 102:71.

HUMAN PLACENTAL MEMBRANE VESICLES AS MODELS FOR INVESTIGATION OF CYSTIC FIBROSIS

A.K. Keenan, M. Cregg, D. Faller, C. Doolan,
C. Somers, M.X. Fitzgerald and M.P. Ryan

Departments of Pharmacology and Medicine
University College, Belfield, Dublin 4, Ireland

The human placenta is a rich source of epithelial material with high transport capabilities containing the same genotype as the foetus. The CF gene is expressed in placenta which constitutes an available source of human biopsy material suitable for investigation of transport defects in CF.

In order to examine and define transport processes located in the apical or basolateral membranes of the placental syncytiotrophoblast, we have isolated and characterized brush border (BBMV) and basolateral membrane vesicles (BLMV) from human term placentas. The studies described in this report have investigated the status of β-adrenoceptor dependent chloride transport at a number of stages in the signal transmission chain.

BBMV and BLMV were prepared by homogenization and differential centrifugation. The purity of individual preparations was assessed (i) using the marker enzymes alkaline phosphatase and gamma-glutamyl transpeptidase and (ii) by scanning and transmission electron miscroscopy. β-adrenoceptor binding was measured by incubating membranes with ^{125}I-cyanopindolol at 37°C for 45 min and non-specific binding was measured in the presence of 10 mM propranolol. Adenylate cyclase activity was measured by monitoring the conversion of -^{32}P ATP to -^{32}P cAMP and separating products from reactants by sequential ion exchange chromatography. cAMP-dependent protein kinase activity was measured as the amount of -^{32}P phosphate of ATP transferred to the synthetic peptide substrate kemptide. Uptake of ^{36}-Cl into BBMV was measured in the presence of an outwardly directed Cl gradient. The latter was generated by exchange of Cl with gluconate, consequent efflux of intravesicular Cl and generation of an intravesicular positivity.

Saturation β-adrenoceptor binding analysis revealed that the Bmax and K_D values in BLMV were 203 ± 80 fmol/mg protein and 40 ± 20 pM respectively. The corresponding values for BBMV were 35 ± 18 fmol/mg and 33 ± 6 pM. While no adenylate cyclase activity was detectable in BBMV, significant stimulation of activity was observed in BLMV with forskolin (at the level of the catalytic unit), with Gpp(NH)p and NaF (at the level of the G protein), but not with isoprenaline at the β-receptor. The low extent of catecholamine stimulation in placenta despite the presence of large numbers of β-receptors has been previously reported (Whitsett et al., 1980).

Stimulation of cAMP-dependent protein kinase activity was demonstrable in crude cytosol (data not shown), BBMV and in BLMV. As can be seen in Table 1, the activity stimulated by 1 mM cAMP was comparable in BBMV and BLMV, as was the extent of inhibition by 2mM Ca. Significant phosphorylation of endogenous substrates (kemptide blank) was evident in BLMV, in contrast to the data shown for the single experiment with BBMV. The dose-dependency of cAMP stimulation was examined in BLMV and activity peaked at 2 μM, declining thereafter with higher concentrations of nucleotide (10-100 μM).

TABLE 1

Stimulation of protein kinase activity in human placental membrane vesicles (a) BBMV and (b) BLMV. Data presented are (a) means of 1 or (b) means ± s.e.m of 3 determinations, both sets performed in duplicate.
+ one determination; ++ basal activity in the absence of kemptide.

Stimulant	Activity (a)	(pmol PO4 transferred/min/mg) (b)
None (basal)	7.5	21 ± 5.2
cAMP (1 μm)	57	58 ± 6.9
$MgCl_2$ (4 mM)	n.d.	35 ± 11.0
$CaCl_2$ (2 mM)	n.d.	8.6 ± 6.9
$CaCl_2$ + cAMP	20	27+ ±
Kemptide blank++	1.2	29 ± 6.4

A time-dependent uptake of ^{36}Cl into BBMV was measured which equilibrated to a value of 17 nmol Cl/mg protein at approximately 12 min, and which was abolished by 5μM valinomycin added either before uptake commenced or 2 min later.

In summary, this study has shown that membrane vesicles prepared from human placenta can be used to investigate β-receptor dependent regulation of chloride transport. These systems are currently being used in our laboratory to investigate modulation of chloride transport via protein kinase activation and to identify putative regulatory proteins thus phosphorylated in normal and CF tissue.

Acknowledgements

This work was supported by the Cystic Fibrosis Association of Ireland.

Reference

Whitsett, J., Johnson, C.L., Noguchi, A., Darovec-Beckerman, C. & Costello, M. (1980) **J. Clin. Endocrinol. and Metabolism**, 50, 27.

THE DELTAF508 MUTATION IN ISRAELI CF FAMILIES

I. Lerer, S. Cohen, M. Chemke*, A. Friedmen, and D. Abeliovich

Department of Human Genetics, Molecular Biology Unit
Hadassah University, Jerusalem
*Clinical Genetics Unit, Kaplan Hospital
Rehovot, Israel

We analyzed DNA from CF patients and their families for linked DNA markers (RFLPs) and for the presence of the DeltaF508 mutation.

The distribution of the haplotypes at the XV2C/taq1 and KM19/Pst1 loci are given in Table 1. The Israeli CF-chromosomes differ from those of northern Europe, especially in their proportion showing the B haplotype (74% instead of 88%).

Table 1

Distribution of KM.19 and XV-2c haplotypes in various Israeli communities

Haplo-	Ashkenazi				Non-Askenazi				Arab				Total	
type	CF	(%)	N	(%)	CF	(%)	N	(%)	CF	(%)	N	(%)	CF	N
A	-		18	(33)	2	(7)	6	(20)	-		8	(32)	2%	31%
B	42	(90)	4	(8)	18	(64)	4	(13)	9	(56)	6	(24)	74%	14%
C	3	(6)	22	(46)	7	(25)	16	(54)	3	(19)	11	(44)	16%	48%
D	2	(4)	4	(8)	1	(4)	4	(13)	4	(25)	-		8%	7%
	47		48		28		30		16		25			

Classification of the Israeli CF families according to their origin revealed that the haplotype distribution among Ashkenazi CF chromosomes was comparable to that of northern Europe, while a marked difference exists in the non-Ashkenazi Jews and in the Arabs. Based on these data, we assumed the existance of one predominant mutation in the Ashkenazi Jews and different mutation(s) in the non-Ashkenazi Jews and the Arab families.

The CF mutation DeltaF508 was found in 32% (36/113) of the CF mutations in Israel. To date this is the lowest frequency published. Correlation of the DeltaF508 mutation according to haplotype and family origin is given in Table 2. The DeltaF508 mutation was found among all ethnic groups with the lowest frequency among the Arabs. It was associated with allele 2 of KM19/Pstl in 28/29 chromosomes and in 79% was associated with haplotype B, but only 35% of CF chromosomes with haplotype B had the DeltaF508 mutation. It seems, therefore, that in our population other mutational event(s) occurred on the same haplotype background as the DeltaF508 in northern Europe.

The implication of the low frequency of DeltaF508 mutation in Israel for heterozygote screening is obvious since only 1/9 of the couples at a one-in-four risk for having an affected child are expected to be identified.

Table 2

Distribution of DeltaF508 mutation according to KM19/XV-2c haplotypes and ethnic origin
Number of CF-bearing chromosomes

Ethnic Origin	A	DeltaF 508	B	DeltaF 508	C	DeltaF 508	D	DeltaF 508	Total CF	DeltaF508	%
Ashkenazi Jews	-	-	35	11	3	-	2	2	40	13	(32%)
Non-Ashkenazi Jews	2	1	18	9	8	-	1	1	29	11	(38%)
Arabs	-	-	12	3	3	-	8	2	23	5	(22%)
Total	2	1	65	23	14	-	11	5	92	29	(31%)

A,B,C,D - KM19/XV-2c haplotypes

THE DELTAF508-DELETION IN 99 CF PATIENTS OF SWITZERLAND

S. Liechti-Gallati, I. Parsai, R. Kraemer,
A. Rudeberg, S. Braga and H. Moser

Medical Genetics Unit, Department of Pediatrics
(Inselspital), University of Bern, Bern, Switzerland

The cystic fibrosis (CF) gene has recently been isolated, and one major mutation has been defined as a 3 bp deletion in exon 10 resulting in the loss of a phenylalanine residue at position 508 (DeltaF508). The DeltaF508-deletion frequency in Europe seems to increase gradually from South-East to North-West.

We have studied 99 CF patients from our area: DeltaF508 was detected by PCR amplification and separation of the products by PAGE. Haplotypes were determined using RFLP's of the probes XV2c, KM19, MP6d-9, J3.11. The Swiss data show 72% of the CF chromosomes carrying the DeltaF508 mutation. 53% of our patients are homozygous for DeltaF508, 37% are heterozygous and 10% were found not to have this deletion on either chromosome. Haplotype analyses showed the allele constellation 1(XV-2c) - 2 (KM19) - 2 (MP6d-9) - 1 (J3.11) in 78% of the CF chromosomes with the DeltaF508 deletion, suggesting strong correlation between this haplotype and the DeltaF508 mutation. The remaining 22% of the DeltaF508 carrying chromosomes demonstrate linkage with the haplotype 1-2-2-2, that may be the result of a recombination event between J3.11 and the CF mutation. No other haplotypes were found to be linked with DeltaF508.

Nine different haplotypes were detected on our non-deleted CF chromosomes. The most frequent haplotypes are represented by the allele constellations 1-1-2-2 (27%) and 2-1-1-1 (22%), followed by 2-1-1-2 (14.5%) and 1-2-2-2 (14.5%), suggesting at least four further mutations in Switzerland that may cause the cystic fibrosis defect. The other five haplotypes may either be due to recombination events between the marker and the CF gene or may represent some rarer mutations.

Four of our CF patients without any DeltaF508 deletion are homozygote for the haplotype constellation 1-1-2-2, strongly suggesting homozygosity for the same mutation. They all show relatively late onset of pseudomonas infection, but high percentage of underweight and low clinical and radiographic scores including physical activity, pulmonary disease, radiologic lung involvement and gastrointestinal manifestations.

The other patients without DeltaF508 show milder clinical course than DeltaF508 compound heterozygotes or DeltaF508 homozygotes.

THE PREVALENCE OF THE DELETION F508 IN A BELGIAN CYSTIC FIBROSIS POPULATION

W. Lissens, M. Bonduelle, A. Malfroot[1], I. Dab[1] and I. Liebaers

Departments of Medical Genetics and [1]Pediatry
A.Z. - Vrije Universiteit Brussel
Brussels, Belgium

We have used the polymerase chain reaction and dot blot analysis (Kerem et al., 1989) to screen for the deletion F508 in 75 Belgian families with at least one CF patient. Haplotype analysis with DNA probes XV-2c and KM19 had already been performed in most of these families (Devoto et al., 1989). The combined results are given in table 1. The frequency of the deletion F508 was 78.7% (118 of 150 CF chromosomes).

TABLE 1

Frequency of the deletion F508 in Belgian CF chromosomes and distribution of haplotypes of XV-2c and KM19 in CF and normal alleles.

	CF ALLELES DELETION F508		NORMAL ALLELES
HAPLOTYPE	+	-	
A	1	7	45
B	103	14	21
C	0	7	41
D	1	1	19
N.D.	13	3	-
	118	32	126

N.D.: haplotype not determined

The Identification of the CF (Cystic Fibrosis) Gene
Edited by L.-C. Tsui *et al.*, Plenum Press, New York, 1991

Haplotypes:	XV-2c	KM19
A	1	1
B	1	2
C	2	1
D	2	2

1 = absence of restriction site
2 = presence of restriction site

Of the CF chromosomes for which haplotypes were available, 98% (103/ 105) of deletion F508 chromosomes have haplotype B. This haplotype also accounts for almost half (14/29) of non-deletion F508 CF chromosomes. A total of 87% (117/134) of CF chromosomes are associated with the B haplotype. In contrast, this haplotype is only represented in 17% of normal parental alleles.

We have also looked for 4 other CF mutations, reported by members of the CF Genetic Analysis Consortium, in the 32 non-deletion F508 chromosomes. Deletion I507 (exon 10) and 3 mutations in codons 533, 549 and 551 in exon 11 of the CFTR gene were not found.

References

1. Devoto, M., L. De Benedetti, M. Seia, L. Piceni Sereni, M. Ferrari, M.L. Bonduelle, A. Malfroot, W. Lissens, A. Balassopoulou, G. Adam, D. Loukopoulos, P. Cochaux, G. Vassart, R. Szibor, J. Hein, K. Grade, W. Berger, B. Wainwright, and G. Romeo, 1989, Haplotypes in cystic fibrosis patients with or without pancreatic insufficiency from four European populations, **Genomics**, 5:894.
2. Kerem, B-S., J.M. Rommens, J.A. Buchanan, D. Markiewicz, T.K. Cox, A. Chakravarti, M. Buchwald, L.-C. Tsui, 1989. Identification of the cystic fibrosis gene : genetic analysis. **Science**, 245:1073.

CYSTIC FIBROSIS DELTA F508 MUTATION IN A FRENCH POPULATION

G. Lucotte and E. Barre

Laboratoire d'Anthropologie Physique
du College de France
Paris 5e, France

The recent identification of the cystic fibrosis (CF) gene (Rommens et al., 1989; Riordan et al., 1989; Kerem et al., 1989) and of the three base pair deletion delta F508 as its most common mutation is of great importance both for basic CF research and for clinical genetics. Genetic evidence established that 68% of CF chromosomes were found to carry delta F508 (whereas this mutation was never seen on normal chromosomes) and that chromosomes with the deletion nearly always had an (XV-2c)1, (KM.19)2 haplotype (previously designated as a "high risk" haplotype for CF).

We tested 122 CF chromosomes (children affected and parents) from French Parisian individuals (Spanish, Italian and North-African surnames were excluded), coming from 61 independent French CF families originated from France from two generations at least, for the presence-absence of delta F508 mutation and the correlation between this mutation and the haplotype distribution defined by XV-2c and KM.19. The detection of the delta F508 mutation was carried out on DNA amplified by polymerase chain reaction (PCR) according to Kerem et al. (1989): amplified DNA was spot blotted and filters hybridized to ^{32}P kinasated allele specific oligomers.

The delta F508 mutation has been observed in 98 out of the 122 CF chromosomes (80%), which is significantly higher than the frequency observed by Kerem et al. (1989): 43 of our patients are homozygous for the delta F508 mutation, 12 are compound heterozygous delta F508/other, and 6 do not show the mutation.

Of the families for which XV-2c, KM.19 were available 92% of delta F508 chromosomes had haplotype 1,2, and 100% of delta F508 chromosomes had haplotype 2,2 (the 1,2 haplotype accounted for 16% of normal parental chromosomes, with 2,1 being the most com-

The Identification of the CF (Cystic Fibrosis) Gene
Edited by L.-C. Tsui *et al.*, Plenum Press, New York, 1991

mon type). Similar results are obtained in some other European regions, with the exception of Spain and Italy (Estivill et al., 1989; McIntosh et al. 1989).

Borderline sweat test values were associated with the presence of delta F508 in one of the two chromosomes of one CF patient. All of our 4 patients classified as CF pancreatic sufficient present a mild phenotype, and none is homozygous for delta F508. Five of our patients had a proven history of meconium ileus; all of them are homozygous for delta F508.

The observed frequency of delta F508 implies that 70% of CF patients in France are homozygous for this mutation, and an equal proportion of prenatal diagnoses in pregnancies at 1:4 risk may now be made by mutation analysis. The identification of the delta F508 mutation allowed us to improve several genetic diagnoses that were only partially informative using haplotype studies : when no DNA of the index case was available, and to confirm a clinical diagnosis (the supposed CF child died before the sweat test could be performed).

None out of 24 non-delta F508 CF chromosomes carried the G551D, G553X or S549N mutations.

References

Estivill, X., Chillon, M., Casals, T., Bosch, A., Morral, N., Nunes, V., Gasparini, P., 1989, Delta F508 gene deletion in cystic fibrosis in southern Europe, **The Lancet**, dec. 9, 1404.

Kerem, B.S., Rommens, J.M., Buchanan, J.A., Markiewicz, D., Cox, T.K., Chakravarti, A., Buchwald, L.-C. Tsui, 1989, Identification of the cystic fibrosis gene : genetic analysis, **Science**, 245:1073.

McIntosh, I., Lorenzo, M.L., Brock, D.J., 1989, Frequency of delta F508 mutation on cystic fibrosis chromosomes in UK, **The Lancet**, Dec. 9, 1404.

Riordan, J.R., Rommens, J.M., Kerem, B.S., Alon, N., Rozmahel, R., Grzelczak, Z., Zielenski, J., Lok, S., Plavsic, N., Chou, J.L., Drumm, M.L., Iannuzzi, M.C., Collins, F.S., Tsui, L.-C., 1989, Identification of the cystic fibrosis gene : cloning and characterization of complementary DNA, **Science**, 245:1066.

Rommens, J.M., Iannuzzi, M.C., Kerem, B.S., Drumm, M.L., Melmer, G., Dean, M., Rozmahel, R., Cole, J.L., Kennedy, D., Zsiga, N.H.M., Buchwald, M., Riordan, J.R., Tsui, L.-C., Collins, F.S., 1989, Identification of the cystic fibrosis gene : chromosome walking and jumping, **Science**, 245:1059.

FREQUENCY OF THE DELTA-F508 MUTATION AND FLANKING MARKER HAPLOTYPES AT THE CYSTIC FIBROSIS LOCUS FROM 167 CZECH FAMILIES

M. Macek, Jr. (1,2), V. Vavrova (3), I. Boehm (2,6), M. Stuhrmann (2), A. Reis (2), M. Macek (3), R. Duspisova (5), E. Jelinkova (5), K. Sperling (2), M. Krawczak (4) and J. Schmidtke (2)

(1) Department of Clinical Biochemistry
Faculty of Medicine, Charles University
Prague, Czechoslovakia
(2) Institute of Human Genetics
Free University Berlin (West)
D1000, Heubnerweg 6, Berlin 19
(3) Institute for Child Development Research
Pediatric Faculty, Charles University
Prague, Czechoslovakia
(4) Institute of Human Genetics
Göttingen University, Federal Republic of Germany
(5) Center of Medical Genetics, Pediatric Faculty
Charles University, Prague, Czechoslovakia
(6) Genetic Laboratory of Dr. Waldenmaier
Munich, Federal Republic of Germany

This study was undertaken in order to compare the predominantly Slavic Czech Cystic Fibrosis (CF) population of Bohemia and Moravia (Czech Republic of Czechoslovakia) with other Caucasian populations in Europe for the deltaF508 CFTR gene mutation (Kerem et al., 1989) and CF-linked marker loci MET, D7S23, D7S399 and D7S8. This geographical region was a crossroad of Central Europe, where in particular Slavic and German groups came into close contact. It was to be expected that deltaF508 and CF-linked haplotype frequencies would reflect the population history of this region.

Furthermore, this population study hoped to identify other frequent CF associated haplotypes on the non-deltaF508 chromosomes. Thirdly, knowledge of population specific allele frequencies is of direct importance for the delivery of an adequate diagnostic service.

The Identification of the CF (Cystic Fibrosis) Gene
Edited by L.-C. Tsui *et al.*, Plenum Press, New York, 1991

DeltaF508 mutation frequencies have been measured in most European derived populations (this workshop). Its frequencies range between 80% in North-West England (Super M. et al. 1990, this meeting) and 27% in Turkish CF population (Hundrieser et al., 1990), suggesting a North-to-Southeast gradient. This is valid also for the closely linked B-haplotype (KM19-XV2c 2-1, Estivill et al., 1988).

167 (116 complete) families with between one and four living CF affected children and up to two non-affected children participated in this study. DNA typing was performed by PCR amplification, restriction analysis and agarose or polyacrylamide gel electrophoresis. Polymorphic restriction sites defined by probes Met-H/MspI, XV2c/TaqI, CS7/HhaI, KM19/PstI, Mp6-d9/MspI and J3.11/MspI were used. Haplotype frequencies were estimated applying iterative likelihood maximisation procedures.

Based on 334 CF chromosomes the frequency of deltaF508 is 68% and the frequency of the B-haplotype is 78%. This "high risk" haplotype is present in 62% on deltaF508 chromosomes and in 16% on non-deltaF508 CF chromosomes. Extreme linkage disequilibrium was found between the D7S23 locus (probes CS7, KM19) and D7S399 (probe Mp6-d9). Moreover linkage disequilibrium was found for deltaF508 and all markers tested. Linked marker haplotype frequencies on CF and normal chromosomes are described in more detail in Macek Jr. et al., 1990.

Our study supports the North-to-Southeast gradient of deltaF508 and B-haplotype frequencies. In the Eastern part of Czechoslovakia (Slovak Republic) deltaF508 is found in 63% on CF chromosomes (Puliti et al. 1990, this meeting). Obtained data further suggest the presence of another frequent CF mutation associated with the B-haplotype.

Acknowledgement

This work was supported by grants from the Deutsche Gesellschaft zur Bekaempfung der Mukoviszidose and Deutsche Forschungsgemeinschaft to M.M. and J.S.

References

1. Estivill, X., Farrall, M., Williamson, R., et al., 1988, Linkage disequilibrium between cystic fibrosis and linked DNA polymorphisms in Italian families: A collaborative study, **Am. J. Hum. Genet.**, 43:23.
2. Hundrieser, J., Bremer, S., Peinemann, F., et al., 1990, Frequency of the deltaF508 deletion in the CFTR gene in Turkish cystic fibrosis patients, **Hum. Genet.** (in press).
3. Kerem, B.-S., Rommens, J.M., Buchanan, J.A., et al., 1989, Identification of the cystic fibrosis gene : Genetic analysis, **Science**, 245:1073.
4. Macek, M. Jr., Vavrova, V., Boehm, I., et al., 1990, Fre-

quency of the deltaF508 mutation and flanking marker haplotypes at the CF locus from 167 Czech families, **Hum. Genet.** (in press).

5. Reis, A., Bremer, S., Schlosser, M., et al., 1990, Distribution patterns of the deltaF508 mutation in the CFTR gene on CF-linked marker haplotypes in German population, **Hum. Genet.** (in press.

THE RESPONSE OF CHLORIDE TRANSPORT TO CYCLIC AMP, CALCIUM AND HYPOTONIC SHOCK IN NORMAL AND CYSTIC FIBROSIS FIBROBLASTS

Teresa Mastrocola and Michela Rugolo

Dip. di Biolcgia Ev.Sp., Universita' di Bologna
Bologna, Italy

Cl^- transport was shown to be altered in skin fibroblasts from cystic fibrosis (CF) patients (Rugolo et al.,1986), and an abnormal Cl^- metabolism was also observed by Mattes et al. (1987). In the present study we have investigated the effect of two cAMP analogs, 8-Bromo cyclic AMP (8-Br-cAMP) and dibutyryl cyclic AMP (dbt-cAMP), of a beta agonist, isoproterenol, on $^{36}Cl^-$ efflux from normal and CF fibroblasts. Cl^- efflux consists of a superposition of two efflux events, operationally defined as two compartments, each exhibiting first order exponential decay. In table 1 it is shown that 15 min preincubation in the presence of cAMP analogs (0.1mM 8-Br-cAMP and 1mM dbt-cAMP) and isoproterenol (5μM) increased the rate constant (k) of Cl^- efflux from the fast compartment of 10%, 5% and 9% respectively, in both normal and CF fibroblasts. These results are in disagreement with those previously reported by Lin and Gruenstein, 1987, who observed a 13% stimulation of Cl^- efflux in normal fibroblasts, but failed to detect any activation in CF cells. However, in our experiments, the extent of activation of Cl^- transport by the cAMP-dependent effectors was weak in control cells, so that no difference could be detected between normal and CF fibroblasts. The addition of A23187 (1-5μM) to the efflux medium did not enhance Cl^- transport in both normal and CF fibroblasts, suggesting that elevation of intracellular Ca^{2+} concentration does not directly activate Cl^- transport. It is noteworthy that the k of Cl^- efflux from the slow compartment was not modified in any experimental conditions (not shown).

A large activation of Cl^- efflux was observed after exposure of human fibroblasts to hypotonic medium. Substitution of Cl^- in the hypotonic efflux medium by gluconate and addition of furosemide did not affect Cl^- efflux, indicating that the anion/anion exchange and the anion/cation cotransport did not play any

TABLE 1

Effect of cAMP analogs, isoproterenol and A23187 on the k of Cl^- efflux from the fast compartment in a normal fibroblast line.

Additions	k, min^{-1}
None	1.487 ± 0.27
Dbt-cAMP 1mM	1.643 ± 0.37
8-Br-cAMP 0.1mM	1.555 ± 0.30
Isoproterenol 5μM	1.620 ± 0.31
A23187 3μM	1.308 ± 0.27

Values are the mean ± SD of at least 4 determinations.

significant role. Conversely, the hypotonic shock-activated Cl^- efflux was inhibited by 0.1mM 5-nitro-2-(3 phenylpropylamino) benzoate (NPPB), a blocker of the cAMP-activated Cl^- channel of respiratory epithelia. This and other evidence (Rugolo et al., 1989) indicate the involvement of a conductive Cl^- pathway in the response of fibroblasts to hypotonic shock. The effect of hypotonicity on Cl^- efflux was similar in normal and CF fibroblasts.

References

Lin, P. and Gruenstein, E., 1987, Identification of a defective cAMP-stimulated Cl^- channel in cystic fibrosis fibro blasts, **J. Biol. Chem.**, 262:15345.

Mattes, P.M., Maloney, P.C. and Littlefield, J.W., 1987, Altered chloride metabolism in cultured cystic fibrosis skin fibroblasts, **Proc. Natl. Acad. Sci. USA**, 84:3009.

Rugolo, M., Romeo, G. and Lenaz, G., 1986, Kinetic analysis of chloride efflux from normal and cystic fibrosis fibroblasts, **Biochem. Biophys. Res. Comm.**,160:1330.

Rugolo, M., Mastrocola, T., Flamigni, A. and Lenaz, G., 1989, Chloride transport in human fibroblasts is activated by hypotonic shock, **Biochem. Biophys. Res. Comm.**, 160:1330.

2D-ELECTROPHORESIS OF MITOCHONDRIAL PROTEINS FROM CYSTIC FIBROSIS PATIENTS

L. Picci, L. Frentagni, G. Mastella*,
E. Scarso, P. Pizzochero, P. Mattiazzo,
L. Chiandetti, F. Anglani and F. Zacchello

Dept. of Pediatrics, University of Padova, Italy
*Cystic Fibrosis Center, Borgo Trento Hospital
Verona, Italy

INTRODUCTION

Experimental data from some laboratories (increased mitochondrial Ca^{++} concentration, increased O_2 consumption, response to inhibitors, altered enzyme kinetics) point to a deranged mitochondrial function in cystic fibrosis (CF) (Burton and Shapiro, 1989).

Whether these findings are related to a primary genetic defect of mitochondrial structure/function, or they reflect a secondary derangement of cellular metabolism, is still a matter of investigation.

Recently, we have developed a method for obtaining mitochondrial preparation, starting from relatively small amounts of peripheral blood mononuclear cells. These preparations showed some degrees of contamination from other subcellular components, but mitochondria retained their functional integrity when they were evaluated by transmembrane potential and oxigen consumption measurements.

The purpose of this study was to analyse the 2D-electrophoresis pattern of such preparations from CF homozygotes and obligate heterozygotes, in order to look for alterations of protein composition, related to the CF genotype.

MATERIAL AND METHODS

Mitochondria Purification

10-15 ml of heparinized venous blood were diluited 1:1 with isotonic Tris/HCl buffer, pH 7.6, layered on Ficoll-Paque and centrifuged. The mononuclear cells were washed and resuspended with 5 mM TES/KOH buffer, pH 7, containing 0.25 M sucrose, 1 mM K-EDTA and 0.05% (w/v) BSA. The cells were disrupted by osmotic lysis at 4°C, and the omogenate centrifuged to remove nuclei and cellular debris. The supernatants were centrifuged twice at 10000 rpm for 15 min. The purity of the resulting mitochondrial fractions was evaluated by measuring the activity of marker enzymes (MAO, LDH, CDH and acid phosphatase), while their functional integrity was studied by measuring the transmembrane potential (TPP ion selective electrode) and oxigen consumption.

Mitochondrial Protein Separation

Mitochondrial proteins were solubilized by three "freezing and thawing" cycles in a solution containing 9 M urea and 2% CHAPS. Protein concentration was measured according to BCA methods (Smith et al., 1985). 2D-electrophoresis was performed according to O'Farrell (1975) with slight modifications, starting from aliquots corresponding to 20 μg protein. Proteins were silver stained according to Merrill (1981) with modifications.

RESULT AND DISCUSSION

We analyzed first 6 CF homozygotes and 6 obligate heterozygotes. The 2D-electrophoretic patterns were compared with those from 6 control subjects.

We could not find any obvious differences between CF homo- and heterozygotes; however it was possible to clearly identify a set of protein spots that seemed peculiar of the CF genotype (homo- and heterozygotes) and could differentiate it from the control ones.

Therefore we performed a blind study to verify the usefulness of our procedure in recognizing the CF genotype. Mitochondrial fractions were prepared from 5 CF homozygotes, 5 obligate heterozygotes and 4 control subjects.

8 electrophoretic runnings, among the 16 performed, were correctly interpreted, and the genotype (CF vs not CF) correctly assigned: homozygotes 4/5, obligate heterozygotes 2/5, control subjects 2/6.

It is possible that: 1) CF mitochondria do not harbour the basic defect, even if our data are not conclusive on this respect; 2) blood mononuclear cells are not the right starting material and further studies should be performed on mitochondria obtained from epithelial cells.

References

Burton, L., Shapiro, B.L., 1989, Evidence for a mitochondrial lesion in cystic fibrosis, **Life Sciences**, 44:1327.

Smith, P.K., Krohn, R.I., Hermanson, G.T., Mallia, A.K., Gartner, F.H., Provenzano, M.D., Fujimoto, E.K., Goeke, N.M., Olson, B.J., Klenk, D.C., 1985, Measurement of protein using Bicinchoninic acid, **Anal. Biochem.**, 150:76.

O'Farrell, P.H., 1975, High resolution two-dimensional electrophoresis of proteins, **J. Biol. Chem.**, 250:4007.

Merril, C.R., Goldman, D., 1981, Ultrasensitive stain for proteins in polyacrylamide gels shows regional variation in cerebrospinal fluid proteins, **Science**, 211:1437.

FREQUENCY OF CYSTIC FIBROSIS MUTATIONS AND ASSOCIATED HAPLOTYPE DISTRIBUTION IN SLOVAK CF PATIENTS

A. Puliti (1), J.J. Telleria Orriols (1), P. Ronchetto (1), L. Fenu (1), M. Devoto (1), G. Romeo (1), L. Kadasi (2), J. Gecz (2), and V. Ferak (2)

1) Laboratorio di Genetica Molecolare
Istituto Giannina Gaslini, Genoa, Italy
2) Institute of Molecular Physiology and Genetics
Slovak Academy of Science, Bratislava, Czechoslovakia

INTRODUCTION

After the identification of the cystic fibrosis (CF) gene (Kerem et al., 1989; Riordan et al., 1989; Rommens et al., 1989) several mutations have been described in different exons (G. Cutting, M. Dean, M. Goossens, L.-C. Tsui, personal communications). The most frequent of them is the deltaF508 mutation (a 3 bp deletion which results in the loss of a phenyl-alanine residue corresponding to codon 508), which is present in 68% of the total CF chromosomes in the original Canadian sample (Kerem et al., 1989). However, the frequency of deltaF508 varies among different populations: in particular in Europe different frequencies have been observed (Estivill et al., 1989; McIntosh et al., 1989) and a South-East to North-West gradient of distribution results from an overall analysis of the population data from Europe (EWGCFG, 1990).

In the present work, a sample of 23 Slovak CF patients has been analysed for the presence of deltaF508 and the association between the CF mutations and the RFLPs revealed by the probe/enzyme combinations KM.19/ PstI, metH/TaqI, and metD/TaqI.

RESULTS AND DISCUSSION

The frequency of the deltaF508 mutation in our sample of Slovak patients is equal to 63% (29/46) (table 1). In particular, 10 out of 23 patients (43%) are deltaF508 homozygotes, while 9 (39%) are compound heterozygotes deltaF508/others and 4

The Identification of the CF (Cystic Fibrosis) Gene
Edited by L.-C. Tsui *et al.*, Plenum Press, New York, 1991

(18%) carry other mutations (table 2). These data are in agreement with those reported for the other European populations, which show that deltaF508 is generally the most frequent CF mutation.

Table 1

Frequency of deltaF508

	%	No.
Total		46
DF508	63	29
Others	37	17

Table 2

Distribution of genotypes

	%	No.
Total		23
DF508/DF508	43	10
DF508/others	39	9
Others/others	17	4

RFLP analysis shows that allele 2 of KM.19 represents about 95% of the deltaF508 chromosomes. The same allele characterizes also 42% of the other CF chromosomes, while it is present in 26% of the normal chromosomes (table 3). The allelic distributions of metH/TaqI and metD/TaqI in CF and normal chromosomes are also reported in table 3. The only significant association found in our sample is that between allele 2 of KM.19 and deltaF508 (standardized linkage disequilibrium coefficient = -0.68, chi-square = 20.67, $p<0.001$).

Table 3

Distribution of RFLPs in CF and normal chromosomes

	KM.19		metH		metD	
	1	2	1	2	1	2
DF508	1 (5)	19 (95)	18 (75)	6 (25)	17 (77)	5 (23)
Non DF508	7 (58)	5 (42)	9 (56)	7 (44)	9 (90)	1 (10)
Normal	23 (74)	8 (26)	27 (68)	13 (32)	27 (84)	5 (16)

References

Estivill, X., Chillon, M., Casals, T., Bosch, A., Morral, N., Nunes, V., Gasparini, P., Seia, A., Pignatti, P.F., Novelli, G., Dallapiccola, B., Fernandez, E., Benitez, J., Williamson, R., 1989, DeltaF508 gene deletion in cystic fibrosis in Southern Europe, **Lancet**, II:1404.

European Working Group on CF Genetics (EWGCFG), 1990, Gradient of distribution in Europe of the major CF mutation and of its associated haplotype (submitted).

Kerem, B.S., Rommens, J., Buchanan, J.A., Markiewicz, D., Cox, T.K., Chakravarti, A., Buchwald, M., Tsui, L.C., 1989, Identification of the cystic fibrosis gene: genetic analysis, **Science**, 245:1073.

McIntosh, I., Lorenzo, M.L., Brock, D.J.H., 1989, Frequency of the deltaF508 mutation on cystic fibrosis chromosomes in UK, **Lancet**, ii:1404.

Riordan, J.R., Rommens, J., Kerem, B., Alon, N., Rozmahel, R., Grzelczak, Z., Zielenski, J., Lok, S., Plavsis, N., Chou, J.-L., Drumm, M.L., Iannuzzi, M.C., Collins, F.S., Tsui, L.-C., 1989, Identification of the cystic fibrosis gene: cloning and characterization of complementary DNA, **Science**, 245:1066.

Rommens, J., Iannuzzi, M.C., Kerem, B., Drumm, M.L., Melmer, G., Dean, M., Rozmahel, R., Cole, J.L., Kennedy, D., Hidaka, N., Zsiga, M., Buchwald, M., Riordan, J.R., Tsui, L.C., Collins, F.S., 1989, Identification of the cystic fibrosis gene: chromosome walking and jumping, **Science**, 245:1059.

Rommens, J., Kerem, B.-S., Greer, W., Chang, P., Tsui, L.C., Ray, P., 1990, Rapid nonradioactive detection of the major CF mutation, **Am. J. Hum. Genet.**, 46:395.

FREQUENCY OF CYSTIC FIBROSIS MUTATIONS AMONG ITALIAN PATIENTS

Patrizia Ronchetto (1), Laura Fenu (1), Aldamaria Puliti (1), Marcella Devoto (1), Luca Romano (2), Giovanni Romeo (1,2), Laura Cremonesi (3), Paola Carrera (3), Luca Ruocco (3), Manuela Seia (4), Silvia Russo (4), Annamaria Giunta (5), and Maurizio Ferrari (3)

1) Laboratorio di Genetica Molecolare
Istituto Giannina Gaslini, Genoa, Italy
2) Clinica Pediatrica I, Universita' di Genova, Italy
3) Laboratorio Centrale, Istituto Scientifico
Ospedale San Raffaele, Milan, Italy
4) Istituti Clinici di Perfezionamento
Universita' di Milano, Italy
5) Clinica Pediatrica II, Universita' di Milano, Italy

INTRODUCTION

The gene responsible for cystic fibrosis (CF) has been recently cloned (Rommens et al., 1989; Riordan et al., 1989; Kerem et al., 1989) and the most common mutation causing CF (deltaF508) has been identified as a three base pair deletion at codon 508 of exon 10 removing a phenylalanine. Since then other mutations have been identified in different exons, and the information necessary for their detection has been distributed through the CF Genetic Analysis Consortium (L.-C. Tsui, personal communications).

In order to investigate the distribution of mutations in the CF gene, a sample of 203 Italian CF patients has been analysed for the presence of the deltaF508 mutation. In addition, the presence of the mutation has been correlated to the haplotype distribution defined by the two polymorphic markers KM.19 and XV-2c, and to the presence or absence of pancreatic insufficiency (PI) and meconium ileus (MI).

MATERIAL AND METHODS

To investigate the proportion of Italian CF patients carrying the deltaF508 deletion, the correspondent segment of exon

10 of the CFTR gene has been amplified with oligonucleotide primers flanking the deletion in a polymerase chain reaction (PCR) (Riordan et al., 1989). The mutation has subsequently been detected by non denaturing polyacrilamide gel electrophoresis (PAGE) (Rommens et al., 1990). In a few cases, results have also been confirmed by allele specific oligonucleotide (ASO) hybridization (Kerem et al., 1989). In addition, two different RFLPs closely linked to the CF gene, KM.19 and XV-2c, have been evaluated (Estivill et al., 1987a; Estivill et al., 1987b).

Presence or absence of pancreatic insufficiency and presence of meconium ileus at birth have been assessed on the basis of clinical and laboratory findings.

RESULTS AND DISCUSSION

The frequency of the deltaF508 mutation among 406 CF chromosomes subdivided in PI and PS (pancreatic sufficient) is reported in Table 1.

Table 1. Frequency of deltaF508 mutation

	No.	(%)		No.	(%)
deltaF508	212	(52)	others	194	(48)
PI	175		PI	135	
PS	15		PS	41	
?	22		?	18	

While the overall frequency of this mutation is 52% in our sample, it is equal only to 27% among PS patients. In particular (Table 2), among PI patients 50 (32%) are homozygotes deltaF508/deltaF508, 75 (48%) are heterozygotes deltaF508/others, and 30 (19%) carry different mutations, whereas none of the PS patients is homozygote deltaF508/deltaF508, in agreement with previously reported data (Kerem et al., 1989). Table 2 also shows the distribution of genotypes in CF patients subdivided according to presence or absence of MI at birth. In our sample, all the MI patients carry at least one copy of the deltaF508 mutation.

Table 2. Distribution of genotypes among 203 Italian CF patients

	DF508/DF508	DF508/others	others/others
PI	50	75	30
PS	-	15	13
?	7	8	5
MI	7	9	-
non MI	48	83	43
?	2	7	5
Tot.	57	98	48

The distribution of haplotypes defined by KM.19 and XV-2c reported in Table 3 shows that the great majority of deltaF508 mutation is associated with haplotype B (XV-2c allele 1, KM.19 allele 2), in agreement with the hypothesis of a single origin for this mutation. DeltaF508 chromosomes associated to other haplotypes (in particular, haplotype D) may be the results of recombination events occurring between the CF locus and the marker loci or between the two marker loci.

Table 3. Distribution of haplotypes for XV-2c and KM.19

	A (1 - 1)		B (1 - 2)		C (2 - 1)		D (2 - 2)	
DF508	-		177	(.94)	1	(.01)	11	(.06)
Others	32	(.18)	85	(.49)	38	(.22)	20	(.11)
PI	18		68		22		13	
PS	12		12		10		4	
MI	-		7		-		1	
non MI	30		75		32		17	

The distribution of the same haplotypes among non deltaF508 chromosomes indicates that haplotype B is still the most frequent CF haplotype in our sample, accounting for 49% of the non deltaF508 CF chromosomes. In particular, it is noteworthy that the frequency of haplotype B rises to 56% among PI chromosomes, whereas it is only 32% among PS chromosomes. In addition, 7 out of 8 MI chromosomes are characterized by a B haplotype.

References

Estivill, X., Farrall, M., Scambler, P.J., Bell, G., Hawley, K.M.F., Lench, N.J., Bates, P., Kruyer, H.C., Frederick, P.A., Stainer, P., Watson, E.K., Williamson, R., Wainwright, B.J., 1987a, A candidate for the cystic fibrosis locus isolated by selection for the metilation-free islands, **Nature**, 326:840.

Estivill, X., Scambler, P.J., Wainwright, B.J., Hawley, K., Frederik, P., Baiget, M., Kere, J., Williamson, R., Farrall, M., 1987b, Patterns of polymorphism and linkage disequilibrium for cystic fibrosis, **Genomics**, 1:257.

Kerem, B.S., Rommens, J., Buchanan, J.A., Markiewicz, D., Cox, T.K., Chakravarti, A., Buchwald, M., Tsui. L.-C., 1989, Identification of the cystic fibrosis gene: genetic analysis, **Science**, 245:1073.

Riordan, J.R., Rommens, J., Kerem, B., Alon, N., Rozmahel, R., Grzelczak, Z., Zielenski, J., Lok, S., Plavsis, N., Chou, J.-L., Drumm, M.L., Iannuzzi, M.C., Collins, F.S., Tsui, L.-C., 1989, Identification of the cystic fibrosis gene: cloning and characterization of complementary DNA, **Science**, 245:1066.

Rommens, J., Iannuzzi, M.C., Kerem, B., Drumm, M.L., Melmer, G., Dean, M., Rozmahel, R., Cole, J.L., Kennedy, D., Hidaka, N., Zsiga, M., Buchwald, M., Riordan, J.R., Tsui, L.-C., Collins, F.S., 1989, Identification of the cystic fibrosis gene: chromosome walking and jumping, **Science**, 245:1059.

Rommens, J., Kerem, B.-S., Greer, W., Chang, P., Tsui, L.-C., Ray, P., 1990, Rapid nonradioactive detection of the major CF mutation, **Am. J. Hum. Genet.**, 46:395.

CULTURED HUMAN PLACENTAL TROPHOBLASTS AS MODELS FOR INVESTIGATING DEFECTIVE REGULATION OF CHLORIDE TRANSPORT AND PROTEIN SECRETION IN CYSTIC FIBROSIS

M.P. Ryan, C. Somers, M.X. Fitzgerald and A.K. Keenan

Departments of Pharmacology and Medicine
University College, Dublin, Ireland

Human placenta provides a major source of epithelial material for investigating basic defects in cystic fibrosis (CF). The placenta has high transport capabilities, contains the same genotype as the foetus and is known to express the CF gene. The present study concerns the establishment of a primary cell culture system for human placental trophoblasts and characterisation of chloride transport and gonadotrophin secretion as models for investigating CF defects.

Placentas were obtained usually within 30 min of delivery, and transported back to the laboratory on ice in sterile phosphate buffered saline: Dulbeccos modified Eagles medium (PBS: DMEM) containing penicillin (100 U/ml), streptomycin (100 μg/ml) and nystatin (100 units/ml). The subsequent steps were based on procedures described by Kliman et al. (1986) and Contractor and Sooranna (1985) and were carried out under a laminar flow hood. Placental trophoblast tissue was washed in DMEM:PBS (1 L) to remove blood and dissected free from chorionic membranes and villi. The trophoblast tissue (30 ml packed volume) was then incubated (30 min at 37°C) with trypsin (5.2 mg/ml) and DNAase (0.2 mg) in Hanks balanced salts solution (containing EDTA 0.02%, HEPES 20 mM, penicillin 100 U/ml, streptomycin 100 μg/ml). The dispersed cells were filtered through surgical gauze, layered onto foetal calf serum (FCS (2.5 ml) and centrifuged (2000 r.p.m.) for 15 min. Fresh trypsin:DNAase solution was added to the residual undigested tissue and the above procedures repeated twice. Cell pellets from the 3 sequential incubations were suspended, and pooled in Hanks solution and filtered through nylon gauze (500 μm). The filtrate was placed on a discontinuous Percoll gradient (5 - 70% Percoll in Hanks solution) and centrifuged at 2000 r.p.m. for 20 min in the presence of 0.25 mg DNAase (to prevent cell flocculation). The gradient

The Identification of the CF (Cystic Fibrosis) Gene
Edited by L.-C. Tsui *et al.*, Plenum Press, New York, 1991

fractions corresponding to density marker beads 1.048 and 1.062 were collected and cells were cultured in collagen coated multi-well plates in DMEM with 20% foetal calf serum.

The total yield of cells per preparation was approximately 10^9 of which over 90% were viable as indicated by trypan blue exclusion. After 1-2 days in culture, cells fused to form syncytiotrophoblasts. The epithelial nature of the cells was confirmed by the finding that cells stained positively to a monoclonal antibody for α keratin. Immunoperoxidase staining for the endothelial cell marker vimentin revealed much weaker staining than that for α keratin. Gonadotrophin secretion was monitored daily using an ELISA technique with the Tandem immuno enzymatic assay kit (Hybritech, Belgium).

Basal gonadotrophin secretion increased until day 5 of culture. Thereafter gonadotrophin secretion declined. Addition of 8-Br cAMP (2 mM) and theophyllin (2 mM) resulted in a 5-fold increase in gonadotrophin production. This stimulation of gonadotrophin secretion was maximal on day 4 of culture.

When ^{36}Cl uptake was investigated, a rapid initial uptake of Cl to a mean value of 521 nmol/mg protein was seen at 1 min. An equilibrium mean value of 480 nmol/mg protein was maintained between 10 and 60 min with a slight decline to 392 nmol/mg protein at 120 min. A time-dependent efflux of ^{36}Cl uptake was measured following loading of cells by incubation in 1 μCi ^{36}Cl for 120 min. A significant increase ($p < 0.01$) in ^{36}Cl efflux was found in the presence of dibutyrl cAMP (2 mM).

These findings demonstrate that primary cultures of placental cells can be established and used to investigate protein (gonadotrophin) secretion and Cl transport. The findings that gonadotrophin secretion and ^{36}Cl efflux were stimulated by cAMP analogues suggest this model may be useful for investigating CF defects. We are currently using the cultured syncytiotrophoblast model to investigate defective regulation of chloride transport at cellular level in CF.

Acknowledgements

This work was supported by the Cystic Fibrosis Association of Ireland.

References

Contractor, S.F. and Sooranna, S.R., 1988, **J. Devl. Physiol.**, 10:47.

Kliman, H.J., Nestler, J.E., Sermasi, E., Sanger, J.M. and Strauss, J.F., 1986, **Endocrinology**, 118:1567.

A DELETION MUTATION OF THE CYSTIC FIBROSIS TRANSMEMBRANE CONDUCTANCE REGULATOR (CFTR) LOCUS: DeltaI507

Martin Schwarz (1), Claire Summers (2), Lesley Heptinstall (2), Clive Newton (2), Alexander Markham (2) and Maurice Super (1)

1 Paediatric Genetics Unit
Royal Manchester Children's Hospital, Manchester, UK
2 ICI Diagnostics, Northwich, Cheshire, UK

SUMMARY

During the development of an amplification refractory mutation system (ARMS)[1] assay for the detection of the DeltaF508 mutation[2,3,4] and corresponding normal locus in cystic fibrosis we discovered a family in which a further variant of the sequence exists. PCR amplification and direct sequencing of a region of exon 10 of the CFTR locus indicated the deletion of the three base pairs encoding isoleucine$_{506}$ or isoleucine$_{507}$, or possibly the presence of a single base substitution in conjunction with the DeltaF508 mutation. The resulting protein has a deletion of an isoleucine residue at position 507 as opposed to the previously described deletion of phenylalanine at position 508. We conclude that the loss of an isoleucine residue at position 507 (DeltaI507) is another defective variant of the Cystic Fibrosis Transmembrane Conductance Regulator (CFTR) gene product.

INTRODUCTION

In the population of the North-West of England approximately 80% of CF genes contain the DeltaF508 deletion[5]. We have evaluated the direct analysis of this deletion in the CFTR gene using two methods - the ARMS technique[1] and Allele-Specific Oligonucleotide probes (ASOs)[4]. The results from both types of assay were consistently in agreement, with the exception of the family reported here.

This family is from the South-West of England, and includes a 12 year old boy with cystic fibrosis. The clinical presentation was one of failure to thrive in early infancy, with proven pancreatic insufficiency (faecal fat 10.9g in 24 hrs) together with diagnostic sweat electrolytes. His chest disease is mild and includes an asthmatic component. On only one occasion, earlier in 1989, has Pseudomonas been isolated from sputum.

The XV2c/Taq I and KM19/Pst I haplotype data[8] for the father were atypical in that the genotype at both loci was 2/2 i.e. homozygous for the presence of the respective Taq I and Pst I restriction sites.

METHOD

For the ASO technique 50ng of genomic DNA were amplified by polymerase chain reaction (PCR) using the primers and conditions previously reported[4]. The upstream amplification primer was modified however, so as to agree with the cDNA sequence[3]. PCR products were separated in duplicate 1.4% agarose gels and transferred onto Hybond N+ (Amersham) according to manufacturer's protocols. The membranes were hybridized to 10ng of 5'-end-labelled allele-specific oligonucleotide, in a solution of 5xSSC, 1xDenhardt's solution, 0.1% SDS, and 1000μg/ml sheared, denatured salmon sperm DNA, at 42°C. Post-hybridization washes were begun in 0.5 x SSC, 0.1% SDS increasing in stringency to 0.1 or 0.05 x SSC. Each wash was at 42°C for 10-15 minutes. Autoradiography was carried out overnight at room temperature.

ARMS analysis was performed on 25ng of genomic DNA and utilised primers specific for the DeltaF508 sequence and for the normal allele in separate reactions. Each reaction also contained a common primer. The primer sequences were dTATATTCATCATAGGAAACACCAATGATAT for DeltaF508, dTATATTCATCATAGGAAACACCAAAGATGA for the normal allele, and the common primer was dGACTTCACTTCTAATGATGATTATGGGAGA (Newton et al., personal communication).

A pair of primers which amplified a 360bp region of exon III of the human alpha-1-antitrypsin (AAT) gene was included in all reactions. The product from these primers served as an internal control for the reaction. The reactions were performed in a volume of 100μl containing 10mM Tris-HCl pH8.3, 50mM KCl, 1.2mM $MgCl_2$, 0.01% gelatin, 2 units Taq polymerase (Cetus), the four dNTPs (100μM each), the appropriate ARMS primers (1μM each) and the control primers (1μM each). Reactions were overlaid with 50μl liquid paraffin. The amplification regime consisted of 40 cycles of denaturation at 92°C for 2 minutes, annealing and extension at 60°C for 2 minutes, and extension at 72°C for 2 minutes.

The amplified products were separated in composite 2% agarose/1% Nusieve agarose (ICN Biomedicals) gels and visualized with ethidium bromide. Gels were scored for the presence or absence of amplified product for each pair of primers (Fig. 1b).

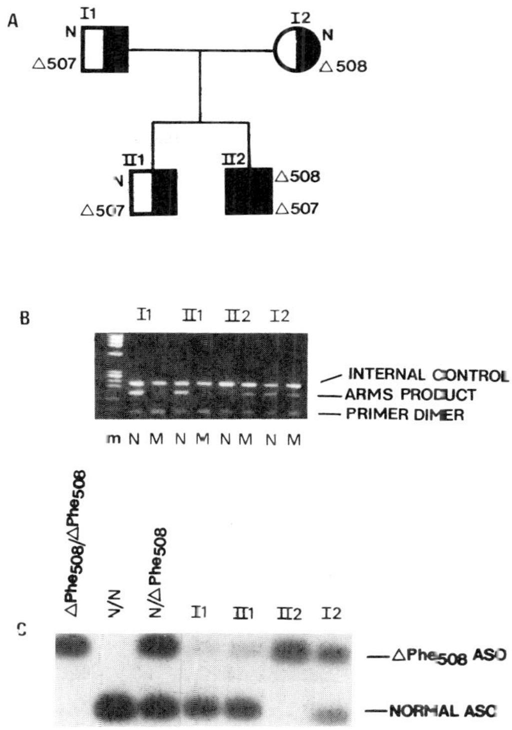

Fig. 1. Panel A: Pedigree of the CF family in which the Delta-I507 deletion mutation was identified. N, Delta507,and Delta508 denote the normal, DeltaI507, and DeltaF508 alleles respectively as determined by direct DNA sequence analysis (Figure 2).
Panel B: Agarose gel of ARMS products from the individuals of the pedigree in Panel A; "m" denotes molecular weight markers; "N" and "M" denote use of normal and DeltaF508 deletion allele-specific ARMS primers respectively.
Panel C: Result from ASO analysis of the individuals from the pedigree in panel A. The first three samples are from characteristic normal, DeltaF508 homozygote and DeltaF508 heterozygote individuals respectively.

RESULTS

After typing the family (Fig 1c) by ASO hybridisation the father (I.1) and the unaffected son (II.1) showed weak hybridisation with the oligonucleotide specific for the DeltaF508 sequence and a strong signal for the normal (non-DeltaF508) sequence. ARMS analysis (Fig 1b) showed these two individuals to be apparently homozygous for the normal allele. Results for the affected boy (II.2) and the mother (I.2) were in agreement for the two methods of ascertainment. Since earlier ARMS analysis showed the ARMS primers to be very specific for their target sequence, it appears that the father and unaffected boy possess a sequence which is weakly detected by the allele-specific oligonucleotides, but which the ARMS primers do not recognise. Such a sequence would have to be similar to the DeltaF508 sequence to allow partial hybridisation to occur.

The DNA from the region around the DeltaF508 deletion was sequenced directly for each family member (Fig 2). Sequence information for the parents shows that the DeltaF508 allele is maternal, while the DeltaI507 is paternal.

DISCUSSION

The sequence data suggest that the DeltaI507 mutation is either a deletion of the three bases 507 encoding isoleucine, or the DeltaF508 deletion in conjunction with an adenine to thymidine substitution in codon 507. We note with interest that

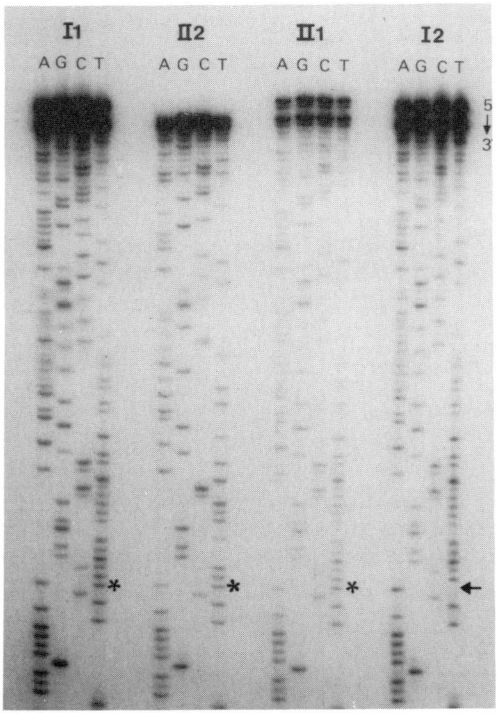

Fig. 2. Direct DNA sequence analysis of the individuals from the pedigree in Figure 1A. The asterisk marks the T residue present in the DeltaI507 allele (note the comigrating A residue contributed by the normal and DeltaF$_{508}$ alleles). The arrow shows the absence of the T residue in the mother who is a carrier of the DeltaF$_{508}$ mutation. Sequencing of PCR products was carried out as previously described.

others reporting a similar finding (Kerem et al., personal communication) favour the interpretation of a deletion of the ATC (codon 506 or 507) triplet, giving rise to a deletion of the isoleucine residue directly. While this explanation fully satisfies the data, we feel that the interpretation of the results as a DeltaF508 deletion in conjunction with an A/T substitution is also valid, even though it occurs on a haplotype in which DeltaF508 is rare. A/T substitutions of this type are not without precedent, eg. the classic GAG to GTG mutation which gives rise to sickle cell anaemia. The latter hypothesis could be vindicated by the finding of the postulated A/T change in normal CFTR genes; this would cause a replacement of the isoleucine at position 507 with a phenylalanine residue, both of which are uncharged amino acids.

The true incidence of the DeltaI507 mutation is difficult to assess, but it is likely that it is rare since it has occurred in conjunction with a relatively rare XV2c/KM19 haplotype, although it appears in a number of French CF chromosomes (Simon-Bouy, personal communication).

These findings demonstrate the high degree of specificity of the ARMS technique, since it gave a result which indicated the absence of the DeltaF508 mutation in the father, while the ASO result was ambiguous. Figure 3 shows the normal and DeltaF508 ARMS primers aligned with the coding strands of the normal, DeltaF508 and DeltaI507 alleles respectively. The results show that even with a mismatched nucleotide 6 bases from the 3' terminus of the DeltaF508 primer, the specificity of the system is retained.

The ARMS method is the subject of European patent application 0 332 435 and corresponding worldwide patents on behalf of ICI plc.

Fig. 3. DNA sequences with their presunptive corresponding amino acid sequences in the region of the DeltaF$_{508}$ mutation.

Codon	501	502	503	504	505	506	507	508	509	510	511
Normal	Thr	Ile	Lys	Glu	Arg	Ile	Ile	Phe	Gly	Val	Ser
	ACC	ATT	AAA	GAA	AAT	ATC	ATC	TTT	GGT	GTT	TCC
DeltaF$_{508}$	Thr	Ile	Lys	Glu	Arg	Ile	Ile	...	Gly	Val	Ser
(maternal)	ACC	ATT	AAA	GAA	AAT	ATC	AT.	..T	GGT	GTT	TCC
DeltaI$_{507}$	Thr	Ile	Lys	Glu	Arg	Ile	...	Phe	Gly	Val	Ser
(paternal)	ACC	ATT	AAA	GAA	AAT	ATC	...	TTT	GGT	GTT	TCC

Acknowledgements

We wish to thank Dr John Osborne for bringing this family to our attention and Dr Lap-Chee Tsui for helpful discussions.

References

1. Newton, C.R., Graham, A., Heptinstall, L.E., (1989), **Nucl. Acids Res.**, 17:2503.
2. Rommens, J.M., Ianuzzi, M.C., Kerem, B., 1989, **Science**, 245: 1059.
3. Riordan, J.R., Rommens, J.M., Kerem, B., 1989, **Science**, 245: 1066.
4. Kerem, B., Rommens, J.M., Buchanan, J.A., 1989, **Science**, 245: 1073.
5. Schwarz, M.J., Super, M., Wallis, C., 1990, **Hum. Genet.** (In press).
6. Newton, C.R., Kalsheker, N., Graham, A., 1988, **Nucleic Acids Res.**, 16:8233.
7. Newton, C.R. and Markham, A.F., 1988, Method for detecting nucleotide sequences, European Patent Application 0 332 435.
8. Estivill, X., (1987) Genomics, 1:257.

DNA TECHNOLOGY FOR PRENATAL DIAGNOSIS OF CYSTIC FIBROSIS IN ITALY

M. Seia°, G. Cozzi°, L. Sereni°, M. Ferrari*, L. Cremonesi*, A.E. Covone+, L. Fenu+, P. Ronchetto+, M. Devoto+, and G. Romeo+

°Lab. Ricerche Cliniche, ICP, Milan Italy
*Ist. Scient. Ospedale San Raffaele
Lab. Centrale, Milan, Italy
+ Lab. Genetica Molecolare
Istituto G.Gaslini, Genoa, Italy

INTRODUCTION

Since 1985, genetic diagnosis of Cystic Fibrosis (CF) by RFLPs analysis (Tsui L.-C. et al., 1985; Wainwright et al., 1985) is available in families with affected children.

The recent identification of the gene and of the major mutation present in CF patients (deletion of 3 bp which results in a loss of Phenilalanine at position 508 of the putative gene product) (Kerem et al., 1989; Riordan et al., 1989) has allowed a new direct approach to genetic diagnosis of the disease.

We report the results of our experience regarding DNA linkage analysis in order to evaluate feasibility of Prenatal Diagnosis (PD). We refer altogether our data on the distribution of F508 mutation in our Italian CF families, which are the basis to discuss feasibility of PD by direct analysis.

MATERIALS AND METHODS

DNA from 295 unrelated Italian families at risk for CF, with at least one affected child, has been analyzed with several markers.

The following probe/enzyme combinations have been used: metD/TaqI, met/MspI, W3D1.4/HindIII, XV-2c/TaqI, KM-19/PstI, E.9/MspI, Mp6D.9/MspI, J3.11/MspI. DNA extraction, restriction digestion, Southern Blotting, hybridization and Polymerase Chain Reaction have been performed according to conventional methods.

Oligonucleotide primer sequences and PCR conditions for the detection of F508 mutation were those indicated by Kerem et al.(1989). The amplified products were analyzed by electrophoresis on polyacrilamide 12% gel and stained with ethidium bromide.

Chorionic villus sampling for PD was done in 30 families by transabdominal route and 1 by transcervical, under real time ultrasound control at 7-12 weeks of gestation.

RESULTS AND DISCUSSION

The data of the feasibility study by RFLPs of CF prenatal diagnosis showed that 289 out of 295 couples (98%) resulted fully informative for genetic counselling. In 6 cases (2%) it was not possible to detect one parental phase and therefore they were only partially informative. The mutation F508 was present on 52.2% of CF chromosomes studied. 57 cases (28%) were homozygous for the deletion; 98 (48%) were heterozygous and 48 cases (24%) carried other mutations. In an attempt to explain the different percentages observed in the same population (Estivill et al., 1990), we have tried to trace regional distribution of the mutation and the data show that in general in the North the F508 deletion is more frequent than in the Center or South and in particular in some regions like Marche, Calabria and Basilicata the mutation is less frequent.

The prenatal diagnosis results are the following: 8 normal, 18 carrier, 4 affected and 1 unaffected. In 25 cases diagnosis was confirmed at birth or after abortion, 6 pregnancies are still in progress.

Our data show that in our CF population the direct method based on the study of F508 mutation can be carried out in 52.2% of CF chromosomes and in the remaining cases RFLP studies must be used. Future identification of the other mutations will increase the possibility of the direct specific diagnosis of CF.

References

Tsui, L.-C., Buchwald, M., Barker, D., Braman, J.C., Knowlton, R., Schumm, J.W., Eiberg. H., Mohr, J., Kennedy, D., Plavsic, N., Zsiga, M., Markiewicz, D., Akots, G., Brown, V., Helms, C., Gravius, T., Parker, C., Rediker, K., Donis Keller, H., 1985, Cystic Fibrosis locus to human chromosome 7cen-q22, **Nature**, 318:384.

Wainwright, B.J., Scambler, P.J., Schmidtke, J., Watson, E.K., Law, H.-Y., Farral, M., Cooke, H.J., Eiberg, H., Williamson, R., 1985, Localization of cystic fibrosis locus to human chromosome 7cen-q22, **Nature**, 318:384.

Kerem, B., Rommens, J.M., Buchanan, J., Marchiewicz, D., Cox, T.K., Chakravarti, A., Buchwald, M., Tsui, L.-C., 1989b, Identification of the cystic fibrosis gene: genetic analysis, **Science**, 245:1073.

Riordan, J.R., Rommens, J.M., Kerem, B., Alon, N., Rozmahel, R., Grzelczak, Z., Zielenski, J., Lok, S., Plavsic, N., Chou, J.-L., Drumm, M.L., Iannuzzi, M.C., Collins, F.S., Tsui, L.-C., 1989, Identification of the Cystic Fibrosis gene: cloning and characterization of complementary DNA, **Scienze**, 245:1066.

INCIDENCE OF CYSTIC FIBROSIS AT THE FAROE ISLANDS

Nicolina Sørensen and Marianne Schwartz

Department of Pediatrics, Section of Clinical Genetics
Rigshopspitalet, Denmark

INTRODUCTION

The Faroe Islands are a small group of islands in the North Atlantic Ocean. The origin of the population is partly Irish, partly Norwegian and Danish. The Faroe Islands are a part of the Danish Kingdom with a population of approx. 50,000. The annual number of births has been about 750 over the last 40 years.

We have investigated the incidence of Cystic Fibrosis (CF) in this relatively isolated society in the period 1957-1989. In this period 13 known patients with cystic fibrosis have been born, which gives a relative incidence of approx. 1:2000. Of these 13 CF patients 6 are still alive, and it was possible to investigate all the families with tightly linked polymorphic DNA markers (XV2c and KM19) and to test for the F508 mutation.

Patients: A total of 100 persons including sibs, parents and their sibs, grandparents and great grandparents were analyzed.

DNA preparation and RFLP analysis with the probes XV2c and KM19 were performed by routine methods. Detection of the F508 mutation was performed by PCR analysis followed by allele specific hybridization (Kerem et al., 1989).

All CF chromosomes analyzed in this study have the most common haplotype (B) as well as the F508 deletion (Table 1 and 2). The direct testing for this mutation made it possible for us to

Table 1

FREQUENCY OF F508 AND HAPLOTYPE B IN CF CHROMOSOMES

	F508	B
The Faroe Islands	100 %	100 %
Denmark	86 %	91 %

Table 2

HAPLOTYPES ON NON-CF CHROMOSOMES

The Faroe Islands		Denmark
A	39 %	33 %
B	14 %	14 %
C	31 %	42 %
D	16 %	11 %

make an unambiguous assessment of the carrier status of the family members, and to follow the mutation through several generations. We have tested 9 healthy sibs to CF patients and 58 sibs to their parents. In these groups we found 8 and 37 carriers respectively. In total we identified 45 carriers compared to the expected 34, which is a clear overrepresentation of carriers (Table 3).

Table 3

FREQUENCY OF CARRIERS

	Analyzed	Expected	Observed
Sibs 2/3 risk	9	6	8
Parent's sibs 1/2 risk	58	29	37
Total	67	35	45

We have been able to follow 6 of the families through 3 generations. By studying the birth register in this small population we were able to show that the origin of the CF gene could be traced to only 4 families, although no first cousin marriages have been detected in the parent or grandparent generations. Due to the small size and isolated character of these Islands it is, however, not unexpected to find such close family relationships.

The population of The Faroe Islands is very similar to the rest of the Danish population in repect to the high frequency of the F508 mutation and haplotype B. However, the incidence of CF in Denmark is only 1:4,700 compared to the apparently much higher incidence in the Faroe Islands. The high rate of consanguinity in The Faroe Islands might explain the apparent higher incidence. The relatively high rate of carriers has to be tested on a larger number of individuals.

Reference

Kerem, B., Rommens, J.R., Buchanan, J.A., Markiewiez, D., Cox, T.K., Chakravarti, A., Buchwald, M., Tsui, L.-C., 1989, Identification of the cystic fibrosis gene: genetic analysis, **Science**, 245:1073.

INDEX